Journey through Mathematics

Enrique A. González-Velasco

Journey through Mathematics

Creative Episodes in Its History

 Springer

Enrique A. González-Velasco
Department of Mathematical Sciences
University of Massachusetts at Lowell
Lowell, MA 01854
USA
Enrique_GonzalezVelasco@uml.edu

ISBN 978-0-387-92153-2   e-ISBN 978-0-387-92154-9
DOI 10.1007/978-0-387-92154-9
Springer New York Dordrecht Heidelberg London

Library of Congress Control Number: 2011934482

Mathematics Subject Classification (2010): 01-01, 01A05

Cover Image: Drawing in the first printed proof of the fundamental theorem of calculus, published by Heir of Paolo Frambotti, Padua in 1668, by James Gregory in GEOMETRIÆ PARS VNIVERSALIS (The Universal Part of Geometry).

Printed on acid-free paper

Springer is part of Springer Science+Business Media (www.springer.com)

To my wife, Donna,
who solved quite a number of riddles for me.
With thanks for this, and for everything else.

TABLE OF CONTENTS

PREFACE

In the fall of 2000, I was assigned to teach history of mathematics on the retirement of the person who usually did it. And this with no more reason than the historical snippets that I had included in my previous book, *Fourier Analysis and Boundary Value Problems*. I was clearly fond of history.

Initially, I was unhappy with this assignment because there were two obvious difficulties from the start: (*i*) how to condense about 6000 years of mathematical activity into a three-month semester? and (*ii*) how to quickly learn all the mathematics created during those 6000 years? These seemed clearly impossible tasks, until I remembered that Joseph LaSalle (chairman of the Division of Applied Mathematics at Brown during my last years as a doctoral student there) once said that the object of a course is not to cover the material but to uncover part of it. Then the solution to both problems was clear to me: select a few topics in the history of mathematics and uncover them sufficiently to make them meaningful and interesting. In the end, I loved this job and I am sorry it has come to an end.

The selection of the topics was based on three criteria. First, there are always students in this course who are or are going to be high-school teachers, so my selection should be useful and interesting to them. Through the years, my original selection has varied, but eventually I applied a second criterion: that there should be a connection, a thread running through the various topics through the semester, one thing leading to another, as it were. This would give the course a cohesiveness that to me was aesthetically necessary. Finally, there is such a thing as personal taste, and I have felt free to let my own interests help in the selection.

This approach solved problem (*i*) and minimized problem (*ii*), but I still had to learn what happened in the past. This brought to the surface another large set of problems. The first time I taught the course, I started with secondary sources, either full histories of mathematics or histories of specific topics. This proved to be largely unsatisfactory. For one thing, coverage was not extensive enough so that I could really learn the history of my chosen topics. There is also the fact that, frequently, historian *A* follows historian *B*, who in turn follows historian *C*, and so on. For example, I have at least four books in my collection that attribute the ratio test for the convergence of infinite series to Edward Waring, but without a reference. I finally traced this partial misinformation back to Moritz Cantor's *Vorlesungen über Geschichte der Mathematik*.[1] This is history by hearsay, and I could not fully put my trust in it. There is also the matter of unclear or insufficient references, with the additional problem that sometimes they are to other secondary sources. Finally, I had to admit that not all secondary sources offer the truth, the whole truth, and nothing but the truth (Rafael Bombelli, the discoverer of complex numbers, has particularly suffered in this respect). The long and the short of it is this: it's a jumble out there.

After my first semester teaching the course, it was obvious to me that I had to learn the essential facts about the work of any major mathematician included here straight from the horse's mouth. I had to find original sources, or translations, or reprints. I enlisted the help of our own library and the Boston Library Consortium, with special thanks to MIT's Hayden Library. Beyond this, I relied on the excellent service of our Interlibrary Loan Department. But even all this would have been insufficient and this book could not have been written in its present form. I purchased a large collection of books, mostly out of print and mostly on line, and scans of old books on CD, all of which were of invaluable help. Special thanks are also due to the *Gottfried Wilhelm Leibniz Bibliothek*, of Hanover, for copies of the relevant manuscripts of Leibniz on his discovery of the calculus. As for the rest, the very large rest, I went on line to several digitized book collections from around the world, too many to cite individually. It is a wonder to me that history of mathematics could be done before the existence of these valuable resources.

Many of the works I have consulted are already translated into English (such as those by Ptolemy, Aryabhata, Regiomontanus, some of Viète's, Napier, Briggs, Newton, and—to a limited and unreliable extent—Leibniz), but in most other cases the documents are available only in the language originally written in or in translations into languages other than English (such as those by Al Tusi, Saint Vincent, Bombelli, most of Gregory's, Fermat, Fourier, da Cunha, and Cauchy). Except by error of omission, the translations in this book that are not credited to a specific source are my own, but I wish to thank my colleague Rida Mirie for his kind help with Arabic spelling and translation.

---

[1] Vol. 4, B. G. Teubner, Leipzig, 1908, p. 275. Cantor gave the reference, but with no page number, and then he put his own misleading interpretation of Waring's statement in quotation marks! For a more detailed explanation of Waring's test, stronger than the one given later by Cauchy, see note 21 in Chapter 6.

As much as I believe that a text on mathematics must include as many proofs as possible at the selected level, for mathematics without proofs is just a story, I also believe that history without complete and accurate references is just a story, a frustrating one for many readers. I have endeavored to give as complete a set of references as I have been able to. Not only to original sources but also to facsimiles, translations into several languages, and reprints, to facilitate the work of the reader who wishes to do additional reading. These details can be found in the bibliography at the end of the book. For easy and immediate access, references are also given in footnotes at each appropriate place, but only by the author's last name, the work title, volume number if applicable, year of publication the first time that a work is cited in a chapter, and relevant page or pages.

I can only hope that readers enjoy this book as much as I have enjoyed writing it.

Dunstable, Massachusetts   *Enrique A. González-Velasco*
June 15, 2010

1

TRIGONOMETRY

1.1 THE HELLENIC PERIOD

Trigonometry as we know it and as we call it today is a product of the seventeenth century, but it has very deep roots. In antiquity, it all started with a stick sunk in the soil perpendicular to the ground, and with the measurement of its shadow in the sun. From repeated observations of the shadow, such things as the length of the day or the length of the year could be determined, as well as the location in the year of the solstices and equinoxes. Because the Greeks would obtain knowledge from the length of the stick's shadow, they called the stick a *gnomon* (γνώμων), a Greek word that has the same root as "to know."

At present there is a wider interest in plane trigonometry, and it is the history of this discipline that we shall outline, but it should not come as a surprise that some of the earliest users were primarily concerned with spherical trigonometry. When humans could find some time for tasks other than procuring food and shelter, when scientific curiosity was possible and started to develop, people looked at the heavens and tried to figure out the mystery of the sun, moon, and stars circling around. Thus, astronomy was one of the earliest sciences and, together with geography, made the development of trigonometry, plane and spherical, a necessity.

Four astronomers were the main known contributors to the development of the first phase of this subject: Aristarchos of Samos, Hipparchos of Nicæa, Menelaos of Alexandria, and Klaudios Ptolemaios.[1]

---

[1] Except in quotations from other sources, I write Greek names with Greek rather than Latinized endings, as in Aristarchos rather than Aristarchus.

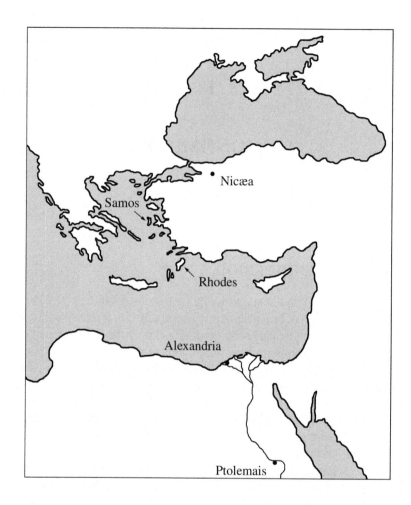

Aristarchos (310–230, BCE) was basically a mathematician, and he was known as such in his own time. Aëtius refers to him as "Aristarchus of Samos, a mathematician and pupil of Strato ... "[2] in his *Doxographi græci*.

Aristarchos' main contribution to astronomy is the proposal of the heliocentric system of the universe, but this work has been lost. We know about it

---

[2] Quoted from Thomas, *Selections illustrating the history of Greek mathematics*, II, 1942, p. 3. Strato of Lampascus was head of the Alexandrian Lyceum from 288/287 to 270/269 BCE [p. 2]. This was a school modeled after Aristotle's Peripatetic Lyceum, so called because Aristotle taught while walking in the garden of the hero Lycos.

from a reference in *The sand-reckoner* of Archimedes [pp. 221–222]:[3]

> Now you are aware ["you" being Gelon, tyrant of Syracuse] that 'universe' is
> the name given by most astronomers to the sphere, the centre of which is the
> centre of the earth, while its radius is equal to the straight line between the
> centre of the sun and the centre of the earth. This is the common account ($\tau\grave{\alpha}$
> $\gamma\rho\alpha\varphi\acute{o}\mu\varepsilon\nu\alpha$), as you have heard from astronomers. But Aristarchus of Samos
> brought out a book consisting of some hypotheses, in which the premises
> lead to the result that the universe is many times greater than that now so
> called. His hypotheses are that the fixed stars and the sun remain unmoved,
> that the earth revolves around the sun in the circumference of a circle, the
> sun lying in the middle of the orbit, and that the sphere of the fixed stars,
> situated about the same centre as the sun, is so great that the circle in which
> he supposes the earth to revolve bears such a proportion to the distance of
> the fixed stars as the centre of the sphere bears to its surface.

As much as the loss of this book and the fact that his contemporaries did not
accept this theory are to be lamented, the heliocentric hypothesis is not central
to our quest.

Aristarchos' only extant book, *On the sizes and distances of the sun and
moon*, is more interesting for our purposes. This is a book on mathematical
astronomy, with little or nothing on the practical side. After stating six basic
hypotheses, starting with[4]

$\bar{\alpha}$ That the moon receives its light from the sun.

and ending with

$\bar{\varsigma}$ That the moon subtends one fifteenth part of a sign of the zodiac.

Aristarchos then gave a set of 18 propositions with proofs. There being twelve
signs of the zodiac in a full circle of $360°$, one sign of the zodiac is $30°$,
which would make the moon subtend an arc of $2°$. Now, this is incorrect,
and Aristarchos himself knew better (although not, apparently, at the time of
writing the preserved manuscript) because Archimedes said that Aristarchos
"discovered that the sun appeared to be about 1/720th part of the circle of

---

[3] This work can be seen in Heath, *The works of Archimedes with The method of
Archimedes*, 1912, pp 221–232. Page references given in brackets are to the Dover Publications edition, 1953.

[4] The stated hypotheses are in Heath, *Aristarchus of Samos*, 1913, p. 302. Page references
given below for additional quotations are to this edition.

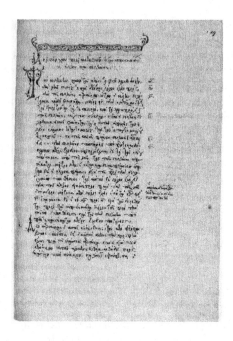

A sixteenth-century manuscript copy of *On the sizes and distances of the sun and moon*.
The header reads: Ἀριστάρχου περὶ μεγεθῶν καὶ ἀποστημάτων ἡλίου καὶ σελήνης.
Reproduced from the virtual exhibition *El legado de las matemáticas*:
de Euclides a Newton, los genios a través de sus libros, Sevilla, 2000.

the zodiac" [*The sand-reckoner*, p. 353]; that is, one half of a degree, and Aristarchos believed that the sun and the moon subtend the same arc (as he stated in Proposition 8 [p. 383]).

His conclusions are erroneous because his sixth hypothesis is, but the proofs are mathematically correct. Most are based on what we now call Euclidean geometry, but Aristarchos also used what we know as trigonometry. For instance, consider [p. 365]

PROPOSITION 4.

The circle which divides the dark and bright portions in the moon is not perceptibly different from a great circle in the moon.

This situation is represented in the next figure (which excludes some of the lines and letters of the original). The circle represents the moon with its center at B, the point A is the eye of an observer on earth, and the sun, eclipsed

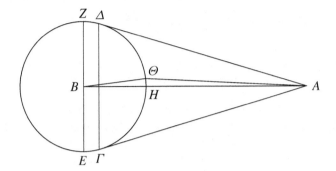

by the moon, is on the left (not shown). The bright portion of the moon is represented by the arc $\Delta ZE\Gamma$ and the dark side by the arc $\Delta H\Gamma$. The circles mentioned in the proposition are shown as the segments $\Delta\Gamma$ and ZE. The point Θ is chosen so that the arc ΘH is equal to half the arc ΔZ.

The proof consists in showing that this last arc is negligible by estimating its maximum possible size if the angle $\Delta A\Gamma$ is $2°$, and it is mostly geometric; but then Aristarchos used the following fact [p. 369]:

And $B\Theta$ has to ΘA a ratio greater than that which the angle $BA\Theta$ has to the angle $AB\Theta$.

To interpret this statement, let Π denote the foot of the perpendicular from Θ to AB (not shown in the figure) and denote the angles $BA\Theta$ and $AB\Theta$ by α and β, respectively. Then, in our usual terminology,

$$\sin\alpha = \frac{\Theta\Pi}{\Theta A} \qquad \text{and} \qquad \sin\beta = \frac{\Theta\Pi}{B\Theta}.$$

This allows us to compute the ratio $B\Theta/\Theta A$ and to restate Aristarchos' statement in the equivalent form

$$\frac{\sin\alpha}{\sin\beta} > \frac{\alpha}{\beta},$$

which shows its true nature as a trigonometric statement.

The remarkable thing is that Aristarchos, who was extremely clear and detailed in the geometric part of the proof, offered no comment or explanation about the preceding statement.

Another instance of the use of trigonometry occurs in the proof of

PROPOSITION 7.

The distance of the sun from the earth is greater than eighteen times, but less than twenty times, the distance of the moon from the earth.

Actually, the sun is about four hundred times farther away than the moon. But this is neither here nor there, for had Aristarchos started with correct values from accurate measurements, he would have arrived at correct results. The relevance of his work is in showing that the power of mathematics can lead to conclusions of the type reached.

In the course of the proof, which is based on a figure some of whose

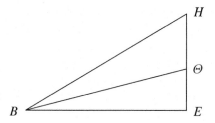

components are reproduced here, Aristarchos asserted the following [p. 377]:

> Now, since HE has to $E\Theta$ a ratio greater than that which the angle HBE has to the angle ΘBE, ...

It is very easy to translate this statement into our present language: if α and β are angles between 0 and $\pi/2$, and if $\alpha > \beta$, then

$$\frac{\tan \alpha}{\tan \beta} > \frac{\alpha}{\beta},$$

and again this is trigonometry. Once more, Aristarchos offered no comment or proof. He simply assumed this as a fact, which suggests that these trigonometric assumptions may have been well-known facts at the time. At any rate, they were known to him.

Very little is known about Hipparchos (190–120, BCE). The main known facts are these: he was born in Nicæa, in Bithynia (now called Iznik in north-western Turkey, about 100 kilometers by straight line southeast from Istanbul); he was famous enough to have his likeness stamped on coins issued by several Roman emperors; and he devoted his scientific life to astronomy, making observations at Alexandria in 146 BCE and at Rhodes, where he probably died, about twenty years later. All but one of his works have been lost, and we have no direct knowledge of his work on trigonometry, even though he is usually regarded to be its founder. We know about his research indirectly, from other sources.

One of Hipparchos' claims to fame is a determination of the solar year that is closer to the real one than the one in use for more than 1600 years after his finding. Comparing the length of the *gnomon*'s shadow at the summer solstice of 135 BCE with that measured by Aristarchos at the summer solstice of 280 BCE, led him to correct the then accepted figure of $365\frac{1}{4}$ days in a now lost tract *On the length of the year*. Ptolemaios provided the following evidence of this fact in his *Mathematike syntaxis*:[5]

> And when he more or less sums up his opinions in his list of his own writings, he [Hipparchos] says: 'I have also composed a work on the length of the year in one book, in which I show that the solar year (by which I mean the time in which the sun goes from a solstice back to the same solstice, or from an equinox back to the same equinox) contains 365 days, plus a fraction which is less than $\frac{1}{4}$ by about $\frac{1}{300}$th of the sum of one day and one night, and not, as the mathematicians suppose, exactly $\frac{1}{4}$-day beyond the above-mentioned number [365] of days.'

This would put the length of the year at 365 days, 5 hours, 55 minutes, and 12 seconds. This length is very close to the true value of 365 days, 5 hours, 48 minutes, and 46 seconds, and represents an opportunity missed by Julius Cæsar when he reformed the calendar in 46 BCE and ignored Hipparchos' value (he was counseled by the professional astronomer Sosigenes).

Trigonometry is supposed to have had its start in a table of chords subtended by arcs of a circle compiled by Hipparchos. We have this on the authority of Theon of Alexandria, who wrote the following in his commentary on Ptolemaios' *Syntaxis*:[6] "An investigation of the chords in a circle is made by Hipparchus in twelve books and again by Menelaus in six."

---

[5] Quoted from Toomer, *Ptolemy's Almagest*, 1984, p. 139.

[6] Quoted from Thomas, *Selections illustrating the history of Greek mathematics*, II, p. 407.

While the exaggerated number of twelve books must be in error, the tables were real.[7] They were based on dividing the circle into 360 parts with each part divided into 60 smaller parts, as the Babylonians had already done. It appears that Hipparchos evaluated the lengths of chords of angles at $7\frac{1}{2}$ parts apart and then interpolated values at intermediate points. But how a table of chords is constructed will be better explained when we cover the work of Ptolemaios, who incorporated that of Hipparchos and provided all necessary details.

There is little to say about Menelaos of Alexandria (c. 70–c. 130) regarding plane trigonometry: writing in Rome, he compiled a table of chords in six books, but it has not survived. He wrote a number of texts, but the only one that has been preserved to our times is an Arabic translation of his *Sphærica* in three books. In it, he made noteworthy contributions to spherical trigonometry, which we shall not discuss.

But the final synthesis of the trigonometric knowledge of antiquity took place in Alexandria, the magnificent city founded in Egypt by Alexander III, son of Philip II of Macedonia, who had set out at the age of 20, in 336 BCE, to conquer the world. He succeeded, indeed, in becoming ruler of most of the known world of his time, totally subduing the Persians. The city of Alexandria was destined to become the center of Hellenic culture for centuries and the undisputed hub of mathematics in the world. On the death of Alexander in 323 BCE, at the age of 33, Egypt was governed by Ptolemy Soter (*Savior*), one of Alexander's generals, who was also his close friend and perhaps a relative, and who became king of Egypt in 305 BCE. Ptolemy I and his successors were enlightened rulers, bringing scholars from all over the world to Alexandria. Here, at the *Museum*, or temple of the Muses, they had a magnificent library at their disposal, botanical and zoological gardens, free room and board in luxurious conditions, exemption from taxes, additional salaries, and plenty of free time to engage in their own scholarly pursuits. Their only duty was to give regular lectures, a situation equivalent to that of a very generously endowed modern university plus quite a few additional perks.

Ptolemy I founded a city that was called Ptolemais, named after him, as a center of Hellenic culture in upper Egypt, and it was here that Klaudios Ptolemaios (c. 87–150) is said to have been born. He is usually known by the name Claudius Ptolemy, and, since this is almost as much of a household name as Euclid's, we shall use it henceforth. The Claudius part of his name suggests

---

[7] These and Hipparchos' methods have been reconstructed by Toomer in "The chord table of Hipparchus and the early history of Greek trigonometry" (1973/74) 6–28.

that he was a Roman citizen, but at some time he moved to Alexandria, where he studied under Theon of Smyrna, and probably remained at the *Museum* for the rest of his life.

Ptolemy wrote books on many subjects, such as optics, harmonics, and geography, but his fame rests on his great work *Mathematike syntaxis* (Mathematical collection) in thirteen books, probably written during the last decade of the reign of Titus Ælius Antoninus (138–161), also known as Antoninus Pius, one of the five good Roman emperors. This work was also called *Megale syntaxis* (Large collection) to distinguish it from smaller or less-important ones. Later, writers in the Arabic language combined their article *al* with the superlative form *megiste* of *megale* to make *al-mjsty*. This is why the renamed Ἐ μέγιστη σύνταξις is usually known as the *Almagest*. This book

The first printed edition of the *Almagest*, Venice, 1515.
Reproduced from the virtual exhibition *El legado de las matemáticas:
de Euclides a Newton, los genios a través de sus libros*, Sevilla.

did for astronomy and trigonometry what Euclid's *Elements* did for geometry; for more than a thousand years it remained the source of most knowledge in those fields, and there were numerous manuscript copies and translations into Syriac, Arabic, and Latin.

Ptolemy described his intention in writing this compilation at the outset, at the end of Article 1:[8]

> For the sake of completeness in our treatment we shall set out everything useful for the theory of the heavens in the proper order, but to avoid undue length we shall merely recount what has been adequately established by the ancients. However, those topics which have not been dealt with [by our predecessors] at all, or not as usefully as they might have been, will be discussed at length, to the best of our ability.

Regarding trigonometry and the table of chords, which are in Book I of the *Almagest*, Ptolemy compiled all the knowledge on this subject up to his time, but did not tell us which parts were due to Hipparchos or to Menelaos. We shall assume that their results are included in Ptolemy's presentation.

1.2 PTOLEMY'S TABLE OF CHORDS

Tables of chords are important for the same reason that today's trigonometric functions are: they allow us to solve triangles, and this is necessary in applications to astronomy and geography. For this reason and because the Almagest represented the definitive state of western trigonometry for the next millennium, the construction of this table of chords will be presented in sufficient detail.

Ptolemy began [Art. 10][9] by dividing the perimeter of a circle into 360 parts, and then the diameter into 120 parts. Although Ptolemy used the same word here for parts ($\tau\mu\acute{\eta}\mu\alpha\tau\alpha$), these two parts are different. Otherwise, the ratio of the perimeter to the diameter would be 3, but Ptolemy determined that this ratio is approximately $3\frac{17}{120}$. We shall call the 360 parts of the perimeter "degrees" (from the Latin *de gradus* = "[one] step away from"), and use the modern notation $360°$.[10] We shall also use "minute" and "second" (derived

---

[8] Quoted from Toomer, *Ptolemy's Almagest*, p. 37.

[9] The more verbal translation by Thomas in *Selections illustrating the history of Greek mathematics*, II, pp. 412–443, is closer than Toomer's to the Greek original style, and, while this may be too tiring for the long haul, it may be quite adequate for a shorter presentation. For this reason I stay closer to this translation here. Quotations are from this edition and page references are to it.

[10] It may not be modern at all. For the origin of the present symbols for degrees, minutes, and seconds, see Cajori, *A history of mathematical notations*, II, 1929, pp. 142–147.

from the Latin *pars minuta prima*, or first small part, and *pars minuta secunda*, or second small part) for the usual concepts and with the usual notation of one or two strokes. We need a different notation for the 120 parts of the diameter, and adopt the widely used superscript ^{P} for this purpose. Finally, before embarking on his construction of a table of chords of central angles in a circle, Ptolemy said that he will [p. 415]

> use the sexagesimal system for the numerical calculations owing to the inconvenience of having fractional parts, especially in multiplications and divisions.[11]

With this notational hurdle out of the way, Ptolemy proceeded to state a few necessary theorems to compute the chords of angles in steps of half a degree. To interpret his results in current notation, refer to the next figure and

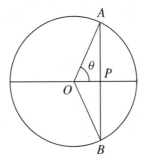

note that if θ is an acute angle and if we denote the length of the chord AB of the angle 2θ by crd 2θ, then we have the basic relation

$$\sin \theta = \frac{AP}{OA} = \frac{2AP}{2OA} = \frac{AB}{\text{diameter}} = \frac{\text{crd}\,2\theta}{120}.$$

We start with Ptolemy's first theorem [p. 415]:

First, let $AB\Gamma$ be a semicircle on the diameter $A\Delta\Gamma$ and with centre Δ, and from Δ let ΔB be drawn perpendicular to $A\Gamma$, and let $A\Gamma$ be bisected at E,

---

[11] Ptolemy was right in this choice: the number 60 is divisible by 2, 3, 4, 5, 6, 10, 12, 15, 20 and 30, and of the fractions with denominators 2 to 9 only 60/7 gives a nonterminating quotient. By contrast, 10—our likely choice for a system base—is divisible by 2 and 5, and has nonterminating quotients when divided by 3, 6, 7 and 9.

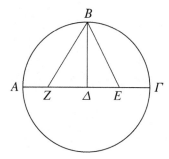

and let EB be joined, and let EZ be placed equal to it, and let ZB be joined. I say that ZΔ is the side of a [regular] decagon, and BZ of a [regular] pentagon.

We omit the proof. Ptolemy's is somewhat involved, since it depends on results from Euclid's *Elements* that, besides depending on previous propositions by Euclid, are unlikely to be available in the mind of the average reader. It is possible to give a short modern proof, which might be jarring and anachronistic at this point, so the best compromise is to give a reference.[12]

This lack of proof notwithstanding, if the preceding theorem is accepted, then it can be used to start the evaluation of some chords. Thus, Ptolemy continued as follows [pp. 417–419]:

> Then since, as I said, we made the diameter consist of 120^P, by what has been stated ΔE, being half of the radius, consists of 30^P and its square of 900^P, and $B\Delta$, being the radius, consists of 60^P and its square of 3600^P, while the square on EB, that is, the square on EZ, consists of 4500^P; therefore EZ is approximately 67^P 4 55, and the remainder ΔZ is 37^P 4 55.

Hold your horses, some reader might be thinking. How did they find, in antiquity, that $\sqrt{4500} = 67^P$ 4 55, and what does this notation mean? The easiest thing to explain is the notation: 67^P 4 55 means

$$67 + \frac{4}{60} + \frac{55}{3600}$$

---

[12] Ptolemy's proof can, of course, be seen in any of the translations of the *Almagest* already mentioned. The best recent presentation that I have seen of all the results in Article 10, with very brief and clear modern proofs, is in Glenn Elert, *Ptolemy's table of chords. Trigonometry in the second century*, www.hypertextbook.com/eworld/chords/shtml, June 1994.

in the sexagesimal system that Ptolemy said he was going to use. As for the extraction of the square root, he did not offer any explanation whatever. Is this is a trivial matter that we should figure out by ourselves? Apparently not, because Theon of Alexandria, writing his *Commentary on Ptolemy's Syntaxis* more than 200 years after its writing, spent some time and effort showing how to evaluate this root.[13] However, an electronic calculator gives $\sqrt{4500} = 67.08203933$, from which value we get the 67^P. Now,

$$0.08203933 = \frac{(0.08203933)(60)}{60} = \frac{4.9223598}{60},$$

from which we get the 4 *pars minuta prima,* so to speak. And then

$$0.9223598 = \frac{(0.9223598)(60)}{60} = \frac{55.341588}{60},$$

which, rounding down, provides the 55 *pars minuta secunda.*

Having obtained $EZ = 67^P 4 \ 55$ (we are writing equalities for convenience, but it should be understood that many of these computations provide only approximate values), Ptolemy then concluded that the side of the regular decagon is

$$\Delta Z = EZ - \Delta E = 67^P 4 \ 55 - 30^P = 37^P 4 \ 55$$

and, since this side subtends an arc of 36°, then, using our notation,

$$\text{crd } 36° = 37^P 4 \ 55.$$

As for the side BZ of the regular pentagon, which subtends an arc of 72°, Ptolemy made the following computation (but only in narrative form) [p. 419]:

$$BZ = \sqrt{B\Delta^2 + \Delta Z^2} = \sqrt{60^2 + (37^P 4 \ 55)^2} = \sqrt{4975^P 4 \ 15},$$

and then

$$\text{crd } 72° = 70^P 4 \ 55.$$

We have computed only two chords, but a few more are easy. Since the side of the regular hexagon is equal to the radius, we have

$$\text{crd } 60° = 60^P;$$

---

[13] His evaluation can be seen in Rome, *Commentaires de Pappus et de Théon d'Alexandrie sur l'Almageste,* II, 1936, pp. 469–473. Also in Thomas, *Selections illustrating the history of Greek mathematics,* I, 1941, pp. 56–61.

the side of the regular square is the square root of twice the square of the radius, so that

$$\text{crd } 90° = \sqrt{60^2 + 60^2} = \sqrt{7200} = 84^p\,51\ 10;$$

and the side of the regular triangle is $\sqrt{3}$ times the radius (using well-known Euclidean geometry the reader can deduce this result in a short time), which gives

$$\text{crd } 120° = 60\sqrt{3} = \sqrt{10800} = 103^p\,55\ 23.$$

Now consider two chords subtending a semicircle in the manner of those shown in the next figure (not included in the *Almagest*). Ptolemy observed

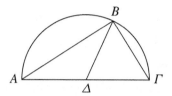

that [p. 421] "the sums of the squares on those chords is equal to the square on the diameter" (since the angle $AB\Gamma = 90°$); that is,

$$AB^2 + B\Gamma^2 = A\Gamma^2.^{14}$$

Then he gave the following example. If $\widehat{B\Gamma}$ is an arc of 36°, then

$$\text{crd } 144° = AB = \sqrt{A\Gamma^2 - B\Gamma^2} = \sqrt{120^2 - (37^p 4\ 55)^2} = 114^p\,7\ 37.$$

Similarly,

$$\text{crd } 108° = \sqrt{120^2 - \text{crd}^2 72°} = 97^p\,4\ 56.$$

At this moment we sum up by stating that we have computed the chords of 36°, 60°, 72°, 90°, 108°, 120°, and 144°. Not a small harvest, but very far from the promised table.

To continue the computation of chords, Ptolemy set out "by way of preface this little lemma ($\lambda\eta\mu\mu\acute{\alpha}\tau\iota o\nu$) which is exceedingly useful for the business at

---

[14] If we denote the angle $B\Delta A$ in the previous figure by 2θ, this equation can be rewritten as $\text{crd}^2 2\theta + \text{crd}^2(180° - 2\theta) = 120^2$. Then, if we recall that $\text{crd } 2\theta = 120 \sin\theta$, it becomes $120^2 \sin^2\theta + 120^2 \sin^2(90° - \theta) = 120^2$ or $\sin^2\theta + \cos^2\theta = 1$.

hand." Since there is no record that this lemma was known before Ptolemy, it is usually known as Ptolemy's Theorem [p. 423].

Let $AB\Gamma\Delta$ be any quadrilateral inscribed in a circle, and let $A\Gamma$ and $B\Delta$ be joined. It is required to prove that the rectangle contained by $A\Gamma$ and $B\Delta$ is equal to the sum of the rectangles contained by AB, $\Delta\Gamma$ and $A\Delta$, $B\Gamma$.

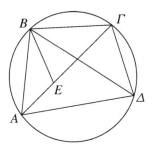

Of course, we would prefer to write the theorem's conclusion as follows:

$$A\Gamma \cdot B\Delta = AB \cdot \Delta\Gamma + A\Delta \cdot B\Gamma,$$

but mathematics was expressed verbally in the ancient world and for a long time after that. The development of our usual mathematical notation was a very slow process that matured only in the middle of the seventeenth century, in the work of Descartes and Newton. Be that as it may, starting with the next equation, we shall frequently use the signs $=$, $+$, and $-$, and replace "the rectangle contained by ..." with a dot to denote a product.

Here is Ptolemy's own proof of the stated theorem. It is based on well-known Euclidean geometry.

For let the angle ABE be placed equal to the angle $\Delta B\Gamma$ [that is, choose E so that this is so]. Then, if we add the angle $EB\Delta$ to both, the angle $AB\Delta$ equals the angle $EB\Gamma$. But the angle $B\Delta A$ equals the angle $B\Gamma E$, for they subtend the same segment [the chord AB]; therefore the triangle $AB\Delta$ is isogonal (ἰσογώνιον) with the triangle $B\Gamma E$. Therefore, the ratio $B\Gamma$ over ΓE equals the ratio $B\Delta$ over ΔA. Therefore

$$B\Gamma \cdot A\Delta = B\Delta \cdot \Gamma E.$$

Again, since the angle ABE is equal to the angle $\Delta B\Gamma$, while the angle BAE is equal to the angle $B\Delta\Gamma$, therefore the triangle ABE is isogonal with the triangle $B\Gamma\Delta$. Analogously, the ratio BA over AE equals the ratio $B\Delta$ over $\Delta\Gamma$. Therefore,

$$BA \cdot \Delta\Gamma = B\Delta \cdot AE.$$

But it was shown that

$$B\Gamma \cdot A\Delta = B\Delta \cdot \Gamma E,$$

and therefore as a whole [meaning $BA \cdot \Delta\Gamma + B\Gamma \cdot A\Delta = B\Delta \cdot AE + B\Delta \cdot \Gamma E = (AE + \Gamma E)B\Delta = A\Gamma \cdot B\Delta$]

$$A\Gamma \cdot B\Delta = AB \cdot \Delta\Gamma + A\Delta \cdot B\Gamma,$$

which was to be proved.

No computation of chords was carried out directly using this theorem; this was done from its corollaries. Although not labeled or numbered by Ptolemy, we shall do so for easy reference [p. 425].

COROLLARY I.

This having first been proved, let $AB\Gamma\Delta$ be a semicircle having $A\Delta$ for its diameter, and from A let the two [chords] AB, $A\Gamma$ be drawn, and let each of them be given in length, in terms of the 120^p in the diameter, and let $B\Gamma$ be joined. I say that this also is given [here "given" means "found"].

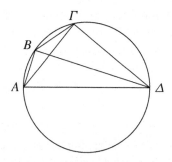

Ptolemy's sketch of proof, for it is no more than that, is very short.

For let $B\Delta$, $\Gamma\Delta$ be joined; then clearly these also are given because they are the chords subtending the remainder of the semicircle. Then since $AB\Gamma\Delta$ is a quadrilateral in a circle,

$$AB \cdot \Gamma\Delta + A\Delta \cdot B\Gamma = A\Gamma \cdot B\Delta$$

[by Ptolemy's theorem]. And $A\Gamma \cdot B\Delta$ is given, and also $AB \cdot \Gamma\Delta$; therefore the remaining term $A\Delta \cdot B\Gamma$ is also given. And $A\Delta$ is the diameter; therefore the straight line $B\Gamma$ is given.

But this corollary gives no formula to find $B\Gamma$. We now fill in the missing steps. From the equation stated in this sketch of proof we obtain

$$B\Gamma = \frac{A\Gamma \cdot B\Delta - AB \cdot \Gamma\Delta}{A\Delta} = \frac{A\Gamma\sqrt{A\Delta^2 - AB^2} - AB\sqrt{A\Delta^2 - A\Gamma^2}}{A\Delta},$$

and $B\Gamma$ can be found from the given chords and the diameter.[15]

Now, returning to the computation of chords, Ptolemy wrote that

by this theorem we can enter [calculate] many other chords subtending the difference between given chords, and in particular we may obtain the chord subtending 12°, since we have that subtending 60° and that subtending 72°.

In fact, using the chord formula just developed and the values already obtained for crd 72° and crd 60°, we would arrive at

$$\text{crd } 12° = \text{crd } (72° - 60°) = 12^P \, 32 \, 36.$$

It is possible to obtain more than this. By successive application of the formula we can calculate crd 18° = crd (108° − 90°) and then crd 6° = crd (18° − 12°), but Ptolemy did not do this, and he had a good reason for it, to be disclosed later. In any event, our table of chords is beginning to flesh out, but is not yet

---

[15] The previous equation can be expressed in terms of the trigonometric function crd if we denote the arcs $\widehat{AB}$ and $\widehat{A\Gamma}$ by 2α and 2θ, respectively. Then it becomes

$$\text{crd } (2\theta - 2\alpha) = \frac{\text{crd } 2\theta \sqrt{120^2 - \text{crd}^2 2\alpha} - \text{crd } 2\alpha \sqrt{120^2 - \text{crd}^2 2\theta}}{120}$$

$$= \text{crd } 2\theta \sqrt{1 - \frac{\text{crd}^2 2\alpha}{120^2}} - \text{crd } 2\alpha \sqrt{1 - \frac{\text{crd}^2 2\theta}{120^2}}.$$

Dividing both sides by 120, using the identity crd $2\theta = 120 \sin\theta$, and noticing that $0° < \alpha$, $\theta < 90°$, this equation can be rewritten as

$$\sin(\theta - \alpha) = \sin\theta \sqrt{1 - \sin^2\alpha} - \sin\alpha \sqrt{1 - \sin^2\theta} = \sin\theta \cos\alpha - \sin\alpha \cos\theta.$$

We see that this familiar trigonometric formula was implicitly contained in Ptolemy's work.

near the promise of finding the chords of all arcs in half degree intervals. So, more trigonometry must be developed [p. 429].

COROLLARY 2.

Again, given any chord in a circle, let it be required to find the chord subtending half the arc subtended by the given chord. Let $AB\Gamma$ be a semicircle upon the

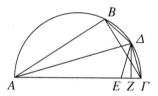

diameter $A\Gamma$ and let the chord ΓB be given, and let the arc ΓB be bisected at Δ, and let AB, $A\Delta$, $B\Delta$, and $\Delta\Gamma$ be joined, and from Δ let ΔZ be drawn perpendicular to $A\Gamma$. I say that $Z\Gamma$ is half of the difference between AB and $A\Gamma$.

We can abbreviate Ptolemy's proof as follows. Choose a point E on $A\Gamma$ so that $AE = AB$. In the triangles $AB\Delta$ and $AE\Delta$, $AE = AB$, $A\Delta$ is a common side, and the angle $BA\Delta$ is equal to the angle $EA\Delta$ (because the arcs $B\Delta$ and $\Delta\Gamma$ are equal by hypothesis); and therefore $B\Delta = \Delta E$. Since $B\Delta = \Delta\Gamma$, we obtain $\Delta\Gamma = \Delta E$, so that $\Delta E\Gamma$ is isosceles, and then

$$Z\Gamma = \tfrac{1}{2}E\Gamma = \tfrac{1}{2}(A\Gamma - AE) = \tfrac{1}{2}(A\Gamma - AB).$$

This finishes the proof of the last statement in the corollary, but we are not done with the required task, which is to find the chord $\Delta\Gamma$. To this end, note that the right triangles $A\Delta\Gamma$ and $\Delta Z\Gamma$ are isogonal, since they have the angle at Γ in common, and then

$$\frac{A\Gamma}{\Gamma\Delta} = \frac{\Gamma\Delta}{\Gamma Z},$$

so that

$$\Gamma\Delta^2 = A\Gamma \cdot \Gamma Z.$$

Ptolemy got to this point but did not provide an equation giving $\Gamma\Delta$. Lacking any type of notation for equations, this is not surprising, but we can do it easily. Replacing ΓZ with its value found above,

$$\Gamma\Delta^2 = \tfrac{1}{2}A\Gamma \cdot (A\Gamma - AB) = \tfrac{1}{2}A\Gamma\left[A\Gamma - \sqrt{A\Gamma^2 - B\Gamma^2}\right].$$

Finding the square root of both sides and recalling that $A\Gamma = 120^P$ yields

$$\Gamma\Delta = \sqrt{\tfrac{1}{2}\left(120^2 - 120\sqrt{120^2 - B\Gamma^2}\right)}.^{16}$$

Ptolemy summed up the usefulness of this result as follows [p. 431]:

And again by this theorem many other chords can be obtained as the halves of known chords, and in particular from the chord subtending 12° can be obtained the chord subtending 6° and that subtending 3° and that subtending $1\frac{1}{2}°$ and that subtending $\frac{1}{2}° + \frac{1}{4}° \left(= \frac{3}{4}°\right)$. We shall find, when we come to make the calculation, that the chord subtending $1\frac{1}{2}°$ is approximately $1^P\ 34\ 15$ (the diameter being 120^P) and that subtending $\frac{3}{4}°$ is $0^P\ 47\ 8$.

This is probably the reason why Ptolemy did not compute the chord of 6° from Corollary 1, namely that it can be done more simply from Corollary 2. His next result was [p. 431]:

COROLLARY 3.

Again, let $AB\Gamma\Delta$ be a circle about the diameter $A\Delta$ and with center Z, and from A let there be cut off in succession two given arcs AB, $B\Gamma$, and let there be joined AB, $B\Gamma$, which, being the chords subtending them, are also given. I say that, if we join $A\Gamma$, it will also be given.

To prove this, construct the diameter BE and the remaining segments shown in the picture. By Ptolemy's theorem, applied to the quadrilateral $B\Gamma\Delta E$,

$$B\Delta \cdot \Gamma E = B\Gamma \cdot \Delta E + BE \cdot \Gamma\Delta.$$

---

[16] If we denote the arc $\widehat{B\Gamma}$ by 2θ (this makes $\theta < 90°$), in which case $\Gamma\Delta = \operatorname{crd}\theta$ and $B\Gamma = \operatorname{crd} 2\theta$, then

$$\operatorname{crd}\theta = \sqrt{\frac{120^2 - 120\sqrt{120^2 - \operatorname{crd}^2 2\theta}}{2}}.$$

Dividing both sides by 120, using the equation $\operatorname{crd} 2\theta = 120\sin\theta$ (but with θ in place of 2θ), and in view of the fact that $\theta < 90°$, the above equation takes the form of the well-known trigonometric identity

$$\sin\tfrac{1}{2}\theta = \sqrt{\frac{1 - \sqrt{1 - \sin^2\theta}}{2}} = \sqrt{\frac{1 - \cos\theta}{2}}.$$

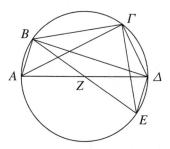

Therefore,

$$\Gamma\Delta = \frac{B\Delta \cdot \Gamma E - B\Gamma \cdot \Delta E}{BE}.^{17}$$

Now, $B\Gamma$ and AB are given, while $B\Delta$ and ΓE can be found from AB and $B\Gamma$, respectively, using the Pythagorean theorem. Thus, since $\Delta E = AB$, $\Gamma\Delta$ can be determined from the preceding equation, and then

$$A\Gamma = \sqrt{A\Delta^2 - \Gamma\Delta^2}.$$

As for the computation of chords, which was Ptolemy's task, this is the way he put it [p. 435]:

> It is clear that, by continually putting next to all known chords a chord subtending $1\frac{1}{2}°$ and calculating the chords joining them, we may compute in a simple manner all chords subtending multiples of $1\frac{1}{2}°$, and there will still be left only those within the $1\frac{1}{2}°$ intervals—two in each case, since we are making the diagram in half degrees.

---

[17] If we denote the arcs $\overset{\frown}{AB}$ and $\overset{\frown}{B\Gamma}$ by 2θ and 2α, respectively, recall that $BE = 120^P$, and observe that $\Delta E = AB$, the equation for $\Gamma\Delta$ can be rewritten as

$$\operatorname{crd}(180° - 2\theta - 2\alpha) = \frac{\operatorname{crd}(180° - 2\theta)\operatorname{crd}(180° - 2\alpha) - \operatorname{crd}2\alpha\operatorname{crd}2\theta}{120}.$$

Using the identity $\operatorname{crd}2\theta = 120\sin\theta$ once more and then dividing by 120, this equation becomes

$$\sin(90° - \theta - \alpha) = \sin(90° - \theta)\sin(90° - \alpha) - \sin\alpha\sin\theta,$$

or equivalently

$$\cos(\theta + \alpha) = \cos\theta\cos\alpha - \sin\alpha\sin\theta,$$

another modern, well-known trigonometric identity.

The next difficulty, in order to complete the table in intervals of half a degree, is precisely the computation of the chord of half a degree. Ptolemy did this from the following "little lemma" (another λημμάτιον) [pp. 435–437].

For let ABΓΔ be a circle, and in it let there be drawn two unequal chords, of

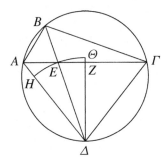

which AB is the lesser and BΓ the greater. I say that

$$\frac{\Gamma B}{BA} < \frac{\text{arc } B\Gamma}{\text{arc } BA}.$$

To the attentive reader this should be a case of *déjà vu*. This is precisely the trigonometric identity that Aristarchos had used in the proof of his Proposition 4, and it is sometimes known as *Aristarchos' inequality*. But while he had made us think that it is a well-known fact by the absence of proof, Ptolemy provided a proof. It depends on some propositions from Euclid's *Elements*, which were well known to Ptolemy and that he mentioned without a reference. Readers who are less familiar with Euclidean geometry may wish to consider the following just a sketch of proof.

Let BΔ be the bisector of the angle ABΓ, and let new points H, E, Z, and Θ be as shown, the first three on an arc of circle with center Δ. It should be clear that

$$\frac{\text{area of triangle } \Delta EZ}{\text{area of triangle } \Delta EA} < \frac{\text{area of sector } \Delta E\Theta}{\text{area of sector } \Delta EH}.$$

Since the two triangles in this inequality have the same height ΔZ and since the two sectors have the same radius, the inequality reduces to

$$\frac{EZ}{EA} < \frac{\angle Z\Delta E}{\angle E\Delta A},$$

where we have used the symbol $\angle$ for angle. Adding 1 to both sides and simplifying[18] gives

$$\frac{ZA}{EA} < \frac{\angle Z\Delta A}{\angle E\Delta A},$$

and then, multiplying both sides by 2 and noting that Z bisects $A\Gamma$ because $B\Delta$ bisects the arc $A\Delta\Gamma$,

$$\frac{\Gamma A}{EA} < \frac{\angle \Gamma\Delta A}{\angle E\Delta A}.$$

Subtracting 1 from both sides and simplifying yields

$$\frac{\Gamma E}{EA} < \frac{\angle \Gamma\Delta E}{\angle E\Delta A} = \frac{\angle \Gamma\Delta B}{\angle B\Delta A}.$$

Using now the well-known facts[19]

$$\frac{\Gamma E}{EA} = \frac{\Gamma B}{BA} \quad \text{and} \quad \frac{\angle \Gamma\Delta B}{\angle B\Delta A} = \frac{\text{arc } \Gamma B}{\text{arc } BA}$$

proves the lemma.

Having concluded the theoretical part of his task, Ptolemy had to complete the computations. First he needed the chord of $\frac{1}{2}^{\circ}$, and to obtain it he used the next figure [p. 441] showing two chords with a common endpoint.[20] Assuming first that AB subtends an angle of $\frac{3}{4}^{\circ}$ and $A\Gamma$ an angle of 1°, and using the inequality in the preceding lemma but with $A\Gamma$ in place of ΓB,

$$A\Gamma < BA\frac{\text{arc } A\Gamma}{\text{arc } BA} = \frac{4}{3}BA.$$

---

[18] This operation on a fraction was well known at least since Euclidean times, and only one word was used to describe it. Ptolemy used συνθέντι (synthenti), meaning "putting together." When these works were later translated into Latin the word *componendo* was used in its place, and this word was still frequently used well into the seventeenth century. Similarly, the operation described below, in which 1 is subtracted rather than added, was called διελόντι (dielonti) by Ptolemy, meaning "having [the fraction] divided," and was translated as *dirimendo*.

[19] They were well known to Ptolemy, since these are Propositions 3 and 33 of Book VI of Euclid's *Elements* [pp. 195 and 273 of the Dover Publications edition, 1956]. Readers with a knowledge of high-school geometry will have no trouble proving these facts using the following hints. For the first, draw a line through A parallel to EB until it intersects the line containing the segment $B\Gamma$ at a point that can be denoted by I, and show first that AB and BI have the same length. For the second, recall that an angle inscribed in a circle, such as $\Gamma\Delta B$, is one-half of the central angle subtended by the same chord.

[20] This new figure is unnecessary, for the previous one shows such chords, and it is unfortunate that Ptolemy decided to change the notation on us for the new figure.

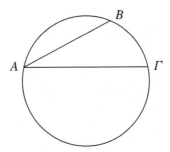

Since $BA = $ crd $\frac{3}{4}^{\circ}$ was shown just before Corollary 3 to be 0^P 47 8, it follows that four-thirds of this value is 1^P 2 50, and then

$$\text{crd } 1^{\circ} = A\Gamma < 1^P\, 2\ 50.$$

Next, use the same chords but assume now that AB subtends an angle of 1° and $A\Gamma$ an angle of $1\frac{1}{2}^{\circ}$. As above, we have

$$BA > A\Gamma \, \frac{\text{arc } BA}{\text{arc } A\Gamma} = \frac{2}{3} A\Gamma.$$

Since we have already found that $A\Gamma = \text{crd } 1\frac{1}{2}^{\circ} = 1^P$ 34 15 and two-thirds of this value is 1^P 2 50, we have

$$\text{crd } 1^{\circ} = BA > 1^P\, 2\ 50.$$

Since crd 1° has been shown to be both smaller and larger than 1^P 2 50, it must "have approximately this identical value 1^P 2 50."[21] Using now Corollary 2 (or, rather, the formula developed after it for the chord of half an arc), we find that

$$\text{crd } \tfrac{1}{2}^{\circ} = 0^P\, 31\ 25.$$

Ptolemy concluded as follows [p. 443]:

> The remaining intervals may be computed, as we said, by means of the chord subtending $1\frac{1}{2}^{\circ}$. In the case of the first interval, for example, by adding $\frac{1}{2}^{\circ}$ we obtain the chord subtending 2°, and from the difference between this and 3° we obtain the chord subtending $2\frac{1}{2}^{\circ}$, and so on for the remainder.

---

[21] Ptolemy's values for crd $1\frac{1}{2}^{\circ}$ and crd $\frac{3}{4}^{\circ}$ are only approximations. More precise calculations would show that

$$1^P\, 2\ 49\tfrac{4}{5} < \text{crd } 1^{\circ} < 1^P\, 2\ 50,$$

validating Ptolemy's assertion.

Then he or his paid calculators evaluated the entries in the table, which made up Article 11. We show in the next table the first six lines as a sample of this arrangement.

Arcs	Chords			Sixtieths			
$\frac{1}{2}^\circ$	0	31	25	0	1	2	50
1°	1	2	50	0	1	2	50
$1\frac{1}{2}^\circ$	1	34	15	0	1	2	50
2°	2	5	40	0	1	2	50
$2\frac{1}{2}^\circ$	2	37	4	0	1	2	48
3°	3	8	28	0	1	2	48

This is Ptolemy's partial description of the table [p. 443]:

> The first section [one column] will contain the magnitude of the arcs increasing by half degrees, the second will contain the lengths of the chords subtending the arcs measured in parts of which the diameter contains 120, and the third will give the thirtieth part of the increase in the chords for each half degree, in order that for every sixtieth part of a degree we may have a mean of approximation differing imperceptibly from the true figure and so be able to readily calculate the lengths corresponding to the fractions between the half degrees.

While the first part of this statement is clear, the second may benefit from an example. There are thirty intervals of $1'$ from $\frac{1}{2}^\circ$ to 1°, and if the difference $1^p 2\ 50 - 0^p 31\ 25 = 0^p 31\ 25$ is divided by 30 we obtain $0^p 1\ 2\ 50$. Thus, we can take

$$\text{crd } \frac{1}{2}^\circ 1' \approx 0^p 31\ 25 + 0^p 1\ 2\ 50 = 0^p 32\ 27\ 50.$$

Within an interval as small as $\frac{1}{2}^\circ$, this provides a good approximation.

With this we conclude our description of only Articles 10 and 11 of Book I of the *Almagest*. The complete work is much larger and, although much of it is a compilation of previous knowledge, the last five books on planetary motion represent Ptolemy's most original contribution. This part of the *Almagest* has been called a masterpiece and remained the standard in astronomy until the sixteenth century.

1.3 THE INDIAN CONTRIBUTION

The Gupta empire of northern India was founded by Chandragupta I, who ruled from Pataliputra, or City of Flowers, the modern Patna on the Ganges

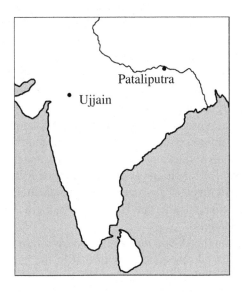

River. Near Pataliputra, but the exact location is unknown, was the town of Kusumapura that would emerge as one of the two major early centers of mathematical knowledge in fifth- and sixth-century India (the other one was at Ujjain). It is not surprising that Kusumapura became a center of learning, being near the capital of the empire and center of the trade routes. Knowledge from and to other parts of the world would naturally flow through Pataliputra in that golden age of classical Sanskrit.

An astronomer called Aryabhata may have worked in Kusumapura in the time of the emperor Budhagupta. Although the exact place of Aryabhata's birth is not known (he is generally thought to be a southerner, possibly from as far as Kerala, on the southern coast of India), the date is. He himself said that he was 23 years of age when he finished a work called the *Aryabhatiya* (Aryabhata's work) in 499. Thus, he was born in 476, and he lived until 550. The *Aryabhatiya* is the only text of Aryabhata that has survived. It is a brief volume written in verse as a mnemonic aid, since some of its portions were meant to be memorized rather than read. It consists of only 123 stanzas (of which five may have been added later), and is divided into four sections, each of them called *pada*: an introduction, a calculations section or *ganitapada* of 33 verses—containing the trigonometry we want to examine—and the rest is on astronomy.

Aryabhata was not the only Indian scholar whose work on trigonometry has survived. Several other works on astronomy, called *Siddhantas* or "established conclusions," were written in what we may call medieval or pre-medieval India. Of these, only the anonymous *Surya Siddhanta* is completely extant. The versions of both books that have survived to the present are from the sixteenth century, and, while it is believed that the copy of the *Aryabhatiya* is identical or almost identical to the original, it is known that that of the *Surya Siddhanta* has undergone many alterations.

Only the facts are presented in these texts, but this knowledge is assumed to be ancient and of divine origin (*Surya* means Sun, and this was supposed to be a book of revelation of the knowledge of the Sun), and no mention is made of either astronomical observations or mathematical deductions that may have led to it, nor is credit given to any preceding mathematicians. This makes it impossible to determine whether the Hindus were original or indebted to the Greeks or the Babylonians.

If Aryabhata and other *Siddhanta* authors were aware of the Greek work on chords, they replaced this concept with that of the half-chord, as shown in the next figure, and this has remained their most important and lasting

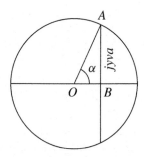

contribution. Actually, Aryabhata talked of the half-chord, *ardha-jya*, and of the chord half, using the name *jya-ardha* as was done in the *Surya Siddhanta* once, but later both documents simply used the word *jya*, also spelled *jyva*, for chord.

To construct a table of *jyvas*—if we are permitted this plural—the Hindus needed to choose a unit of length. Departing from Hellenic use, they replaced the two Ptolemaic units—one for arcs and one for parts of the diameter—by a single unit. They divided the length of the circumference into 21,600 parts and then used any one of these parts as their only unit of length. We can call these parts minutes, since there are 21,600′ minutes in a complete circumference of 360°.

How can the *jyva* be measured in this way? Very simply, and it is well to remember that we do something similar routinely today: first we select the radius as the unit of length, and then we use this length to obtain the radian measure of an arc. To do this we use our knowledge of the value of π, which is defined as the ratio of the length of the circumference, C, to the diameter. For us $\pi = C/2$, since the length of the diameter is two radians, and then $C = 2\pi$ radians. For the Hindus, $C = 21,600'$, and if R denotes the length of the radius,

$$R = \frac{21,600}{2\pi} = \frac{10,800}{\pi}.$$

In either case, one needs the value of π.

The *Surya Siddhanta* gave the approximation $\pi \approx \sqrt{10}$ in stanza 59 of Chapter I, but this is not the value used later to compute the *jyvas*. Aryabhata was a little more careful in the tenth stanza of the *ganitapada*:[22]

10. One hundred plus four, multiplied by eight, and added to sixty-two thousand: this is the nearly approximate measure of the circumference of a circle whose diameter is twenty thousand.

He did not include an explanation, but the result of Aryabhata's calculation is

$$62,000 + 8(100 + 4) = 62,832,$$

which divided by a diameter of 20,000 gives a quotient of 3.1416 as an approximation of π, the best approximation to his day. Thus, Aryabhata could find the length of the radius to be

$$R = \frac{10,800}{3.1416} \approx 3438'.$$

This is, then, the *jyva* of 90°.[23]

---

[22] Shukla and Sarma, *Aryabhatiya of Aryabhata*, 1976, p. 45. The translations from the *Aryabhatiya* included here are from this source, but I have consistently replaced their word "Rsine" with *jyva*.

[23] If Aryabhata used his value of π, he rounded up to obtain R, so his value is not precise. An electronic calculator will give the quotient $10,800/\pi$ as 3437.746771′ or 57° 17′ 44.8″, which is closer to the true radian.

From this and a few elementary mathematical facts the Hindu mathematicians were able to construct a table of *jyvas*. They are given in the *Surya Siddhanta* in stanzas 15 to 22 of the second chapter, beginning as follows (an explanation will be given after each quotation): [24]

> 15. The eighth part of the minutes of a sign is called the first half-chord (*jyârdha*); that, increased by the remainder left after subtracting from it the quotient arising from dividing it by itself, is the second half-chord.

Once more, since there are 12 signs of the zodiac in a full circle of 360°, a sign of the zodiac is equivalent to 30°, or 1800′, and the eighth part of this is $3\frac{3}{4}^{\circ}$, or 225′. This is the value of first half-chord, the *jyva* of $3\frac{3}{4}^{\circ}$ (the Hindus knew, as we do, that for small angular values the arc subtended by the angle and the half-chord are approximately equal). The quotient arising from dividing it by itself is 1, and then the remainder left after subtracting this quotient from the first *jyva* is 224. Increasing the first *jyva* by this amount, we obtain the second *jyva*: $225 + 224 = 449$.

Stanza 16 then gives the general rule to obtain an arbitrary *jyva* from the preceding ones. It is understood, although not so stated in the rule, that we are computing the *jyvas* of equally spaced angles in intervals of $3\frac{3}{4}^{\circ}$.

> 16. Thus, dividing the tabular half-chords in succession by the first, and adding to them, in each case, what is left after subtracting the quotients from the first, the result is twenty-four tabular half-chords (*jyârdhapinda*),[25] in order, as follows:

Thus, each *jyva* is obtained by adding to the preceding *jyva* what is left after subtracting from the first *jyva* the quotients obtained by dividing all preceding *jyvas* by the first.

To express the rule for obtaining *jyvas* in familiar mathematical terms rather than in narrative form, we shall coin a temporary trigonometric function, denoted by jya. Then, if we denote the angle $3\frac{3}{4}^{\circ}$ by θ, the first *jyva* is just jya θ and the nth, where n is a positive integer between 1 and 24, is jya $n\theta$.

---

[24] Burgess, "Surya-Siddhanta. A text-book of Hindu astronomy," 1860, p. 196. The translations from the *Surya-Siddhanta* included here are from this source, but I have consistently replaced the word "sine" with "half-chord" (Burgess gave transliterations of the Sanskrit terms in parentheses).

[25] According to Burgess, p. 201, *pinda* means "the quantity corresponding to."

Then the stated rule, to obtain each *jyva* by adding to the previous one "what is left after subtracting the quotients from the first," can be written as

$$\text{jya}\,n\theta = \text{jya}\,(n-1)\theta + \text{jya}\,\theta - \sum_{k=0}^{n-1} \frac{\text{jya}\,k\theta}{\text{jya}\,\theta},$$

$n = 1, \ldots, 24$. The inclusion of $\text{jya}\,0° = 0$ in the sum on the right, which was not explicitly done by the Hindu author, does not alter the sum. For instance, for $n = 3$, the right-hand side becomes

$$449 + 225 - \frac{225}{225} - \frac{449}{225} \approx 449 + 225 - 1 - 2 = 671,$$

which is jya 3θ.

Using the *jyva* rule, the *jyvas* of the arcs in the first quadrant in intervals of $3\frac{3}{4}°$ can be found to be (approximately) 225, 449, 671, ..., 3431, and 3438. These are the values stated in Sanskrit verse in stanzas 17 to 22 of the *Surya Siddhanta*, but it is easier for us to read them in table form.[26]

Arc	Jiva	Arc	Jiva	Arc	Jiva
$3\frac{3}{4}°$	225′	$33\frac{3}{4}°$	1910′	$63\frac{3}{4}°$	3084′
$7\frac{1}{2}°$	449′	$37\frac{1}{2}°$	2093′	$67\frac{1}{2}°$	3177′
$11\frac{1}{4}°$	671′	$41\frac{1}{4}°$	2267′	$71\frac{1}{4}°$	3256′
$15°$	890′	$45°$	2431′	$75°$	3321′
$18\frac{3}{4}°$	1105′	$48\frac{3}{4}°$	2585′	$78\frac{3}{4}°$	3372′
$22\frac{1}{2}°$	1315′	$52\frac{1}{2}°$	2728′	$82\frac{1}{2}°$	3409′
$26\frac{1}{4}°$	1520′	$56\frac{1}{4}°$	2859′	$86\frac{1}{4}°$	3431′
$30°$	1719′	$60°$	2978′	$90°$	3438′

Aryabhata did not express the *jyvas* directly, but rather, in the twelfth stanza

---

[26] A reproduction of the table of half-chords, from a Sanskrit edition of the *Surya Siddhanta* published at Meerut, India, in 1867, can be seen in Smith, *History of mathematics*, II, 1925, p. 625.

of his introduction, or *gitikapada*,[27] to the *Aryabhatiya* he gave a sequence of differences between the values of the *jyvas* shown here [p. 29]:

12. 225, 224, 222, 219, 215, 210, 205, 199, 191, 183, 174, 164, 154, 143, 131, 119, 106, 93, 79, 65, 51, 37, 22, 7. These are the *jyva* differences in terms of minutes of arc.

The reason for giving differences is related to the rule that he used to compute the stated values, in the twelfth stanza of the *ganitapada*:[28]

प्रथमाच्चापज्याधाद्यैरूनं खण्डितं द्वितीयार्धम् ।
तत्प्रथमज्याधाँशैस्तैस्तैरूनानि   शेषाणि   ॥ १२ ॥

which translates as follows:[29]

12. The first *jyva* divided by itself and then diminished by the quotient will give the second *jyva* difference. For computing any other difference, [the sum of] all the preceding differences is divided by the first *jyva* and the quotient is subtracted from the preceding difference. Thus all the remaining differences [can be calculated].

The interpretation of this statement is straightforward. The quotient of the first *jyva* difference, $\text{jya}\,\theta - \text{jya}\,0 = 225$, by itself is 1, which subtracted from $\text{jya}\,\theta$ gives the second *jyva* difference: $\text{jya}\,2\theta - \text{jya}\,\theta = 224$. To compute any other difference, $\text{jya}\,(n+1)\theta - \text{jya}\,n\theta, n = 1, 2, \ldots, 23$, the sum of all the preceding differences,

$$\sum_{k=1}^{n} [\,\text{jya}\,k\theta - \text{jya}\,(k-1)\theta] = \text{jya}\,n\theta,$$

---

[27] The introduction includes ten aphorisms in the *gitika* meter, called *dasagitika-sutra*, which Aryabhata had previously written as an independent tract. The stanza quoted here is the tenth stanza in the *dasagitika-sutra*, and is numbered in this manner in Clark, *The Aryabhatiya of Aryabhata*, 1930, p. 19.

[28] It shows the use of the word ज्याध = ज्या (*jya*), also spelled जीवा (*jiva*) in stanza 23 of the *Golapada*, the fourth part of the *Aryabhatiya*. The Sanskrit version reproduced here is from Shukla and Sarma, *Aryabhatiya of Aryabhata*, p. 51.

[29] From the available translations I have chosen the one given by Bibhutibhushan Datta and Avadesh Narayan Singh in *History of Hindu Mathematics*, Part III. This part remains unpublished, but this translation is reproduced in Shukla and Sarma, *Aryabhatiya of Aryabhata*, p. 52.

is divided by the first *jyva* and the quotient is subtracted from the preceding difference:

$$\text{jya}\,(n+1)\theta - \text{jya}\,n\theta = \text{jya}\,n\theta - \text{jya}\,(n-1)\theta - \frac{\text{jya}\,n\theta}{\text{jya}\,\theta}.$$

That is,

$$\text{jya}\,(n+1)\theta = 2\,\text{jya}\,n\theta - \text{jya}\,(n-1)\theta - \frac{\text{jya}\,n\theta}{\text{jya}\,\theta}.$$

Taking $n = 2$ as an example to find the *jyva* of $11\frac{1}{4}°$, we have

$$\text{jya}\,3\theta = 2\,\text{jya}\,2\theta - \text{jya}\,\theta - \frac{\text{jya}\,2\theta}{\text{jya}\,\theta} = 898 - 225 - \frac{449}{225} = 673 - 2 = 671,$$

where the quotient on the right has been rounded to the nearest whole number.

Aryabhata's rule is different from the one in the *Surya Siddhanta* only in appearance. Replacing n with $n + 1$ in the *Surya Siddhanta* rule on page 29 yields

$$\text{jya}\,(n+1)\theta = \text{jya}\,n\theta + \text{jya}\,\theta - \sum_{k=0}^{n}\frac{\text{jya}\,k\theta}{\text{jya}\,\theta},$$

$n = 1, \ldots, 23$. Subtracting now the first form from this one, we have

$$\text{jya}\,(n+1)\theta - \text{jya}\,n\theta = \text{jya}\,n\theta - \text{jya}\,(n-1)\theta - \frac{\text{jya}\,n\theta}{\text{jya}\,\theta},$$

which is Aryabhata's rule.

This rule, or its simplified form, is easier to handle than the one in the *Surya Siddhanta* because it does not have a long sum on the right. We assume that the rule is only approximate, but may reasonably ask whether the last term on the right can be replaced by another term $T(\theta)$, to be determined, such that

$$\text{jya}\,(n+1)\theta = 2\,\text{jya}\,n\theta - \text{jya}\,(n-1)\theta - T(\theta)$$

exactly. This will be a simpler task if we switch from *jyvas* to sines using the equation $\text{jya}\,\alpha = R\sin\alpha$, where α is any angle between $0°$ and $90°$ and R is the length of the radius given in minutes. Then we can rewrite the previous equation as

$$\sin(n+1)\theta = 2\sin n\theta - \sin(n-1)\theta - \frac{T(\theta)}{R}.$$

Using the formulas for the sine of a sum and a difference of two angles—already implicit in Ptolemy's work for chords, and quite possibly known to the Hindus for *jyvas*—and simplifying reduces this equation to

$$2(\cos\theta - 1)\sin n\theta = -\frac{T(\theta)}{R},$$

or

$$T(\theta) = 2R(1 - \cos\theta)\sin n\theta = 2(1 - \cos\theta)\,\text{jya}\,n\theta.$$

Using a hand-held calculator to find $\cos\theta = \cos 3.75°$, we obtain

$$T(\theta) = 0.004282153\,\text{jya}\,n\theta,$$

and when this is compared with the term

$$\frac{\text{jya}\,n\theta}{\text{jya}\,\theta} = \frac{\text{jya}\,n\theta}{225} = 0.004444444\,\text{jya}\,n\theta,$$

it follows that

$$\frac{\text{jya}\,n\theta}{\text{jya}\,\theta} = 1.037899497\,T(\theta).$$

Thus we have found the exact formula for the computation of *jyva* differences and have shown to what extent the Hindu formula is an approximation.

The Hindu astronomers of the fifth century could not have given an exact formula for $T(\theta)$ because the cosine was unknown to them. However, there is another trigonometric length that the author of the *Surya Siddhanta* could have used: the versed half-chord. It is defined, with the name *utkramajya* (reverse-order *jyva*), in Chapter II, in the second part of stanza 22, as follows [p. 196]:

> 22. ... Subtracting these [the *jyvas*], in reverse order from the half-diameter, gives the tabular versed half-chords (*utkramajyârdhapindaka*).

To understand this definition, note that if $\alpha = n\theta$ for some n between 0 and 24, then to subtract its *jyva* from the radius in reverse order means to subtract the *jyva* of $(24 - n)\theta = 90° - \alpha$. For instance, if R is the length of the radius in minutes, the *utkramajya* of $\theta = 3\frac{3}{4}°$ is

$$R - \text{jya}\left(90° - 3\tfrac{3}{4}°\right) = 3438 - 3431 = 7.$$

The values of the *utkramajyas* are then given in stanzas 23 to 27 of the *Surya Siddhanta* from 7 for $\alpha = \theta$ to 3438 for $\alpha = 90°$.

In general, if we denote the *utkramajya* of α by uya α, then

$$\text{uya}\,\alpha = R - \text{jya}\,(90° - \alpha) = R - R\sin(90° - \alpha) = R(1 - \cos\alpha),$$

and, in particular,

$$T(\theta) = \frac{2\,\text{jya}\,n\theta\,\text{uya}\,\theta}{R}.$$

Thus, the author of the *Surya-Siddhanta* missed an opportunity to give the following exact equation for the computation of *jyvas*:

$$\text{jya}\,(n+1)\theta = 2\,\text{jya}\,n\theta - \text{jya}\,(n-1)\theta - \frac{2\,\text{jya}\,n\theta\,\text{uya}\,\theta}{R},$$

$n = 1, \ldots, 23$, instead of an approximation. Of course this formula would not give correct values using the approximations jya $\theta \approx 225$ and uya $\theta \approx 7$ instead of the closer values 224.8393963 and 7.360479721, which present-day technology can produce in an instant.

The origin of the *jyva*-producing rule has also been a subject for speculation. Of course, the answer is not known because the astronomers who used it did not include a derivation or an explanation of its origin. This is all Aryabhata had to say about the computation of *jyvas* in the *ganitapada* [p. 45]:

> 11. Divide a quadrant of the circumference of a circle (into as many parts as desired). Then, from (right) triangles and quadrilaterals, one can find as many *jyvas* of equal arcs as one likes, for any given radius.

He did not, however, explain how it is done. Several possible explanations have been proposed since the nineteenth century, but the truth is unknown.[30]

More than a century after Aryabhata, Brahmagupta (598–670), working at Ujjain, incorporated trigonometry in his work *Brahmasphuta Siddhanta*.[31] After changing Aryabhata's value of the radius from 3438 to 3270, he included some interpolation procedures to find the *jyvas* of arcs that are closer together than 225′. His work may have had a profound influence on the development of trigonometry, since it was later studied by astronomers outside of India. Another mathematical astronomer, called Bhaskara (c. 600–c. 680), gave an algebraic formula, in a work called *Mahabhaskariya*, to compute an approximation to the *jyva* of an arbitrary angle that does not rely on the 225′ differences [pp. 207–208 of the included translation]:

---

[30] The explanation proposed by the French astronomer Delambre (1749–1822) in *Histoire de l'astronomie ancienne*, I, 1815, pp. 457–458 deserves consideration.

[31] The word *sphuta* means "corrected." Thus, this is the corrected *Brahma Siddhanta*, an earlier work, now lost. See Burgess, "Surya-Siddhanta. A text-book of Hindu astronomy," pp. 419 and 421.

Subtract the degrees of the *bhuja* [arc] from the degrees of the half-circle. Then multiply the remainder by the degrees of the *bhuja* and put down the result at two places. At one place subtract the result from 40500. By one-fourth of the remainder [thus obtained] divide the result at the other place as multiplied by the *antyaphala* [radius]. Thus is obtained the [*jyva* to that radius].

In other words, if the angle in question is denoted by θ, then

$$\mathrm{jya}\,\theta = \frac{R\theta(180 - \theta)}{\frac{1}{4}[40{,}500 - \theta(180 - \theta)]}.$$

Bhaskara did not bother to tell us how he obtained this formula, but it provides a good approximation.

1.4 TRIGONOMETRY IN THE ISLAMIC WORLD

The Hellenic civilization of Egypt slowly declined through the centuries, the Roman Empire itself was gone with the wind of the barbarian invasions, plunging western Europe in the dark ages, and only the Byzantine Empire in the east managed to hold on, although without much luster. But the void would be filled soon, after Mohammed ibn Abdallah (c. 570–632) founded in 612 a new religion, Islam (meaning "submission" to Allah). Its adherents, called Muslims, set out to the conquest of infidel lands shortly after the death of Mohammed. It was thus that a new empire was born, and by 642 it extended from northern Africa to Persia.

This success notwithstanding, there was disagreement over the matter of Mohammed's successor or caliph (from *khalaf*, to succeed) from the very first moment after his death. Mohammed's cousin and son-in-law, Ali ibn Abu Talib, thought he was the intended successor, but a group of Muslim leaders chose Mohammed's father-in-law to become the first caliph. Eventually, Ali managed to become the fourth caliph in 656, on the assassination of Uthman ibn Affan, the third caliph, by rebel factions. In turn, members of the Umayyad family, to which Uthman belonged, ousted Ali—who would end up assassinated by some of his ex-followers—and proclaimed one of their own, Muawiya ibn Abu Sufyan, as caliph in 661.[32]

---

[32] The followers of Ali, or *Shi'at ul-Ali*, became known later as Shi'ites. Those who believed that the prophet had left the choice of successors entirely to the people were known as Sunnis.

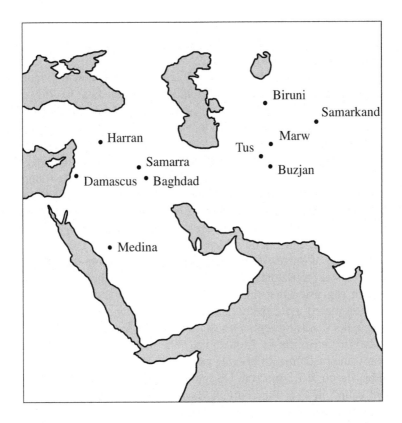

The Umayyads, who turned the caliphate into a dynasty and moved their capital from Medina (more fully, al-Madinah an Nabi, meaning "the City of the Prophet") to Damascus, greatly expanded the Islamic empire, building an efficient government structure in the process, with a large number of Christian Byzantine administrators. By the time the Muslims were checked at Poitiers, near Tours, by Charles, natural son of Pepin of Herstal, and his Franks in 732—Charles would be called *Martel* (the Hammer) after this victory—they dominated a large portion of the world, from Hispania in the west to the Indus river in the east. The court structure that they built may have clashed with the simplicity of early Islam, but it was a source of culture in the arts and sciences.

Arts and sciences notwithstanding, the Umayyads were not popular in some quarters, definitely not in the mind of those who thought they had betrayed the spirit of Islam. A revolt led by Abu al-Abbas (722–754), a descendant of a paternal uncle of Mohammed, ended the Umayyad dynasty in Syria with the massacre of most of the Umayyads on June 25, 750. This earned Abu al-Abbas the name extension *al-Saffah* (the Blood-shedder).

Abu Jafar Abdullah ibn Mohammed al-Mansur (the Victorious) (712–775), the second caliph of the Abbasid dynasty, as it would be known, founded the city of Baghdad (or "God-given" in Middle Persian) on July 30, 762, on the recommendation of the court astrologer, and transferred the capital there. This was a city built in the shape of a circle, surrounded by a moat, and with the mosque in the center. Here Al-Mansur established a program of translation from foreign languages, with a large mathematical and astronomical component, that lasted for well over two hundred years and was the foundation of much scholarly work in the caliphate.[33] The historian Abu'l Hasan Ali ibn Husayn ibn Ali al-Masudi (c. 895–957) stated that al-Mansur sponsored Arabic translations of Ptolemy's *Almagest*, the *Arithmetike eisagoge* of Nicomachos of Gerasa, and Euclid's *Elements* (not extant).[34] There was also a work referred to as the *Sindhind* that had been brought from the land of Sind (modern Pakistan) to Baghdad about 766 by an Indian scholar named Kanka. It has been suggested that this work may have been the *Surya Siddhanta*, but it is considered more likely that it was the *Brahmasphuta Siddhanta* of Brahmagupta. It was through a translation of this document, commissioned by al-Mansur and made by Mohammed ibn Ibrahim al-Fazari, that Indian astronomy and mathematics became known in the Islamic empire.

Under Harun al-Rashid (the Upright) (764–809), who became the fifth Abbasid caliph in 786, the court of Baghdad reached a peak of splendor, and he himself is known as the main character in the *The thousand and one nights*. He commissioned a translation of Euclid's *Elements*, prepared by al-Hajjaj ibn Yusuf ibn Matar, a manuscript copy of which still exists. After a disastrous civil war, it was with Harun's son Abdullah al-Mamun ibn Harun abu Jafar (786–833), the seventh of the Abbasid caliphs, who welcomed all kind of scholars to his court, that intellectual life flourished. In mathematics,

---

[33] Al-Mansur initiated this activity for political reasons. When Alexander the Great conquered Persia and killed king Darius III, he had all the documents in the archives of Istahr (Persepolis) translated into Greek and Coptic, and then destroyed the originals. The neo-Persian empire of the Sasanians (225–651) then attempted to recover their cultural heritage through a program of translations from the Greek into Pahlavi (Middle Persian). Al-Mansur's decision to continue the translation program was part of his policy to adopt and incorporate the ideology of the large Persian component of the population in this part of the Islamic empire. For a complete account of the translation movement see Gutas, *Greek thought, Arabic culture*, 1998.

[34] In his book *Muruj al-zahab wa al-maadin al-jawahir* (Meadows of gold and mines of gems), 947, §3458

not only were additional translations of the *Elements* and the *Almagest* made
during Al-Mamum's reign, but also significant original work was produced,
such as the treatise *Kitab al-jabr wa'l-muqabala* (Treatise of restoring and
balancing), dedicated to al-Mamun by Mohammed ibn Musa al-Khwarizmi
(c. 780–850), the greatest mathematician of medieval Islam.[35] It was on the
translation of Kanka's *Siddhanta* that al-Khwarizmi based the construction of
his astronomical tables.

As for trigonometry, it was still subservient to astronomy in the Islamic
world, mainly through the elaboration of tables of half-chords and other
lengths. For this is one of the main contributions in Arabic to the development
of trigonometry: the introduction of all six trigonometric lengths (they were
still lengths and not functions as they are today). The Hindu *jyva* was adopted
over the Greek chord, but since there is no *v* sound in Arabic it became *jyba*
or simply *jyb* (جيب). The cosine of an arc smaller than 90° was known and
used as the *jyba* of the complement of the arc. In this way it was used by
the famous astronomer Mohammed ibn Jabir al-Battani al-Harrani (c. 858–
929)—working in Samarra, now in modern Iraq—in determining the altitude
of the sun. It appears, together with a table of its values, in his book *On the
motion of the stars*, written about 920. What we now know as the tangent
and cotangent appeared first as shadows of the gnomon, or *miqyas* (مقياس) in
Arabic, as shown in the next figure.

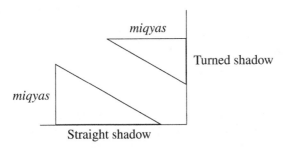

A table of sines and tangents—to use the modern terms—was compiled by
Ahmad ibn Abdallah al-Marwazi al-Baghdadi (c. 770–870), a native of Marw,
in present-day Turkmenistan. Known as Habash al-Hasib (the Calculator),

---

[35] It should be mentioned, in passing, that the title of this book was the origin of the word
algebra through its Latin "translation" *Liber algebræ et almucabola*. A work on arithmetic
was translated as *Liber Algoritmi de numero Indorum* (Al-Khwarizmi's book on Hindu
numbering), and it was thus that the name of the region Khwarazm, around the present-day
town of Biruni, in Uzbekistan, originated the word algorithm.

working in the court of al-Mamun and that of the next caliph Abu Ishaq al-Mu'tasim ibn Harun, he may have computed the first table of tangents ever. His tables were done in intervals of $1°$ and are accurate to three sexagesimal places (parts, minutes, and seconds) as opposed to Ptolemy's, which are accurate only to the minutes. Be all this as it may, al-Hasib did not contribute to trigonometry in the usual sense.

The work of Mohammed ibn Mohammed abu'l-Wafa al-Buzjani (940–998)[36] had a greater impact on trigonometry. In 945 Baghdad was captured by Ahmad Buyeh, of north Persian descent, and the caliph remained a figurehead. Four years later his son Adud al-Dawlah became caliph and founded the Buyid dynasty. He was a great supporter of mathematics, science, and the arts, and attracted Abu'l-Wafa to Baghdad, where he wrote several texts in mathematics, in 959. In 983 Adud al-Dawlah was succeeded by his son Sharaf al-Dawlah, who continued to support the sciences, and Abu'l-Wafa remained in Baghdad.

In addition to elaborating trigonometric tables in smaller steps and with greater accuracy than his predecessors, he started a systematic presentation of the theorems and proofs of trigonometry as a separate discipline in mathematics. In Chapter 5 of his *Almjsty* (المجسطى), which is not a translation of Ptolemy's *Almagest* but a separate work probably written after 987,[37] he considered only the chord, the *jyba*, and the *jyba* of the complement, giving at once formulas for the *jyba* of a double arc and half an arc. In Chapter 6 he defined the tangent or shadow in this manner:[38]

> The shadow of an arc is the line [segment] drawn from the end of this arc parallel to the *jyba*, in the interval included between this end of the arc and a line drawn from the center of the circle through the other end of the same arc.

---

[36] This may be a good place to explain Arabic names. They usually start with a given name, such as Mohammed, followed by *ibn* (son of) and the name of the father. The *i* in *ibn* is silent, so that this word sounds more like *bin*, which is another frequent spelling. Further names of the grandfather and great-grandfather may be added next. Usually, the name ends with the town or region of origin, such as al-Buzjani, which means "the one from Buzjan," located in present-day Iran. Sometimes there is a nickname at the end, such as al-Rashid (pronounced ar-Rashid because the *l* in *al* tends to acquire the sound of the next consonant). In some cases, the person being named had a famous son, whose name is added following the word *abu* (father of). Fate determines how a particular person should be known to posterity, and Mohammed ibn Mohammed ibn Yahya ibn Ismail ibn al-Abbas abu'l-Wafa al-Buzjani is usually known as Abu'l-Wafa.

[37] For this date see Woepcke, "Recherches sur l'histoire des sciences mathématiques chez les orientaux, d'après des traités inédits arabes et persans. Deuxième article," 1855, p. 256.

[38] The definitions quoted here are from Delambre, *Histoire de l'astronomie du moyen age*, 1919, p. 157.

Abu'l-Wafa's statement is represented in the figure. He called this trigono-

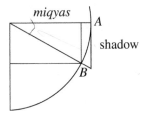

metric line the first shadow or turned shadow. Then the secant, or

> diameter of the shadow, is the line drawn from the center to the end of the shadow. The straight shadow [cotangent] is the shadow of the complement of the arc.

Next Abu'l-Wafa stated, in verbal form, four equations relating these and the former trigonometric lines, the first of which, if R denotes the radius, can be written as

$$\frac{\text{shadow}}{R} = \frac{\text{jyb } \overset{\frown}{AB}}{\text{jyb}\,(90° - \overset{\frown}{AB})}.$$

This is clear from the similarity of the triangles shown in the previous figure. Then Abu'l-Wafa chose $R = 1$ for the following reason [p. 420]:[39]

> Thus it is evident that, if we take unity for the radius, the ratio of the *jyba* of an arc to the *jyba* of its complement is the shadow, and the ratio of the *jyba* of the complement to the *jyba* of the arc is the level shadow.

This choice for the radius was a stroke of modernity, which he applied in the construction of his sine and tangent tables, but later writers did not adopt it.

    As an example of Abu'l-Wafa's style in trigonometry, we return to Chapter 5 of the *Almjsty* and present his proof of the formulas now written as

$$\sin(\alpha \pm \theta) = \sin\alpha\cos\theta \pm \cos\alpha\sin\theta.$$

    He provided two calculations to obtain the sine of a sum or difference, and the first one is as follows [pp. 416–417]:

---

[39] The remaining quotations from Abu'l-Wafa included here are English translations from a French translation of the relevant passages on trigonometry in Carra de Vaux, "L'Almageste d'Abû'lwéfa Albûzdjâni," 1892, pp. 416–420. I have changed the words "sine" and "cosine" to *jyba* and *jyba* of the complement, and have mostly used "segment" as a translation of *ligne*.

Consider the two arcs AB, BC of a circle $ABCD$ (fig. 1) [these arcs are represented below in two possible arrangements]. The *jybas* of each of them

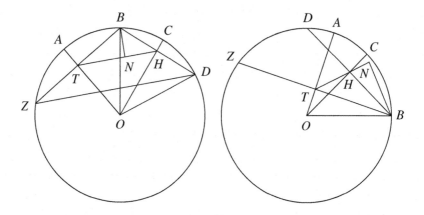

are known. I say that the *jyba* of their sum and that of their difference are also known [can be found]. Join the three points A, B, C to the center O; from the point B drop to the radii OA, OC the perpendiculars BT, BH, and join HT. I say that [in the left figure] HT is equal to the *jyba* of AC. In fact, prolong BH, BT down to D and Z and join DZ; HT will be equal to half of DZ, because the two segments BD, BZ are respectively divided into two equal parts at the points H and T. Therefore the arc DBZ is double the arc ABC, and the segment HT is equal to the *jyba* of the arc CA.

To restate this in current notation we use the central angles $\alpha = \angle AOB$ and $\theta = \angle BOC$ rather than the arcs AB and BC that Abu'l-Wafa used. Then, by the bisecting nature of OA and OC,

$$\angle ZOD = \angle ZOB + \angle BOD = 2\alpha + 2\theta$$

in the left figure, and

$$HT = \tfrac{1}{2}DZ = \tfrac{1}{2}\text{crd } ZOD = \text{jyb}\,(\alpha + \theta) = \sin(\alpha + \theta).$$

The last equation is a consequence of the choice $R = 1$. Then, in order to deal with the sine of a difference, Abu'l-Wafa continued as follows [pp. 417–418]:

In a second figure (fig. 2) [the one on the right], the arc BDZ is double the arc BCA, and the arc BCD, double the arc BC; as a difference the arc DZ is double the arc AC. From the point B drop a perpendicular BN on the line HT. The two angles BHO, BTO being right angles and built on the segment BO, the quadrilateral $BOTH$ can be inscribed in a circle [see Proposition 31 of Book III of Euclid's *Elements*]. The two triangles BOH, BTN, are similar since the angles at H and N are right angles, and the angles at O and T are equal, because these two angles are built on the segment BH [a chord in the circle $BOTH$]. Therefore the other two angles are equal to each other, and we have:

$$\frac{NB}{BT} = \frac{HB}{BO};$$

but the lengths BT, BO, BH are known; thus we can infer BN, and the angle at N being right, we can obtain the segments NH, NT, and finally HT, which we wanted to prove.

This proves, indeed, that HT can be obtained from the given *jybas*: $BT = $ jyb α and $BH = $ jyb θ, and the proof stated here is just as valid for the left figure. But in this case,

$$\angle ZOD = \angle ZOB - \angle BOD = 2\alpha - 2\theta,$$

and therefore

$$HT = \tfrac{1}{2}DZ = \tfrac{1}{2}\mathrm{crd}\, ZOD = \mathrm{jyb}\,(\alpha - \theta) = \sin(\alpha - \theta).$$

This concludes the calculation of HT, that is, the calculation of the sine of a sum or difference, but it does not give the promised and expected familiar formulas. However, Abu'l-Wafa would return to this topic, expressing himself in the following manner [p. 418]:

Let us return to the two figures that we have drawn. The two triangles HNB, BOT are similar, because the angles BNH, BTO are right and the angles at H and O are built on the segment BT [in the right figure it is the angles BHT, BOT that are built on BT, and since they are inscribed in the circle $BOTH$ on opposite sides of the same chord they are supplementary]. Therefore we have:

$$\frac{BH}{HN} = \frac{BO}{OT};$$

but the lengths BH, BO, OT are known; thus HN will be known. Furthermore, the two triangles BNT, OBH are also similar, because the two angles at H and N are right and the two angles at O and T are built on the segment BH. Then we have:

$$\frac{BT}{TN} = \frac{BO}{OH};$$

but the lengths BT, BO and OH are known; thus TN will be known. Hence the segment HT will be known which is, as we have already seen, the *jyba* of the arc AC.

From these equations, and recalling that $BO = 1$, we immediately obtain

$$HN = OT \cdot BH = \cos\alpha \sin\theta,$$
$$TN = BT \cdot OH = \sin\alpha \cos\theta.$$

Then, in the left figure,

$$\sin(\alpha + \theta) = TH = TN + NH = \sin\alpha \cos\theta + \cos\alpha \sin\theta,$$

and in the right figure

$$\sin(\alpha - \theta) = TH = TN - NH = \sin\alpha \cos\theta - \cos\alpha \sin\theta.$$

These are the well-known formulas in today's notation. Abu'l-Wafa stated them in narrative form as follows:[40]

حساب جيب بمجموع القوسين وجيب تفاضلهما اذا كانت
كل واحد منهما معلوما إذا أردنا ذلك ضربنا جيب كل واحد
منهما ڡ جيب تمام الاخر دقايق ڡا حصل جمعناها ان اردنا
جيب بمجموع القوسين واخذنا تفاضلهما ان اردنا جيب
تفاضلهما

which translates as

> Calculation of the *jyba* of the sum of two arcs and the *jyba* of their difference, when each of them is known. We multiply the *jyba* of each of these arcs by the *jyba* of the complement of the other, expressed in sexagesimal minutes, and we add the two products if we want to know the *jyba* of the sum; we subtract them if we seek the *jyba* of the difference.

We saw before that the first of the last two identities can be obtained from an interpretation of Corollary 1 of Ptolemy's theorem. Abu'l-Wafa's proof, by contrast, gives the two formulas in one single move, and, because of his

---

[40] Reproduced from Carra de Vaux, "L'Almageste d'Abû'lwéfa Albûzdjâni," p. 419.

choice of the radius, they are given in modern terms at once.[41]

The historian and general scholar Muhammad ibn Ahmad abu'l-Rayhan al-Biruni (973–1048), from Khwarazm, in today's Uzbekistan, wrote an *Exhaustive treatise on shadows* about 1021, in which he gave several relationships among the various trigonometric lengths. But a comprehensive treatment of trigonometry in the Islamic world would appear only in the thirteenth century. While, if we were to talk in cinematographic terms, mathematics developed in slow motion—very slow motion in the tenth century—empires and dynasties evolve, in comparison, like the life of a flower in those films in which its birth, bloom, and death are viewed in the span of one minute, as if speed had been invented for the sole purpose of making mincemeat of history.

Mincemeat is what Baghdad became in 1258 after being conquered by the Mongol leader Hulagu (or Hulegu), for he did not believe in mercy toward the defeated. He was the grandson of Temujin, also called *Genghis Khan* (possibly from the Chinese *cheng-ji*, "successful-lucky," and the Turkish *khan*, "lord"), who had managed to unite various Mongol tribes and began a large campaign of conquests in 1206. Hulagu expanded the Mongol empire to the west, and on his way to Ain Julat, near Nazareth, where he was defeated for the first time, he accepted the surrender of the *Hashashin* of Alamut in 1256. This was a fortress in the mountains of central Iran (*aluh amut* means "eagle's nest") where this order of Nizari Ismailis had its center of operations, sending killers wherever they were deemed necessary to eliminte enemies and creating terror and unrest in the Islamic world.[42]

That this event was important in the development of trigonometry is due to the fact that Mohammed ibn Mohammed ibn al-Hasan al-Tusi (1201–1274), a gifted astronomer and mathematician and a native of Tus—a town right next to the location of modern Meshed in Iran—was at the Alamut fortress at that time. Upon its destruction by Hulagu, he joined his forces and was with them when they sacked Baghdad.

Hulagu was interested in science and made al-Tusi his advisor in such

---

[41] However, Abu'l-Wafa's proof could have been shortened by using Ptolemy's theorem: Applying it to the quadrilateral $BOTH$ in the left figure gives

$$TH \cdot BO = TO \cdot BH + BT \cdot HO,$$

and in the right figure

$$BT \cdot HO = TO \cdot BH + BO \cdot TH.$$

With $BO = 1$ and the appropriate values of TH this gives the well-known formulas.

[42] *Hashashin* is a word of uncertain origin, about which much has been written without proof, from which the modern word "assassin" derives. For a history of this sect see Hodgson, *The secret order of assassins*, 1955. For the word assassin, pp. 133–137.

MOHAMMED AL-TUSI

matters and also put him in charge of religious affairs.[43] In 1262 he completed the construction of an observatory in his new capital, Maraga, near Azerbaijan, for which Mohammed al-Tusi designed several instruments and where he wrote tables and other works. One of these was the *Kitab shakl al-qatta* (Treatise on the transversal figure), in five books,[44] in which he gave the first systematic presentation of trigonometry as independent of astronomy. In this treatise he stated and proved the law of sines for plane triangles and used it to solve triangles (this is the name for the process of determining the unknown angles and sides of a triangle when some are given). The law of sines was not entirely new, but not until the sixteenth century was originality one-tenth as valued as it is today. Compilation was more the name of the game, so that others could have a handy reference when applying a science to the practical world. The law of sines, in particular, was already implied in Ptolemy's work, was proved by Abu'l-Wafa for spherical triangles, and was also stated and proved by al-Biruni. In fact, al-Tusi used and borrowed from al-Biruni's

---

[43] It is possible that Mohammed al-Tusi's connection with religious affairs is the reason for the name Nasir al-Din (frequently spelled Nasir Eddin), by which he is very often known. It means "Defender of the Faith."

[44] كتاب شكل القطاع, an Arabic translation of a book that al-Tusi had originally written in Persian a few years before. Page references below are to the French translation by Carathéodory, *Traité du quadrilatère attribué a Nassiruddin-el Toussy*, 1891.

treatise *The key to the knowledge of spherical figures and other figures* while
writing his own treatise.

Mohammed al-Tusi ignored Abu'l-Wafa's choice of a unit radius [45] and
returned to $R = 60$. Then in Chapter 2 of Book III, entitled *On the way to
calculate the sides and the angles of a triangle the ones from the others* [p. 66],
he proposed two ways to do just that, the method of "arcs and chords" and the
method of "arcs and *jybas*." He explained first the method of arcs and chords,
starting with a right triangle and considering three subcases. Then he turned
his attention to arbitrary triangles, dividing his study into four subcases, of
which we shall need the last two [p. 69]: [46]

> III.– Two sides and one angle are given. ... That if the given angle
> is between the two given sides, as the angle A is between the two sides

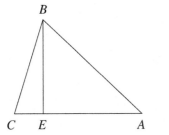 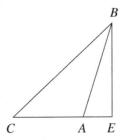

> AB AC, lower from B to AC the perpendicular BE. You will thus have the
> right triangle AEB [47] of which we know the side AB and the angle A; we
> get BE, EA, and thus we fall back into one of the preceding cases; that is in
> the case in which BE, CE are known; then from these we shall know BC
> and the angle C, as we have explained.

In other words, $\operatorname{crd} 2A$ is obtained from a table of chords, BE from the
equation

$$\frac{BE}{BA} = \frac{\frac{1}{2}\operatorname{crd} 2A}{60},$$

---

[45] Al-Tusi stated that Al-Biruni was the first to choose unity as the radius [p. 212].

[46] Since I am translating from a translation, I will do it as literally as possible to minimize the divergence from the original text. However, I have not been able to avoid some
punctuation changes.

[47] The French translation says BEC, but the Arabic version has the correct triangle $\left(\overline{\text{اعب}}\right)$
on page ٥٣ (53). However, the figure for the second triangle is mislabeled in Arabic.

EA from the Pythagorean theorem, $CE = CA \mp EA$, BC is also obtained from the Pythagorean theorem, and crd $2C$ is found from the equation

$$\frac{BE}{BC} = \frac{\frac{1}{2}\text{crd}\,2C}{60}.$$

Then $2C$ is found from a table of chords, and C is one-half of this value.

IV.– The three sides of the triangle ABC are given.[48] We would calculate

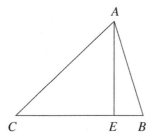 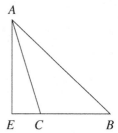

the perpendicular according to the common rule; taking the excess of the two squares of BA, BC over the square of AC, which we shall divide by the double of BC; the quotient will be BE,[49] and then the root of the excess of the square of AB over the square of BE will give the perpendicular. Thus we shall have two right triangles whose angles can be determined, by means of which those of the triangle ABC can then be determined.

This concludes al-Tusi's presentation of the method of arcs and chords, and then, switching to the method of arcs and *jybas*, he stated the following [p. 70]:

---

[48] Note that the labeling of the vertices in the new figure for this case is different from the labeling in Case III.

[49] No proof of this fact is provided because this is "common" knowledge. However, for those without that knowledge it can be very simply proved if we denote the sides opposite the angles A, B, and C by a, b, and c, respectively. Then, if the perpendicular AE is denoted by h and if we put $x = BE$, we have $c^2 = x^2 + h^2$ and

$$b^2 = (a-x)^2 + h^2 = a^2 - 2ax + x^2 + h^2 = a^2 - 2ax + c^2.$$

Solving for x gives

$$x = \frac{a^2 + c^2 - b^2}{2a},$$

which is al-Tusi's statement.

As for that of the arcs and the *jybas*, the fundamental notion is that the ratio of the sides is equal to the ratio of the *jybas* of the angles opposite those sides.

Next he restated it in a more formal manner, which we now call the law of sines.

Let there be a triangle ABC, I say that the segment AB over the segment AC equals [the *jyba* of] *the angle ACB over* [the *jyba* of] *the angle ABC.*[50]

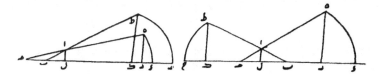

Al-Tusi's law of sines and his first proof.
From page 54 (٥٤) of the Arabic translation of the Persian original,
bound back to back with the French translation by Carathéodory:
Traité du quadrilatère attribué a Nassiruddin-el Toussy.

To prove it he considered three cases [p. 70], in which the angle at *B* (ﺏ) is obtuse, right, or acute. The first and the third cases are illustrated in the next figure, and, following al-Tusi's lead, we shall leave the case of the right angle at *B* to the reader.

---

[50] Strangely enough, al-Tusi did not use the word *jyba* (ﺟﻴﺐ) in this statement, although it appears prominently in the preliminary statement and in the proof.

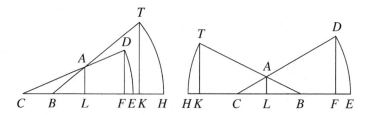

Demonstration.[51] Prolong CB to $CE = 60$ [this may involve a contraction rather than a prolongation, but the argument is the same]. From C with a radius equal to CE describe ED and prolong CA until it meets this arc in D. From D lower the perpendicular DF over CE, that will be the *jyba* of the angle ACB. Prolong BC equally to $BH = 60$. From B with a radius $= BH$ describe HT which will be cut at point T, by the prolongation of AB. Lower the perpendicular TK, this will be the *jyba* of the angle ABC. From A lower AL perpendicular over BC, and because of the similarity of the triangles ABL, TBK you will have:

$$\frac{AB}{AL} = \frac{TB\ (\text{radius})}{TK};$$

in the same manner because of the similarity of the triangles ALC, DFC,

$$\frac{AL}{AC} = \frac{DF}{DC\ (\text{radius})};$$

from which by the established proportion [multiplying the two preceding equations]:

$$\frac{AB}{AC} = \frac{DF\,(\text{jyb}\,ACB)}{TK\,(\text{jyb}\,ABC)},$$

Q.E.D.[52]

Next al-Tusi gave a second proof, presumably his own, that is shorter than the first. It is based on the next two figures, which represent the two cases in which the angle at B of the basic triangle ABC is either obtuse or acute [p. 71].

---

[51] This is al-Biruni's original proof, which can be seen translated into English and in current notation in Zeller, *The development of trigonometry from Regiomontanus to Pitiscus*, 1946, p. 9. A different translation of this proof and the preceding statements of the law of sines can be found in Berggren, "Mathematics in medieval Islam," in *The mathematics of Egypt, Mesopotamia, China, India, and Islam*, 2007, p. 642.

[52] I have used, and will continue using, modern fractions rather than the symbol : which the French translation uses for division. The stated equations are just part of the narrative in both the French and the Arabic versions.

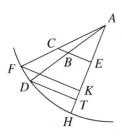 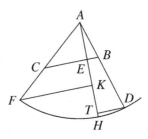

Lower AE perpendicular to BC; prolong AB, AC just so that $AF = AD = 60 = R$. Describe the arc DH. Bring DT, FK perpendicular onto AH. In the triangle ABE the angle E being a right angle, B will be the complement of A; DT is the *jyba* of the angle A; TA is the *jyba* of the angle B. Similarly in the triangle AEC, the angle C being the complement of the angle A,[53] KF is the *jyba* of the angle A; KA is the *jyba* of the angle C. And because of the similarity of the two triangles ABE, ADT;

$$\frac{AB}{AE} = \frac{AD \text{ (radius)}}{AT \text{ (jyb } B)};$$

in the same manner, due to the similarity of the two triangles AEC, AKF,

$$\frac{AE}{AC} = \frac{AK \text{ (jyb } C)}{AF \text{ (radius)}},$$

which gives us [multiplying the last two equations]

$$\frac{AB}{AC} = \frac{KA(\text{jyb } C)}{AT(\text{jyb } B)},$$

Q.E.D.

Then al-Tusi used this theorem to solve triangles, starting with the case of right triangles [pp. 71–72].

Thus if we are dealing with a right triangle, if we know its sides, by the method of the *jybas* [the law of sines], the hypotenuse will be to one of the sides in the ratio of the radius [= jyb 90°] to the *jyba* of the angle opposite that side, and through the *jyba* one finds the angle. And if that which is given is one angle and one side we shall know from those the angles of the right

---

[53] The French translation reads $A + C = $ *arc de demi-circonférence.*

و بوجه اخر يخرج عمود اء على ضلع بــ ه ويخرج اب اه الى ان
يصيرا عند نقطتى ٠ر ستين ستين اعنى بقدر نصفالقطر وترسم قوس ر٠ح
ويخرج عمود اء الى ح ويخرج عمودى ٠ط رك على خط اح ونقول
لماكانت فى مثلث اب ء زاوية ء قائمة كانت زاوية بـ تمام زاوية أ من
قائمة وعمود ٠ط جيب زاوية أ لان جيب تمام ٠ح من الربع اعنى تمام
آمن قائمة فخط طا جيب زاوية بـ وايضا فى مثلث اء ح تكون زاوية
ح تمام زاوية أ من قائمة وعمود كر جيب زاوية أ وخط كا
جيب زاوية ح ولتشابه مثلى اب ء ا ٠ط تكون نسبة اب الى اء
كنسبة ا٠ نصف القطر الى اط ايضاً ولتشابه مثلى اء ح اكر تكون
نسبة اء الى اح كنسبة اك الى ا ر كنسبة اك الى ا ر نصف القطر بل ا٠ فبالمساواة المغتطربة
نسبة ضلع اب الى اح كنسبة كا الذى هو جيب زاوية ح الى ا ط
الذى هو جيب زاوية بـ وذلك مااردناه

Al-Tusi's second proof of the law of sines.
From page 55 (٥٥) of the Arabic translation of the Persian original,
bound back to back with the French translation by Carathéodory:
Traité du quadrilatère attribué a Nassiruddin-el Toussy.

triangle, and the ratio of the *jyba* of the angle opposite the known side to the
jyba of the other angle will be equal to the ratio of the given side to the other
side. Through that the sides will be known.

While al-Tusi dealt with arbitrary triangles in one single condensed paragraph [p. 72], we shall interrupt it three times to insert explanations using
today's mathematical writing. To this end, consider an arbitrary triangle *ABC*
and denote the sides opposite the angles *A*, *B*, and *C* by *a*, *b*, and *c*, respectively.

As for the other triangles, if one knows two angles and one side, the other two
sides will be found according to what we have said about the right triangle.

That is, if A and B are the given angles, then $C = 180° - (A + B)$ is found. Now assume that the known side is a. According to the law of sines,

$$\frac{b}{a} = \frac{\text{jyb } B}{\text{jyb } A} \quad \text{and} \quad \frac{c}{a} = \frac{\text{jyb } C}{\text{jyb } A}.$$

In each equation three of the four quantities are known, and then b and c can be found.

> If what is given is two sides and one angle (not between the sides) the ratio of the side opposite the known angle to the other side will be equal to the ratio of the *jyba* of the known angle to the *jyba* of the angle opposite the other side. And when you have known the angles you would also know the remaining side.

That is, assume that A, a, and b are the given items. Then B is determined from

$$\frac{b}{a} = \frac{\text{jyb } B}{\text{jyb } A}.$$

Now two angles and one side are given, and the triangle is solved as in the previous case.

> If the angle is between the two given sides one would proceed as it has been said.

Now, this is really a short explanation, but al-Tusi is right. This is just Case III of his method of arcs and chords, shown on page 45. Then he concludes with the last of the present cases:

> If the given consists of the three sides, find first the perpendicular as it has been said, after which you will be able to know the angles as you would find them in the case of a right triangle.

We need to add to this explanation by saying that in Case IV of his solution of triangles by the method of arcs and chords (page 46), and with reference to the figure reproduced there, al-Tusi had already evaluated BE and AE. Then EC is easily obtained from BC and BE, so that in each of the right triangles ABE and AEC two sides and the right angle between them are known, a case that has already been covered.

Then al-Tusi concluded Chapter 2 with the following words: "Here ends what we were going to say about triangles." But this is not the end of his application of the law of sines, for in Chapter 3 of Book III, entitled *Rules whose knowledge is very useful in the theory of the plane quadrilateral,* he also used it to prove several of these rules or lemmas. Of these we present only the first as a sample [p. 73].

Let there be in a circle ABC, two arcs AB, AC whose ends touch at the point A: their sum $BAC < \frac{1}{2}$ the circumference is given, as well as the ratio of jyb AB *over* jyb AC. *I say that AB, AC can be determined.*

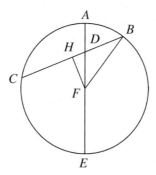

Demonstration.[54] Draw the chord BC and the diameter AE which intersect in D. Drop from the center, FH perpendicular over CB, join BF. The arc BAC being known the chord BC is known too [from a table of chords];

$$\frac{\text{jyb } AB}{\text{jyb } AC}$$

being known [the ratio] DB/DC ... is also known.[55]

It might be appropriate to interrupt the proof to show why this is true. Visualize two perpendiculars from B and from C to AE and denote their intersections with AE by B_p and C_p, respectively. Note that $BB_p = $ jyb $\angle AFB = $ jyb AB and that $CC_p = $ jyb $\angle AFC = $ jyb AC. This and the similarity of the triangles

[54] A different translation of this lemma and its proof can be found in Berggren, "Mathematics in medieval Islam," in *The mathematics of Egypt, Mesopotamia, China, India, and Islam,* p. 643.

[55] The omitted short passages in this quotation and the next one refer to some segments, external to the circle in question, that may be relevant in the rest of al-Tusi's Chapter 3 but are not relevant to this proof.

BDB_p and CDC_p gives

$$\frac{DC}{DB} = \frac{BB_p}{CC_p} = \frac{\text{jyb } AB}{\text{jyb } AC},$$

which shows that the ratio on the left is also known. Then al-Tusi continued as follows:

> We deduce:
> $$\frac{DB + DC = BC}{DB}$$
>
> ... [is known too]. DB, BC, BH ($BC/2$) and consequently DH and FH (*jyba* of the complement of half the arc BAC)[56] are thus known segments [$DH = BH - DB$ and $FH = \sqrt{60^2 - BH^2}$]. Now in the right triangle DHF the two sides of the right angle are known and consequently the angle DFH. Moreover BFH (which corresponds to half of the arc BC) is known. Thus we come to determine the angle BFA, which corresponds to the arc BA, and hence the arc AB itself, Q.E.D.

Thus we see that al-Tusi's treatise prominently displays trigonometry and its applications, although the only trigonometric length that he used in the work described to this point is the *jyba*. However, in Chapter 6 of Book V, entitled *Of the figure called shadow, of its consequences and of its accessories*, al-Tusi defined all six trigonometric lengths and gave some of their properties. Before going on to prove some results in spherical trigonometry, he stated the following [p. 164]:

> Let there be the circle $ABCE$ with center at D and the arc AB. Draw the diameters passing through the points A and B, and from the point A raise the perpendicular AF to AC, which will meet the diameter BE at F;[57] AF is the *shadow* of the arc AB, parallel to BH which is the *jyba*. Similarly, raise on AC the perpendicular DT through the center. The perpendicular TK will be the shadow of the arc TB, while BL is the *jyba*; in other words, TK and BL will be the shadow and the *jyba* of the complement of the arc AB. What we have called shadow, astronomers call the *first shadow or turned shadow* of the arc AB. ... they call TK, the *second shadow* of the arc AB, or the *straight shadow*; FD is for them *the diameter of the 1st shadow*, and KD *the diameter of the second shadow* ...

---

[56] Half the arc BAC is equivalent to the angle BFH, and its complement is the angle FBH.

[57] For some reason, this point is not labeled in the Arabic manuscript.

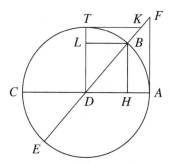

That makes six trigonometric lengths, that last two of which we now call the secant and the cosecant.

Then al-Tusi stated the following relationships for the new trigonometric lengths [pp. 164–165]:

> The first shadow of every arc is the second shadow of its complement and vice versa; the ratio of the shadow to the radius is equal to ratio of the *jyba* of the arc to the *jyba* of its complement; the ratio of the shadow to the diameter of the shadow is equal to ratio of the *jyba* to the radius, because of the similarity of the triangles FAD, BHD; but the triangles AFD, KTD being also similar one has
>
> $$\frac{FA}{AD} = \frac{DT = AD}{TK}; \text{ [58]}$$
>
> hence the radius is the mean proportional between the shadow of an arc and the shadow of its complement, and the shadows of two arcs are as the reciprocals of the shadows of their complements.

In other words, we have some familiar trigonometric identities, starting with

$$R \tan \alpha = R \cot(90° - \alpha)$$

and ending with

$$\frac{\tan \alpha}{\tan \theta} = \frac{\cot \theta}{\cot \alpha}.$$

While some of these identities may be considered unexciting today, the fact is that the introduction of the six trigonometric functions and the relationships between them gave an agility to the applications of trigonometry that it did not have in the times of just arcs and chords or arcs and *jybas*.

---

[58] In the French translation the last denominator appears as KD, which contradicts the text following this equation, but it is correct in the Arabic version on page ١٢٧ (127).

After this, little progress was made in trigonometry in the Islamic world except in the elaboration of finer tables. Worthy of mention are those completed by Ghiyath al-Din Jamshid al-Kashi (of Kashan) in 1414 as a revision of earlier ones by Mohammed al-Tusi. They were dedicated to the Great Khan Ulug Beg, a generous patron of the arts and sciences, whose capital was at Samarkand. Of great merit is al-Kashi's interpolation algorithm based on the approximate solution of a cubic equation.[59]

1.5 TRIGONOMETRY IN EUROPE

Abd al-Rahman, one of the Umayyads to escape the massacre by Abu al-Abbas al-Saffah in 750, managed to reach the Iberian peninsula. In 756 he proclaimed himself emir in Córdoba and in 773 he became independent of Baghdad. Abd al-Rahman III, also of the Umayyad dynasty in Córdoba, who ascended to the throne in 912, turned the emirate into an independent caliphate in 929. Islamic culture had already started to flourish under the emir Abd al-Rahman II. When in the ninth century the largest library in Christian Europe, at the Swiss monastery of Saint Gallen, had a grand total of 36 volumes, that of Córdoba housed 500,000, one per inhabitant.

Abd al-Rahman III was determined in his resolution that the splendor of his court at Córdoba would outshine that of the eastern Abbasid caliphate. He brought artists and scientists to Córdoba, and it was in this way that a large part of Islamic knowledge reached Europe, to be augmented by some of the local scientists. Contributions to trigonometry were only minor, and they took place after the decomposition of the Córdoba caliphate into smaller independent kingdoms. But because of their later influence we should cite the *Zij tulaytulah* (Toledan tables) of Abu Ishaq Ibrahim ibn Yahya al-Naqqas (the Chiseler) al-Zarqel (the Blue-eyed, referring to his father) (1029–1100),[60] a practical astronomer who made observations in Toledo and devoted several pages to the use of the cotangent. Abu Mohammed Jabir ibn Aflah al-Ishbili (from Seville) (c. 1100–c. 1160), frequently known as Geber, the Latinized form of his name, wrote the *Islah al-mjsty* or Correction of the Almagest, which would be influential some centuries later.

But the accumulation of knowledge in Muslim Spain would have been

---

[59] For details see Berggren, *Episodes in the mathematics of medieval Islam*, 1986, pp. 151–154.

[60] Many historians use the form al-Zarqali, but this cannot be considered correct for the reasons given by Millás Vallicrosa in *Estudios sobre Azarquiel*, 1943, pp. 13 and 21–22.

useless if such a wealth of books had remained unread by future generations. However, in 1085 Alfonso VI of Castile, the Valiant, conquered Toledo by the civilized method of convincing King Ismail ibn Yahya al-Qadir that resistance was futile, and entering into an agreement for a peaceful take-over. Less than a century later, the archbishop Raimundo de Sauvetat was pondering what to do with so many old volumes at the Cathedral of Toledo, and decided to make them available to scholars. This was the origin of the *Escuela de Traductores de Toledo* (Toledo School of Translators), which became one of the main avenues for the dissemination of old knowledge to Europe.

Many scholars flocked to Toledo to translate those works, among them Adelard of Bath (1075–1160), who translated the astronomical tables of al-Khwarizmi—revised by Maslama ibn Ahmed al-Madjriti (of Madrid) in the tenth century—in 1126 and Euclid's *Elements* in 1142. In 1145 Robert of Chester (1110–1160) translated al-Khwarizmi's *Kitab al-muqhtasar fi hisab al-jabr wa'l-muqabala* (The compendious treatise on calculation by restoration and reduction) with the title *Liber algebræ et almucabala*, which gave us the word algebra. Gherardo de Cremona (1114–1187) translated more than 80 texts from Arabic to Latin, including Ptolemy's *Almagest*—for which purpose he went to Toledo in 1167 and remained there for life—Euclid's *Elements*, and al-Zarqel's Toledan tables. Iohannes Hispalensis, or Juan de Sevilla (died 1180), translated al-Khwarizmi's book on Hindu numbering.

Many other translators, local and foreign, contributed to this endeavor: Petrus Alphonsi (Moshé Sefardí), Rudolf of Bruges, Dominicus Gundisalvi, Michael Scot, Marcos de Toledo, Alfred of Sareshel, Plato Tiburtinus also known as Plato of Tivoli, Abraham Bar Hiyya, Hugo de Santalla, Hermann of Carinthia (Hernán Alemán), Abraham ben Ezra . . . the list is endless. This translation activity was an essential component of what has been called the twelfth-century Renaissance, and it certainly paved the way for the full-fledged Renaissance of the fifteenth century.[61]

Some of these translations are particularly noteworthy for linguistic reasons. The translators, unable to make any sense of the word *jyb* and taking it to be *jayb*, one of whose meanings is bosom or any object similarly shaped, translated it as *sinus*, which has the same meaning in Latin. We know that the word appears in Gherardo de Cremona's translation of al-Zarqel's Toledan

---

[61] Although Toledo was the main translation center, other translations were made in Barcelona, León, Pamplona, Segovia, and Tarazona in Spain; Béziers, Marseilles, Narbonne, and Toulouse in France; and in the Norman court of Roger II of Sicily. For full details see Chapter IX in Haskins, *The Renaissance of the twelfth century*, 1927.

tables, *Canones sive regulæ super tabulas toletanas*, as follows: "Sinus cuius libet portionis circuli est dimidium corde duplicis portionis illius" (The sine of any portion of a circle is one half the chord of double the portion).[62] It is, however, possible that the same rendering of *jayb* into *sinus* was made earlier by other translators. The names of Plato Tiburtinus and Robert of Chester have been mentioned in this connection.[63]

With this preparatory work, Europe was ready to embrace and improve on trigonometry. The translators had rendered the turned shadow and the straight shadow into *umbra versa* and *umbra recta*, and these trigonometric lengths were used in Europe. At Oxford University, John Manduith (c. 1310) used them in his *Small tract*, and Richard of Wallingford (c. 1292–1335) in his *Quadripartitum de sinibis demonstratis*. In Paris, Jean de Linières (1300–1350), a follower of al-Zarqel, clearly defined the *umbra recta* and *umbra versa* in his *Canones super tabulas primi mobilis*, astronomical tables for the Paris meridian.

The stage was set for the first systematic treatment of trigonometry in Europe. It was written by Johannes Müller (1436–1476), also known as Regiomontanus. A true man of the Renaissance, he adopted the Latin version, Regius Mons, of the name of his hometown of Königsberg (King's Mountain), becoming Joannes de Regio Monte. Actually, he was born at Unfinden, near Königsberg. He went to the Academy of Vienna as a student, where he met Georg Peurbach in 1451 or 1452, and they became like father and son. In 1461 he was appointed professor of astronomy at the University of Vienna, but he did not enjoy this position for long. At some time, Peurbach had introduced Regiomontanus to Cardinal Basilios Bessarion, who, like Peurbach, was interested in a definitive translation of Ptolemy's *Almagest* (previous Latin translations of bad Arabic translations were shunned). When Peurbach died at 37, in 1461, it was up to Regiomontanus to finish the task.

He and Bessarion left for Italy in search of reliable documents in 1462. There Regiomontanus learned Greek and Arabic, finished Peurbach's work as *Epytoma Joannis de Monte Regio in almagestum Ptolemei* in 1463, and wrote his own compilation of trigonometry, which he finished in June 1464. The title of this work, a systematic treatise on trigonometry, as independent from astronomy and even more inclusive than al-Tusi's, is *De triangvlis omnimodis* (On triangles of all kinds). It consists of five books, the first two of which

---

[62] From Smith, *History of mathematics*, II, p. 616.

[63] The first by Cajori, *A history of mathematics*, 1919, p. 105. The second by Boyer, *A history of mathematics*, 1968, p. 278.

JOHANNES MÜLLER
From Ludwig Bechstein, ed., *Zweihundert deutsche Männer in Bildnissen
und Lebensbeschreibungen*, Georg Wigands Verlag, Leipzig, 1854.

are devoted to plane trigonometry and the rest to spherical trigonometry (in
the fourth book he borrowed heavily from Jabir ibn Aflah without giving him
credit, for which he was to be chastised later by Cardano). It was eventually
printed in 1533 and reprinted in 1561 and 1967.[64]

Book I of *De triangvlis* contains fifty-seven propositions, of which numbers
20 to 57 deal with the solution of triangles. The only trigonometric function
used in this book is the sine, which appears in Propositions 20, 27, and 28.
Here are the statement and proof of this last proposition [pp. 66–67]:

XXVIII.

*When the ratio of two sides of a right triangle is given, its angles can be
ascertained.*

---

[64] Hughes, *Regiomontanus on triangles*, 1967. Page references are to this edition, and,
except for very minor changes, the translations included here are by Hughes.

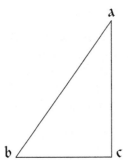

One of the two sides is opposite the right angle or else none is. First, if side ab, whose ratio to side ac is known, is opposite the right angle acb, I say that the angles of this triangle become known. In fact, ac is the sine of the arc of angle abc by the above,[65] provided that ab is the semidiameter; that is, the whole sine [*sinus totus*] of the circle. Therefore, the ratio of the whole sine to the sine of angle abc is known. Hence the latter sine can be found, and finally the angle abc will be known. However, if the ratio of the two sides bc and ac is given, then the ratio of their squares will be known. And by addition [of unity to the last ratio and simplification] the ratio of the sum of the square of bc plus the square of ac—that sum being the square of ab because of the right angle c—to the square of ac will be known, whence the ratio of the lines is known. And what remains is as before.

To rephrase this using current notation, we consider two cases:

1. If the ratio ab/ac is given and if ab is the semidiameter, the equation

$$\frac{\text{ab}}{\text{ac}} = \frac{\text{sinus totus}}{\text{sinus abc}}$$

(where we have written sinus instead of sin as a reminder of the fact that the sinus is the length of a segment and not our usual trigonometric function), allows us to determine sinus abc and then the angle abc from a table of sines.

2. If the ratio bc/ac is given, then

$$\frac{\text{ab}^2}{\text{ac}^2} = \frac{\text{bc}^2 + \text{ac}^2}{\text{ac}^2} = \left(\frac{\text{bc}}{\text{ac}}\right)^2 + 1,$$

---

[65] As Hughes explains, Regiomontanus was referring to his previous Proposition 20, which just stated the definition of the sine of an angle between 0° and 90° as the sine of its arc.

from which the ratio ab/ac becomes known, and then we finish as in case 1.

After the proof, Regiomontanus included a guide to the "operation" of the theorem; that is, computational instructions, starting with: "If one of the two sides is opposite the right angle, multiply the smaller term ..." He concluded with an example in which the ratio of ab to ac is 9 to 7. This was a typical arrangement that he employed throughout the book.

Book II, which contains an organized treatment of trigonometry in 33 propositions, starts with the law of sines [pp. 108–109]:

I.

In every rectilinear [as opposed to spherical] *triangle, the ratio of one side to another is as that of the right sine of the angle opposite the one to the right sine of the angle opposite the other.*

Regiomontanus explained, with reference to the enclosed figure, what the

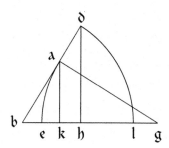

proposition says. Using our terminology, this is

$$\frac{ab}{ag} = \frac{\text{sinus } agb}{\text{sinus } abg} \quad \text{and} \quad \frac{ab}{bg} = \frac{\text{sinus } agb}{\text{sinus } bag}.$$

Then the proof proper begins with the statement that if abg is a right triangle, this is [Proposition] 28 above. If not, but if ab and ag are equal, then the angles opposite these sides are equal and hence their sines are equal. But if one of the sides is longer than the other, say ag is longer than ab, then prolong ba to bð so that the length of bð equals that of ag. The he drew the arcs in the figure, probably because he was following al-Tusi's first proof, which has the arcs. Drawing perpendiculars ak and ðh to the base bg, it is evident that ak is to ðh as the sinus of angle agb is to the sinus of angle abg. Moreover, in view of the similarity of the triangles ᴧbk and ᴆbᏏ,

$$\frac{ᴧb}{ᴧg} = \frac{ᴧb}{bᴆ} = \frac{ᴧk}{ᴆᏏ} = \frac{\text{sinus } ᴧgb}{\text{sinus } ᴧbg},$$

"which is certain, and that is what the proposition asserted."

As a sample of his use of this theorem in the solution of triangles we present his seventh proposition [pp. 112–115].

VII.

If the perimeter of a triangle is given with two of its angles, any one of its sides will be known.

Clearly, if two angles are known then all three are known, and (a shorter proof will be given after the quotation)

therefore, by the reasoning often cited [the law of sines], the ratio of ᴧb to ᴧg will be known, and then by addition [of 1] the sum of bᴧ plus ᴧg will have a known ratio to line ᴧg. Similarly, the ratio of ᴧg to gb will be known, and the ratio of bᴧ plus ᴧg to gb will be found [as the product of the two previous ratios]. Thus, by addition [of 1] the total perimeter of triangle ᴧbg has a known ratio to line bg. And since the hypothesis gave the perimeter itself, line bg will be known. Hence the other two sides will also be declared known.

Or, more simply, since the ratio of any two sides is known by the law of sines, then

$$\frac{ᴧb + bg + gᴧ}{gb}$$

is known, so that line bg will be known. Similarly for the other two sides.

Two features of this second book are remarkable. The first is the use of algebra, specifically quadratic equations, in the proofs of Propositions 12 and 25. The second is the fact that Regiomontanus implicitly used the trigonometric formula for the area of a triangle in the proof of Proposition 26, reproduced next. This book was extremely influential in Europe and highly praised by Tÿge Brage (a name that he Latinized as Tycho Brahe),[66] a Danish astronomer

---

[66] This spelling of this name is from what appears to be his signature (but the fourth letter of his last name is less than clear). It is reproduced below his portrait by Hans Peter Hansen in Frederik Reinhold Friis, *Tyge Brahe: en historisk fremstilling. Efter trykte og utrykte kilder*, Gyldendalske Boghandel, Copenhagen, 1871.

x x v i.
Data area trianguli cum eo,quod fub duobus lateribus continetur
rectangulo,angulus quem bafis refpicit,aut cognitus emerget,a ut cū
angulo cognito duobus rectis æquipollebit.

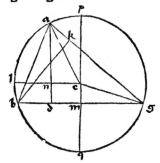

Refumptis figurationibus præcedētis,
fi perpendicularis b k uerfus lineā a g ,p
cedens,extra triangulum ceciderit,erit per
ea,quæ in præcedenti commemorauimus,
proportio b k ad b a nota,& ideo per
primi huius angulum b a k notum acci∗
piemus,fic angulus b a g cum angulo b
a k noto duobus rectis æquiualebūt. Si ue
ro perpendicularis b k intra triangulum
ceciderit,quemadmodū in tertia figuratio
ne præcedentis cernitur, erit ut prius a b
ad b k notā habens proportionē , & ideo
angulus b a k fiue b a g notus conclude
tur.At fi perpendicularis b k coinciderit lateri a b, neceffe eft angulum b a g
fuiffe rectum,& ideo cognitum,quod quidem accidit,quando area trianguli pro∗
pofiti æquatur ei,quod fub duobus lateribus eius continetur rectangulo.

Reproduced from Regiomontanus, *De triangvlis omnimodis*, p. 58.

who studied a supernova that became visible in the sixteenth century. He stated
that he "found the longitude and latitude of this new star with the help of the
infallible method of the doctrine of triangles," and that a good part of the
propositions that he used were from the fourth book of Regiomontanus.[67]

The second edition of *De triangvlis* contains some additions. There are
two tables of sines, one based on a *sinus totus* of 6,000,000 parts and another
in which it is subdivided into 10,000,000 parts. Those were times in which
decimal numbers had not been introduced and it was convenient to compute
the sines as whole numbers. To obtain precision, the radius or *sinus totus* must
then be a huge number. A shorter table of only 95 sines was also included
based on a radius of 600,000,000 parts. In spite of the amount of effort needed
to compute them, they would soon become obsolete.

However, both editions of this book represented a setback in one direction:
that Regiomontanus used only the sine function, thereby limiting the agility
with which trigonometry could be applied. But he used the *umbra versa* from
1465, the year after he finished *De triangvlis*.

---

[67] Tychonis Brahe, Dani *De nova et nvllivs ævi memoria privs visa stella, iam pridem
anno à nato Christo 1572. mense Nouembrj primùm conspecta, contemplatio mathematica*,
1573, f. B4^v = *Opera omnia*, 1648, p. 357.

About 1467 Regiomontanus left Italy and found employment managing the manuscript collection of King Matthias I Corvinus of Hungary. In 1467 he finished the *Tabule directionum profectionumque*, possibly printed privately in 1475, which includes 31 astronomical problems but is mostly devoted to tables. One of these, which he called *tabula fecunda*, is a table of tangents. It was used in solving his tenth problem, after which he stated that he chose the name *fecunda* because "like a fertile tree it produces many marvelous and useful [results]." [68]

Notable among those who worked in trigonometry after Regiomontanus was the astronomer Mikołaj Kopernik (Latinized as Nicolaus Copernicus) (1473–1543). Some trigonometry appeared in his great work *De revolvtionibvs orbium cœlestium* of 1543, but it had been published separately the year before under the title *De lateribvs et angvlis triangulorum*.[69]

Georg Joachim von Lauchen (1514–1576),[70] usually known as Rheticus because he was born in the former Roman province of Rhætia, was a young professor of mathematics at Wittenberg in 1539 when he resigned his appointment to become a student of Copernicus in Poland. He learned that Copernicus had trouble completing the demonstrations of his trigonometry, and that he even thought of discontinuing his work, but Rheticus encouraged him to persevere until he eventually found the necessary demonstrations.

Rheticus returned to Wittenberg in 1541 to supervise the printing of *De revolutionibvs*, which may never have happened otherwise. He added a table of cosines of his own to the *De lateribvs*, the first table of cosines ever. From Wittenberg he moved to Nuremberg, then to the University of Leipzig, and then to Kraków, where he became a practicing doctor of medicine for the next twenty years. He still found time for mathematics, and even received funding from Emperor Maximilian II, which enabled him to employ six research assistants at some time. With their help, he elaborated a table of all trigonometric lines based on a radius of 10,000,000 in intervals of ten seconds. Then he began a table of what we now call tangents and secants based, for extra precision, on a radius of 1,00000,00000,00000, but he died in 1576 before he was able to finish. The tables were completed by his pupil Lucivs Valentinvs Otho

---

[68] Translated from page 13 of the 1490 edition, which has unnumbered pages.

[69] The plane trigonometry appears in *De revolvtionibvs* in Chapter 13, pp. 19$_b$–21, with minor differences from that in *De lateribvs*. The tables are in Chapter 12, pp. 12–19, and have one fewer place of accuracy than those in *De lateribvs*.

[70] When his father was executed as a convicted sorcerer, the law forbade his family the use of his last name: Iserin. Then Georg Joachim used his Italian mother's last name de Porris, which means "of the leeks." Subsequently, he translated it into German as "von Lauchen."

Tabula Secunda

Numerus		Numerus		Numerus	
δ		δ		δ	
0	00000	31	60086	61	180402
1	1745	32	62486	62	188075
2	3492	33	64940	63	196263
3	5240	34	67452	64	205034
4	6992	35	70022	65	214450
5	8748	36	72654	66	224607
6	10511	37	75356	67	235583
7	12273	38	78129	68	247513
8	14053	39	80978	69	260511
9	15838	40	83909	70	274753
10	17633	41	86929	71	290422
11	19439	42	90040	72	307767
12	21256	43	93254	73	327088
13	23087	44	96571	74	348748
14	24932	45	100000	75	373211
15	26794	46	103551	76	401089
16	28674	47	107236	77	433148
17	30573	48	111062	78	470453
18	32492	49	115037	79	514438
19	34433	50	119177	80	567118
20	36396	51	123491	81	631377
21	38387	52	127994	82	711569
22	40402	53	132704	83	814456
23	42448	54	137639	84	951387
24	44522	55	142813	85	1143131
25	46631	56	148253	86	1430203
26	48772	57	153987	87	1908217
27	50952	58	160035	88	2863563
28	53170	59	166429	89	5729796
29	55432	60	173207	90	Infiniti
30	57734				

Reproduced from page 57 (unnumbered) of *Tabule directionū profectionūq[ue] famosissimi viri Magistri Joannis Germani de Regiomonte in natiuitatibus multum vtiles*, 1490.

(c. 1550–1605) and published in the second volume of the *Opvs palatinvm de triangvlis* in 1596. The first volume contains four trigonometric treatises, of which one is due to Otho. Rheticus is chiefly remembered for his magnificent trigonometric tables. In addition to those in the *Opvs palatinvm*, he began a table of sines based on a radius of 1,00000,00000,00000, which was also completed by Otho in 1598 and published as the *Thesaurus mathematicus*.

1.6 FROM VIÈTE TO PITISCUS

Most notable among the many people who worked on trigonometry in the sixteenth century was François Viète (1540–1603), also known as Vieta after the Latin form of his name. He was a significant contributor and left us quite a

FRANÇOIS VIÈTE
From Smith, *Portraits of Eminent Mathematicians*, II

number of well known and not so well known trigonometric identities. Viète was a lawyer in the *Parlement de Paris* and then a personal advisor to kings Henri III and Henri IV of France, which did not prevent him from becoming, in his spare time, the best mathematician of the sixteenth century. Viète was the first to understand the generalities underlying the particulars and, in that sense, the first man in the new Europe to really deserve the name mathematician. He wrote extensively on trigonometry, starting with his first published work: *Canon mathematicvs sev ad triangvla cum adpendicibus* (Mathematical table, or about triangles, with appendices), written in four books, of which only the first two were ever published. In the first of these he gave tables of minute-by-minute values of all six trigonometric lengths.

The next illustration shows the second page of the first table. In the top

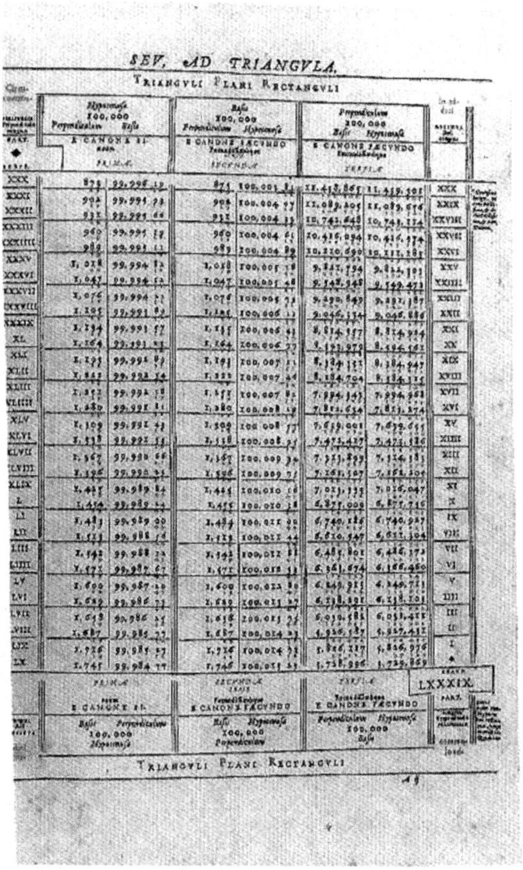

Second page of the table in Viète's *Canon mathematicvs*.
Reproduced from the virtual exhibition *El legado de las matemáticas*:
de Euclides a Newton, los genios a través de sus libros, Sevilla.

part we read the words *Hypotenusa, Basis*, and *Perpendiculum*. These refer to
a right triangle one of whose legs is horizontal, which he called the *base*, and
the other vertical, which he called the *perpendicular*. These names already
appear in the works of Rheticus, who used *perpendiculum* for what we now
call the sine and the tangent, *basis* for the cosine, and *hypotenusa* for the

secant. Below we read *canone sinuum* and *canone fæcundo*. They mean sine table and tangent table. He may have taken this name for the tangent from the *tabula fecunda* of Regiomontanus.

But we have stated that he used all six trigonometric lengths. He referred to the cosine, the cotangent, and the cosecant as the sine, the tangent, and the secant of the rest (the "rest" meaning the complement, or 90° minus the original acute angle). In Latin, "of the rest" is *residuæ*, and thus *fæcundus residuæ* means cotangent and *hypotenvsam residuæ* means cosecant. The second book of the *Canon mathematicvs*, individually titled *Vniversalivm inspectionvm ad canonem mathematicvm, liber singularis* (General examination of the mathematical tables, single book), is devoted to the solution of triangles from a number of trigonometric formulas, such as the following [p. 11]:

> DIFFERENTIA inter Fæcundum & Hypotenvsam Residuæ, est Fæcundus Dimidiæ peripheriæ.

If we accept that the qualifier *Residuæ* applies to both *Fæcundum & Hypotenvsam*, and if *peripheriæ*, meaning arc, is denoted by θ, this statement is equivalent to the trigonometric identity

$$\cot\theta - \csc\theta = \tan\tfrac{1}{2}\theta.$$

The next work of Viète's containing a result that can be considered trigonometric is *Ad logisticem speciosam, notæ priores.*[71] In the last section of this paper, *On the genesis of triangles*, we find the following result [p. 72]:

PROPOSITION XLVIII.

From two equal and equiangular right triangles, to construct a third right triangle.

First the notation is established as follows [p. 72]:

> Let there be two right triangles with these common sides: *A* the hypotenuse, *B* the perpendicular, and *D* the base.

---

[71] English translation by Witmer as "Preliminary notes on symbolic logistic," in *The analytic art by François Viète*, 1983, pp. 33–82. Page references are to and quotations from this translation.

36   Ad Logisticen Spec.
fis. æquatur quadrato adgregati ex bafe primi, & perpendiculo fecundi, plus quadrato
differentiæ inter perpendiculum primi, & bafin fecundi, vel etiam, æquatur quadrato adgregati ex perpendiculo primi & bafe fecundi, plus quadrato differentiæ inter
bafin primi & perpendiculum fecundi.

                     PROPOSITIO XLVIII.

A Duobus triangulis rectangulis æqualibus & æquiangulis , tertium
triangulum rectangulum, conftituere.

   Sunto duo triangula rectangula, quorum communia latera hypotenufa quidem **A**,
perpendiculum B, bafis D. Oporteat ab illis duobus tertium triangulum rectangulum
conftituere. Fiat deductio ficut docuit Propofitio 46. cafu primo. Deduci enim tantum
poteft fynærefeos via , non autem diærefeos. Fit hypotenufa fimilis A quadrato. Bafis, D
quadrato ═ B quadrato. Perpendiculum B in D 2. Tertium autem illud, vocetur triangulum anguli dupli, & ejus refpectu primum vel fecundum dicetur anguli fimpli ob cau-
fas * fuo ponendas loco.

   * *Cauffa eft quod angulus acutus trianguli re-*   Triangulum anguli dupli.
ctanguli à duobus triangulis via fynarefeos effecti æ-
quetur angulis acutis horum triangulorum fimul
adgregatis , cujus theorematis converfum dæmon-
ftravit Anderfonus theoremate fecundo fectionum
angularium. Porro acuti voce intelligitor is angulus, D in B2
cui Perpendiculum fubtenditur.   A *q*.

                     PROPOSITIO XLIX.

A Triangulo rectangulo fimpli , &   D*q* ═ B *q*.
triangulo rectangulo anguli du-
pli , triangulum rectangulum effingere. Vocetur autem tertium illud,
triangulum anguli tripli.

Neither the statement of the proposition nor the notation clarifies what the task
really is, but from explanations previously provided by Viète and a look at the
figure that he provided, shown above, the third rectangle is to be constructed
such that its hypotenuse is A^2 ($Aq.$ in Viète's notation, as an abbreviation of
A *quadrato*, or A square) and the acute angle on the right is twice that of the
original angle (*anguli dupli*).

The proof, based on a previous Proposition 46, is not sufficiently interesting
for inclusion here, but the result is, and it is contained in the already mentioned
figure: the new base and perpendicular are $Dq = Bq$ and D in $B2$, meaning
$D^2 - B^2$ and $2BD$, respectively. That this is a trigonometric result follows from
the following interpretation in current notation and terminology. If the angle
formed by A and D in the original triangle is denoted by θ, then $D = A \cos \theta$
and $B = A \sin \theta$. Then, in the new triangle we have

$$A^2 \cos 2\theta = D^2 - B^2 = A^2(\cos^2 \theta - \sin^2 \theta)$$

and

$$A^2 \sin 2\theta = 2BD = 2A^2 \sin \theta \cos \theta.$$

This is the way in which Viète discovered the double angle formulas, previously found by Abu'l-Wafa.

In the next three propositions he considered the cases of the triple, quadru-ple, and quintuple angles, and from these particular cases, Viète noticed a pattern for the expressions giving the base and the perpendicular (what we would now call $\cos n\theta$ and $\sin n\theta$ for $n = 3, 4, 5$, if θ is as defined above). He stated what the pattern was and realized that it could be continued "in infinite progression" [p. 74].

Viète also made this discovery in another paper, *Ad angvlarivm sectionvm analyticen. Theoremata. καθολικώτερα*, possibly written about 1590 but also published posthumously, in 1615, by Alexander Anderson.[72] Since Viète was more explicit in this second paper, we shall present his general result in this context. This work contains ten theorems, and we are interested in the first three. The first states the following (an explanation in today's terminology follows) [p. 418]:

THEOREM I.

If there are three right triangles the acute angle of the first of which differs from the acute angle of the second by the acute angle of the third, the first being the largest of these, the sides of the third will have these likenesses:

The hypotenuse will be analogous to the product of the hypotenuses of the first and second.

The perpendicular will be analogous to the product of the perpendicular of the first and the base of the second minus the product of the perpendicular of the second and the base of the first.

The base [will be analogous] *to the product of the bases of the first and second plus the product of their perpendiculars.*

To interpret this theorem refer to the next figure, of a half-circle resting on a diameter AB, which was used for the demonstration (actually, we have omitted two segments and three letters that are not useful for our purposes). The first, second, and third triangles are ABE, ABD, and ABC, respectively, and the corresponding acute angles mentioned in the hypothesis above are $\angle EAB$, $\angle DAB$, and $\angle CAB$. The hypothesis itself—in spite of the possible confusion created by the word "differs" (*differat*)—is that

$$\angle EAB = \angle DAB + \angle CAB.$$

---

[72] English translation of *Ad angvlarium* by Witmer as "Universal theorems on the analysis of angular sections with demonstrations by Alexander Anderson," in *The analytic art by François Viète*, pp. 418–450. Page references are to and quotations from this translation.

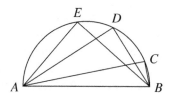

It is clear from the first conclusion that the word "analogous" (*simile*) cannot be taken to mean "equal" for these triangles except in the particular case in which $AB = 1$. If we do this, the equations that were actually proved in the demonstration—in narrative form, of course—become

$$CB = EB \times AD - DB \times AE$$

and

$$CA = AE \times AD + EB \times DB$$

(otherwise, each of these lengths must be divided by AB). Then, if we interpret that the sides AE, AD, and AC are the "bases" of the stated triangles and that the sides EB, DB, and CB are the "perpendiculars," these equations represent the last two conclusions of the theorem, and the word "analogous" (*simile*) can be taken to mean "equal." If we now define $\alpha = \angle EAB$ and $\theta = \angle DAB$ and if we note that $\angle CAB = \angle EAB - \angle DAB = \alpha - \theta$, we see that the first equation is just Abu'l-Wafa's formula for the sine of $\alpha - \theta$ and the second is the companion formula for the cosine of $\alpha - \theta$. So this is not new (although it probably was to Viète), and neither is the second theorem, containing equations that can be turned into similar formulas for $\alpha + \theta$.

To present Viète's third theorem, which is our goal, we shall consider first a different particular case, that in which $D = C$, so that $\angle EAB = 2\angle CAB$. But at this point in the original paper there is a change in notation and the letters D and B are used to represent the lengths AC and CB, respectively. Any subsequent explanations using the letters D and B with two different meanings would be confusing. Some sort of compromise is in order, and in this we shall follow the lead of a master. When Newton was studying Viète's paper as a young man of 21 and writing his own understanding of it,[73] he used lowercase letters for all items previously written in uppercase. Thus, we now write $D = ad = ac$ and $B = db = cb$. Then the two equations of the first theorem's demonstration can be rewritten as

$$B = eb \times D - B \times ae \qquad \text{and} \qquad D = ae \times D + eb \times B$$

---

[73] Whiteside, ed., *The mathematical papers of Isaac Newton*, I, 1967, p. 78.

(for convenience, we continue taking $ab = 1$, which Newton did not). Then solving them for ae and eb and using the fact that $D^2 + B^2 = 1$ gives

$$ae = D^2 - B^2 \qquad \text{and} \qquad eb = 2DB.$$

These are then the base and the perpendicular of the triangle abe, whose acute angle $\angle eab$ is double the acute angle $\angle cab$. If this angle is denoted by θ, then, writing things in present-day notation, $D = \cos\theta$, $B = \sin\theta$, $ae = \cos 2\theta$, and $eb = \sin 2\theta$, so that the preceding equations can be rewritten as

$$\cos 2\theta = \cos^2\theta - \sin^2\theta \qquad \text{and} \qquad \sin 2\theta = 2\sin\theta\cos\theta.$$

Next we start a similar procedure, but this time with $\angle cab = \theta$, $\angle dab = 2\theta$, and $\angle eab = 3\theta$. Replacing e with d in the equations $ae = D^2 - B^2$ and $eb = 2DB$, we obtain

$$ad = D^2 - B^2 \qquad \text{and} \qquad db = 2DB,$$

in which we still have $D = ac$ and $B = cb$. Taking these values to the equations of Theorem I, they become

$$B = eb \times (D^2 - B^2) - 2DB \times ae$$

and

$$D = ae \times (D^2 - B^2) + eb \times 2DB.$$

Solving these equations for ae and eb gives

$$ae = D^3 - 3DB^2 \qquad \text{and} \qquad eb = 3D^2B - B^3,$$

which are the base and the perpendicular of the triangle abe, whose acute angle $\angle eab$ is triple the acute angle $\angle cab = \theta$. In present-day trigonometric notation, these equations are equivalent to

$$\cos 3\theta = \cos^3\theta - 3\cos\theta\sin^2\theta$$

and

$$\sin 3\theta = 3\cos^2\theta\sin\theta - \sin^3\theta.$$

Viète did not include the preceding calculations in his paper, just the results, including also the cases of the quadruple and quintuple angles. That is, he gave the base and the perpendicular of the triangle abe in the cases $\angle eab = n \angle cab$ for $n = 2, 3, 4,$ and 5, as summed up in the table below.

Angle	Base	Perpendicular
Simple	D	B
Double	$D^2 - B^2$	$2DB$
Triple	$D^3 - 3DB^2$	$3D^2B - B^3$
Quadruple	$D^4 - 6D^2B^2 + B^4$	$4D^3B - 4DB^3$
Quintuple	$D^5 - 10D^3B^2 + 5DB^4$	$5D^4B - 10D^2B^3 + B^5$

It is reproduced in the next page from Viète's original publication, on the right of the central vertical double line. In this table, q. represents a square, *cub.* or c. a cube, qq. a fourth power, qc. a fifth power, and *in* indicates multiplication.

The fact is that the computations involved in solving the equations of Theorem I for ae and eb are sufficiently tedious even for these few values of n (they were omitted above for $n = 4, 5$), so this procedure cannot be comfortably continued indefinitely. But Viète noticed a pattern in what he already had. The terms in each line of the table are those obtained on raising the binomial $D + B$ to the nth power:

$$(D + B)^2 = D^2 + 2DB + B^2,$$
$$(D + B)^3 = D^3 + 3D^2B + 3DB^2 + B^3,$$
$$(D + B)^4 = D^4 + 4D^3B + 6D^2B^2 + 4DB^3 + B^4,$$
$$(D + B)^5 = D^5 + 5D^4B + 10D^3B^2 + 10D^2B^3 + 5DB^4 + B^5,$$

but different terms are allotted to the base and the perpendicular in a certain pattern, with alternating plus and minus signs.

Generalizing from these five cases, Viète proposed the following conjecture: if the resulting terms are separated into two groups (the odd terms and the even terms), and if the terms in each group are alternately positive and negative, the base ae will be the first group of terms and the perpendicular eb the second. For instance, for $n = 4$, taking first the odd and then the even terms of the expansion of $(D + B)^4$ and making them alternately positive and negative, we have

$$ae = D^4 - 6D^2B^2 + B^4 \quad \text{and} \quad eb = 4D^3B - 4DB^3,$$

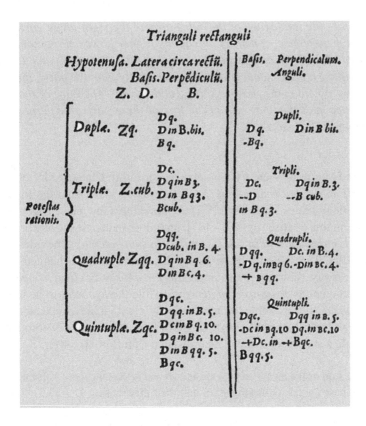

Reproduced from Viète, *Ad angvlarivm sectionvm analyticen*, p. 14

There are some errors in this table, which were corrected in his *Opera mathematica*.

as given in the table. Since the expansion of $(D + B)^n$ was well known in Viète's time for any positive integer n, it provided a general way to evaluate $ae = \cos n\theta$ and $eb = \sin n\theta$ from $B = \cos\theta$ and $D = \sin\theta$. Here is an extract of his very long statement [pp. 422–423]:

THEOREM III.

If there are two right triangles the acute angle of the first of which is a fraction of the acute angle of the second, the sides of the second are like this:

. . .

[*To find*] *the analogues of the sides around the right angle correspond-ing to the hypotenuse, construct an equally higher power of a binomial root composed of the base and perpendicular of the first* [*triangle*] *and distribute the individual homogeneous products successively into two parts, the first of each being positive, the next negative,* [*and so on*]. *The base of the second triangle becomes the analogue of the first of these parts, the perpendicular the analogue of the second.*

. . .

There is more of Viète's trigonometry starting in Chapter XIX of his *Variorvm de rebvs mathematicis responsorvm, liber VIII* (Answers to various mathematical things, book VIII).[74] There are quite a few plane trigonometric identities in this treatise [pp. 401–403], including of course the law of sines in the form *Latera sunt similia sinibus, quibus ea subtendentur* (The sides are like the sines that they subtend) [p. 402]. By this time Viète had changed some of his terminology, referring to the tangent as *prosinus* rather than *fæcundus* and to the cotangent as *prosinus complementi* (while the *hypotenusa* had become *transsinuosa*). For example, imagine a triangle with a horizontal side, called the base, and refer to the other two sides as the legs. Using his new terminology, Viète made a statement that can be translated as follows [p. 402]:[75]

As the sum of the legs is to their difference, so is the tangent of half the sum of the base angles to the tangent of half their difference.

If the legs are denoted by a and b, if A and B are the angles opposite a and b (thus, they are the base angles), and if A is assumed to be larger than B, this statement can be rendered in current terminology as

$$\frac{a+b}{a-b} = \frac{\tan \frac{1}{2}(A+B)}{\tan \frac{1}{2}(A-B)},$$

and is known as the *law of tangents*.

Viète's success in trigonometry was based on his superior command of algebra, the crucial step being his consistent use of consonants for known quantities and vowels for the unknowns. His literal expressions, which he introduced in the *Canon mathematicvs* of 1579, were instrumental in freeing

---

[74] Page references are to Schooten, *Francisci Vietæ opera mathematica*, 1646.

[75] The original reads: *Vt adgregatum crurum ad differentiam eorundem, ita prosinus dimidiæ summæ angulorum ad basin ad prosinum dimidiæ differentiæ.*

algebra from the necessity of considering just specific examples, of which we shall see plenty from earlier times in Chapter 3.

But the recently mentioned law of tangents was not discovered first by Viète. It appeared ten years earlier in a book intended as a text: *Thomæ Finkii flenspurgensis geometriæ rotundi* [circular geometry] *libri XIIII* by Thomas Fincke (1561–1656), a physician and mathematician from the town of Flensburg, Denmark (now Germany). He stated it in a rather complicated verbal form that with the help of a figure,[76] if the notation is the same as for Viète's statement, and if A is assumed to be larger than B, can be rendered into modern terminology as follows:

$$\frac{\frac{1}{2}(a+b)}{\frac{1}{2}(a+b)-b} = \frac{\tan\frac{1}{2}(A+B)}{\tan\left[\frac{1}{2}(A+B)-B\right]}.$$

Of course, this simplifies to Viète's equation.

But Fincke's fame does not rest on the law of tangents, which is not even included in today's trigonometry books, since it is easily derived from the law of sines, but rather on his having coined the present names for the tangent and the secant. He did use the tangent to describe the right-hand side above as *sic tangens semissis anguli crurum exteriores, ad tangemtem* ... (as the tangent of the semi[sum] of the angles opposite the legs is to the tangent of ...), but he had already introduced them by the statement: *Recta sinibus connexa est tangens peripheriæ, aut eam secans* (A straight line associated with the sine is the tangent of the arc, or also the secant), on page 73 of his book, expressing also the hope that these terms be used in this connection. Viète was opposed to them because they could be confused with the same terms used in geometry, but history made Fincke the winner.

The major contributions to trigonometry up to the end of the sixteenth century have already been presented, but there is the embarrassing matter of terminology. It is clear that, with the exception of the word *sinus* used by Gherardo de Cremona and Regiomontanus, neither the name for the subject nor the familiar notation and terminology existed for most of this time. Viète himself did not adopt the use of *sinus* until 1593, and it was not until the 1590s that the words *tangens* and *secans* started gaining adherents.

The reader has had no choice but to have the subject described in this chapter referred to as trigonometry, its modern name. This is because this

---

[76] Both of them can be seen in Zeller, *The development of trigonometry from Regiomontanus to Pitiscus*, p. 89.

subject remained unnamed almost to the end of the sixteenth century. Finally, Bartholomäus Pitiscus (1561–1613) provided a name for it with the publication of his book *Trigonometria: sive de solvtione triangvlorvm tractatus breuis et perspicuus* (Trigonometry: or a brief and clear treatise on the solution of

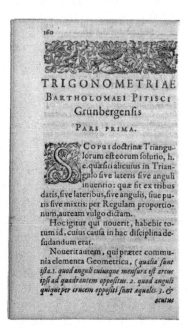

Title page and first page of Pitiscus, *Trigonometria*, 1595.
From Eidgenössische Technische Hochschule Bibliothek, Zürich.

triangles) in 1595. It was that brevity and clarity that made the *Trigonometria* a very popular book, to the extent that it was reissued three times in Latin and was soon translated into English.[77]

As for the usual abbreviations that are so familiar today, it was Fincke who introduced "sin.", "tan.", and "sec." in Book 14 of his *Geometriæ rotundi*, in connection with spherical trigonometry. While these were adopted by Tycho Brahe in a manuscript of 1591 and by Pitiscus in the second edition of his *Trigonometriæ* of 1600, they were not universally accepted at that time. It

---

[77] The word "trigonometry" first appeared in English in 1614 in the translation of the same work as *Trigonometry, or, the doctrine of triangles* by Raph Handson.

was not until 1633 that William Oughtred (1574–1660) used "sin" (without the period) for sine, in *An addition vnto the vse of the instrument called the circles of proportion for the working of nauticall questions* (an addition to *The circles of proportion*).

The word cosine was a late arrival in comparison with tangent and secant. After all, it was not needed, since it was very easy to avoid inventing cos A by referring to $\sin(90° - A)$. In this manner, it was called the complement of the sine or, as Regiomontanus and Viète later wrote, *sinus complementi*. But then, from this description, the Englishman Edmund Gunter (1581–1626) coined the term *co.sinus* in his *Canon triangulorum* of 1620, as well as the term *co.tangens*. They did not gain general acceptance, but John Newton (1622–1678), a teacher of mathematics and textbook author, did adopt them, while changing *co.sinus* into *cosinus*, in his *Trigonometria Britanica* of 1658. This term fared better, and Sir Jonas Moore (1627–1679) in his *A mathematical compendium* of 1674, abbreviated it to "Cos." and *co.tangens* to "Cot.".

However, there was a reversal of usage during the eighteenth century, when three-letter abbreviations ending with a dot became commonplace on the Continent. In the main, the English reverted to two-letter abbreviations without a dot. It was probably Euler's use of "sin.", "tan.", "sec.", and "cot." in Articles 242 to 249 of his *Introductio in analysin infinitorum* of 1748 that sealed the use of this notation. But Euler used "tang." and "cosec.", and on the use of abbreviations for these functions there are variations to this day.

2

LOGARITHMS

2.1 NAPIER'S FIRST THREE TABLES

The mists of sixteenth-century Scotland frequently engulfed Merchiston Tower, near Edinburgh. But today Merchiston Tower is in downtown Edinburgh, right at the heart of the Merchiston Campus of Napier University. In the sixteenth

MERCHISTON TOWER IN THE EIGHTEENTH CENTURY
The lower building with slanted roof and the chimney are eighteenth-century additions.
From Knott, *Napier tercentenary memorial volume*, 1915, p. 54.

century its seventh laird was Archibald Napier (Neper, Nepier, Nepair, Napeir, Nappier, Napper?). John Napier (Nepero in Latin, as we shall see later) was born in 1550 to Sir Archibald and his first wife, Janet Bothwell. At that time, Mary Tudor (later Stuart) was queen of the Scots and Edward VI king of the English. Three years earlier—to put things in perspective—Henry VIII had died, Cervantes was born, and Ivan IV (the Terrible) crowned himself tsar of Russia, the first with that title.

John Napier, an amateur Calvinist theologian who had studied religion at Saint Andrews, predicted that the end of the world would occur in the years between 1688 and 1700, in his book *A plaine discouery of the whole Reuelation of Saint Iohn* of 1593. Not a huge success at prediction, it must be pointed out. Napier's reputation may not have reached the present time either based on his theological studies or on his activities as a designer of weapons of mass destruction, intended for use against the enemies of the true religion. His lasting reputation is due to his third hobby: mathematical computation. As a landowner, he never held a job, and his time and energy could be entirely

POSSIBLY JOHN NAPIER, COMPUTING
Portrait by Francesco Delarame, engraving by Robert Cooper (c. 1810).

devoted to intellectual pursuits.

He developed an interest in reducing the labor required by the many tedious computations that were necessary in astronomical work, involving operations with the very large values of the trigonometric lengths given in the tables of his times. His interest in these matters might have been rekindled in 1590 on the occasion of a visit by Dr. John Craig, the king's physician. James VI of Scotland had set out to fulfill an important royal duty, to find a wife, and had selected Princess Anne of Denmark. In 1590, on the consent of her brother Christian IV, James set sail to Copenhagen to pick her up. Bad weather, however, caused his ship to land first on the island of Hveen near Copenhagen, which was the location of *Uraniborg*, the astronomical observatory of Tycho Brahe.

On his return from Denmark, Dr. John Craig paid the aforementioned visit to Napier to inform him of his findings while visiting Tycho Brahe. There were new, ingenious ways to perform some of those tedious computations required by astronomical calculations, and foremost was the use of *prosthaphæresis*. This imposing name, coined from the Greek words for addition ($\pi\rho o\sigma\theta\acute{\epsilon}\sigma\iota\varsigma$) and subtraction ($\acute{\alpha}\varphi\alpha\acute{\iota}\rho\epsilon\sigma\iota\varsigma$), refers to the use of the trigonometric identity

$$\sin A \cdot \sin B = \tfrac{1}{2}[\cos(A - B) - \cos(A + B)],$$

brought to Hveen by Paul Wittich (c. 1546–1586), an itinerant mathematician who visited *Uraniborg* for four months in 1580.[1] So, why should this be all the rage in computation? Let us say that we want to evaluate the product of two 10-digit numbers a and b. We know that tables of sine and cosine values were available at that time, and it was not a difficult matter to find numbers A and B, at least approximately, such that $a = \sin A$ and $b = \sin B$ (it may be necessary to divide or multiply a and b by a power of 10, but this can easily be managed). Then perform the simple operations $A - B$ and $A + B$, find their cosines in the table, subtract them as indicated in the previous formula, and divide by 2, and that is the product ab (except for an easy adjustment by a power of 10). The big problem of multiplying many-digit numbers was thus reduced to the simpler one of addition and subtraction.

But what about quotients, exponentiations, and roots? Napier sought a general method to deal with these computations, eventually found it, and

---

[1] Its European origin goes back to Johannes Werner, who probably discovered it about 1510, but a similar formula for the product of cosines is usually credited to the Egyptian Abu'l-Hasan Ali ibn Abd al-Rahman ibn Ahmad ibn Yunus al-Sadafi al-Misri (c. 950–1009). Wittich's contribution was his realization that it can be used for multiplication of any two numbers. See Thoren, *The Lord of Uraniborg*, 1990, p. 237.

gave it to the world in his book *Mirifici logarithmorum canonis descriptio* (Description of the admirable table of logarithms) of 1614 (he had become eighth laird of Merchiston in 1608). The English translation's author's preface starts with a statement of his purpose [p. A_5^2]:

> Seeing there is nothing (right well beloued Students in the Mathematickes) that is so troublesome to Mathematicall practise, nor that doth more molest and hinder Calculators, then the Multiplications, Diuisions, square and cubical Extractions of great numbers, which besides the tedious expence of time, are for the most part subject to many slippery errors. I began therefore to consider in my minde, by what certaine and ready Art I might remoue those hindrances.[2]

This is a small volume of 147 pages, 90 of which are devoted to mathematical tables containing a list of numbers, mysteriously called *logarithms*, whose use would facilitate all kinds of computations. Before the tables, some geometrical theorems are given in the *Descriptio* about their properties, and examples are provided of their usefulness. About the use of logarithms we shall talk presently, but first we should say that no explanation was given in this book about how they were computed. Instead, there is the following disclaimer in an *Admonition* in Chapter 2 of *The first Booke* [pp. 9–10]:

> Now by what kinde of account or method of calculating they may be had, it should here bee shewed. But because we do here set down the whole Tables, ..., we make haste to the vse of them: that the vse and profit of the thing being first conceiued, the rest may please the more, being set forth hereafter, or else displease the lesse, being buried in silence.

In short, if his Tables were well received Napier would be happy to explain how the *logarithms* were constructed; otherwise, let them go into oblivion. It happened that the *Descriptio* was a huge editorial success; a book well received and frequently used by scientists all over the world. Then an explanatory book became necessary but, although it was probably written before the *Descriptio*, it was published only posthumously (Napier died in 1617, one year after Shakespeare and Cervantes) under the title *Mirifici logarithmorvm canonis constrvctio*, when his son Robert included it with the 1619 edition of the *Descriptio*.[3]

---

[2] From Wright's translation as *A description of the admirable table of logarithmes* ("all perused and approued by the Author"), 1616. Page references are to this edition.

[3] Page references are given by article number so that any of the available sources (see the bibliography) can be used to locate them. Quotations are from Macdonald's translation.

MIRIFICI

Logarithmorum
Canonis deſcriptio,

Ejuſque uſus, in utraque
Trigonometria; ut etiam in
omni Logiſtica Mathematica,
Ampliſsimi, Faciĺimi, &
expeditiſsimi explicatio.

Authore ac Inventore,
IOANNE NEPERO,
Barone Merchiſtonii,
&c. Scoto.

EDINBURGI,
Ex officinâ ANDREÆ HART
Bibliopolæ, cIↃ. DC. XIV.

From Knott, *Napier tercentenary memorial volume*, Plate IX facing page 181.

The *Constrvctio* gives a glimpse into Napier's mind and his possible sources of inspiration. It starts as follows:

1. A logarithmic table [*tabula artificialis*] is a small table by the use of which we can obtain a knowledge of all geometrical dimensions and motions in space, by a very easy calculation.

... very easy, because by it all multiplications, divisions and the more difficult extractions of roots are avoided ...

It is picked out from numbers progressing in continuous proportion.

2. Of continuous progressions, an arithmetical is one which proceeds by equal intervals; a geometrical, one which advances by unequal and proportionately increasing or decreasing intervals.

It will be easier to interpret what Napier may have had in mind using modern notation. To us, a geometric progression is one of the form $a, ar, ar^2, ar^3, \ldots$, where a and r are numbers, and it is clear that the exponents of r form an arithmetic progression with step 1. It was common knowledge in Napier's time that the product or quotient of two terms of a geometric progression contains a power of r whose exponent is the sum or difference of the original exponents.[4]

To take a specific instance, Michael Stifel (1487–1567) had published a list of powers of 2 and their exponents in his *Arithmetica integra* of 1544, with the exponents in the top row (*in superiore ordine*) and the powers in the

0.	1.	2.	3.	4.	5.	6.	7.	8.
1.	2.	4.	8.	16.	32.	64.	128.	256.

bottom row (*in inferiore ordine*), and right below he taught his readers how to use the table to multiply and divide (translated from *Liber* III, p. 237):

> As the addition (in the top row) of 3 and 5 makes 8, thus (in the bottom row) the multiplication of 8 and 32 makes 256. Also 3 is the exponent of eight itself, & 5 is the exponent of the number 32. & 8 is the exponent of the number 256. Similarly just in the top row, from of the subtraction of 3 from 7, remains 4, thus in the bottom row the division of 128 by 8, makes 16.

That is, to multiply $8 = 2^3$ and $32 = 2^5$ just add the exponents to obtain $2^8 = 256$, and to divide $128 = 2^7$ by $8 = 2^3$ just subtract the exponents to obtain $2^4 = 16$. Thus multiplication and division are easily reduced to additions and subtractions of exponents.

---

[4] When writing r^n I am using current notation, which did not exist in Napier's time, for it was introduced by Descartes in *La Géométrie*, 1637. On page 299 of the original edition (as an appendix to the *Discours de la méthode*, *La Géométrie* starts on page 297) Descartes stated: *Et aa, ou a², pour multiplier a par soy mesme*; *Et a³, pour le multiplier encore vne fois par a*, & *ainsi a l'infini* (And *aa*, or a^2, to multiply a by itself; And a^3, to multiply it once more by a, and so on to infinity). Quoted from Smith and Latham, *The geometry of René Descartes*, 1954, p. 7. This statement can also be seen, in the English translation by Smith and Latham only, in Hawking, ed., *God created the integers, The mathematical breakthroughs that changed history*, 2005, p. 293.

But what about 183/11? There are no such table entries, not even approximately. The trouble is that the values of 2^n are very far apart. For values of r^n to be close to each other, r must be close to 1. Napier realized at this point that the computations of powers of such an r may not be easy (think of, say, $r = 1.00193786$), expressing himself as follows in the *Constrvctio*:

> 13. The construction of every arithmetical progression is easy; not so, however, of every geometrical progression. ... Those geometrical progressions alone are carried on easily which arise by subtraction of an easy part of the number from the whole number.

What Napier meant is that if r is cleverly chosen, multiplication by r can be reduced to subtraction. In fact, he proposed—in his own words in Article 14—to choose r of the form $r = 1 - 10^{-k}$. Then multiplication of a term of the progression by r to obtain the next one is equivalent to a simple subtraction:

$$ar^{n+1} = ar^n(1 - 10^{-k}) = ar^n - ar^n 10^{-k},$$

and the last term on the right is easily obtained by a shifting of the decimal point. With such an r, it is not necessary to evaluate its successive powers by multiplication to obtain the terms of the progression.

On this basis Napier made his choice of a and r. First he chose $a = 10,000,000$ because at this point in his life he was mainly interested in computing with sines, and the best tables at that time (c. 1590), those of Regiomontanus and Rhaeticus, took the whole sine (*sinus totus*) to be 10,000,000. Then, in order for the terms of his geometric progression to be sines (that is, parts of the whole sine), he had to choose $r < 1$; and for ease of computation he selected

$$r = 1 - 0.0000001.$$

We have talked about shifting the decimal point as a simple operation. Was such a procedure, or even the decimal point, available in Napier's time? The use of decimal fractions can be found in China, the Islamic world, and Renaissance Europe. It eventually replaced the use of sexagesimal fractions, then in vogue, after Simon Stevin (1548–1620), of Bruges, explained the system clearly in his book in Flemish *De thiende* ("The tenth") of 1585. Their use had already been urged in the strongest terms by François Viète in his *Vniversalivm inspectionvm* of 1579 [p. 17]:

> Finally, sexagesimals & sixties are to be used sparingly or not at all in Mathematics, but thousandths & thousands, hundredths & hundreds, tenths & tens,

and their true families in Arithmetic, increasing & decreasing, are to be used frequently or always.[5]

Not only was Napier aware of decimal fractions and points, but, through his continued use of decimal notation in the *Constrvctio*, he may have been instrumental in making this notation popular in the world of mathematics. He explained it clearly at the beginning of the *Constrvctio*:

5. In numbers distinguished thus by a period in their midst, whatever is written after the period is a fraction, the denominator of which is unity with as many ciphers [zeros] after it as there are figures after the period.

Thus 10000000·04 is the same as $10000000\frac{4}{100}$; also 25·803 is the same as $25\frac{803}{1000}$; also 9999998·0005021 is the same as $9999998\frac{5021}{10000000}$ and so of others.

With his use of decimal notation thus clearly settled and his choice of a and r, Napier then continued his construction in Article 16 as follows:

Thus from the radius, with seven ciphers added for greater accuracy [that is, with seven zeros after the decimal point], namely, 10000000.0000000, subtract 1.0000000, you get 9999999.0000000; from this subtract .9999999, you get 9999998.0000001; and proceed in this way until you create a hundred proportionals, the last of which, if you have computed rightly, will be 9999900.0004950.

To sum up in present-day notation, Napier had applied the formula

$$10^7 r^{n+1} = 10^7 r^n (1 - 10^{-7}) = 10^7 r^n - r^n$$

---

[5] But Viète did not use the decimal point. Instead, he wrote the decimal part of a number as a fraction with an empty denominator, giving the semicircumference of a circle whose *sinus totus* is 100,000 as 314,159,$\frac{265,36}{}$ [p. 15]. Viète used this notation through most of the book, but later he modified it, using first a vertical bar to separate the decimal part, which was also written in smaller or lighter type. For instance, he gave the *sinus* of 60° as 86,602|540,37 [p. 64]. Later, he replaced the vertical bar with another comma, writing now the semicircumference of the given circle as 314,159,265,36 [p. 69].

with $n = 0, \ldots, 100$ to perform the following computations:[6]

$$10000000 \times 0.9999999^0 = 10000000$$
$$10000000 \times 0.9999999^1 = 9999999$$
$$10000000 \times 0.9999999^2 = 9999998.0000001$$
$$10000000 \times 0.9999999^3 = 9999997.0000003$$

$$\cdot \quad \cdot \quad \cdot \quad \cdot \quad \cdot \quad \cdot \quad \cdot \quad \cdot \quad \cdot \quad \cdot \quad \cdot$$

$$10000000 \times 0.9999999^{100} = 9999900.0004950.$$

This completes his *First table*, which, writing in Latin, he called "canon." Of course, the entries on the right-hand sides of the fourth line and those below are only approximations, as shown by the fact that we have only subtracted 0.9999998 from the third entry to obtain the fourth, but Napier had already remarked in Article 6: *When the tables are computed, the fractions following the period may then be rejected without any sensible error.*

This first table is just the start of Napier's work, and these entries do not even appear in the *Descriptio*. The use of this table in computation is limited in at least two ways (we will mention a third later on). The first is that it contains very few entries, and the second is that all the numbers on the right-hand sides are very close to 10,000,000 due to the choice of r. But these problems are easy to deal with by computing more entries and choosing a different value of r.

Thus Napier started another table, but this time, to move a little further from 10,000,000, he chose r to be the last number in the first canon (omitting the decimals, for dealing with such large numbers one need not trouble with them) over the first one,

$$r = \frac{9999900}{10000000} = 0.99999 = 1 - \frac{1}{100000},$$

---

[6] At the risk of delving into tedium by harping on mathematical notation, the $\times$ sign for multiplication did not exist in Napier's time. It was introduced for the first time in an anonymous appendix to the 1618 edition of Wright's translation of the *Descriptio* entitled "An appendix to the logarithmes," pp. 1–16, probably written by William Oughtred. It is reprinted in Glaisher, "The earliest use of the radix method for calculating logarithms, with historical notices relating to the contributions of Oughtred and others to mathematical notation," 1915. It also appears in Oughtred, *Clavis mathematicæ* (The key to mathematics), 1631. However, the $=$ sign that we use below did exist; it had been introduced by Robert Recorde (1510–1558) in his algebra book *The whetstone of witte* of 1557. The original symbol was longer than the present one, representing two parallel lines "bicause noe. 2. thynges, can be moare equalle." This book has unnumbered pages, but is divided into named parts. This quotation is from the third page of "The rule of equation, commonly called Algebers Rule." This is page 222 of the text proper, not counting the front matter.

and computed 51 entries by the same method as before, but now using the
equation $10^7 r^{n+1} = 10^7 r^n - 100 r^n$ to obtain

$$10000000 \times 0.99999^0 = 10000000$$
$$10000000 \times 0.99999^1 = 9999900$$

$$\cdot \quad \cdot \quad \cdot \quad \cdot \quad \cdot \quad \cdot \quad \cdot \quad \cdot$$

$$10000000 \times 0.99999^{50} = 9995001.222927$$

(he erred: the last entry should be 9995001.224804). This is his second table.
Then he started again with the ratio

$$r = \frac{9995000}{10000000} = 0.9995 = 1 - \frac{1}{2000}$$

and computed 21 entries:

$$10000000 \times 0.9995^0 = 10000000$$

$$\cdot \quad \cdot \quad \cdot \quad \cdot \quad \cdot \quad \cdot \quad \cdot \quad \cdot$$

$$10000000 \times 0.9995^{20} = 9900473.57808.$$

He placed the numbers so obtained in the first column of his third canon.
Finally, from each of these values he computed 68 additional entries using the
ratio

$$0.99 \approx \frac{9900473}{10000000},$$

and these he located horizontally from each first value. Thus, his third table had
69 columns, and the entries in each column are 0.99 of those in the preceding
column. The last entry of the last column is

$$9900473.57808 \times 0.99^{68} \approx 4998609.4034,$$

"roughly half the original number" (again, the last two decimal digits are in
error).

From Napier's point of view, interested as he was at the time in computations with sines and applications to astronomy, this was enough, since—other
than the magnification by 10,000,000—the right-hand sides above are the
sines of angles from 90° to 30°. He would easily deal with smaller angles in
Articles 51 and 52 of the *Constrvctio*.

First Column.	Second Column.		69th Column.
10000000.0000	9900000.0000		5048858.8900
9995000.0000	9895050.0000		5046334.4605
9990002.5000	9890102.4750		5043811.2932
9985007.4987	9885157.4237		5041289.3879
9980014.9950	9880214.8451	$\cdots$	5038768.7435
.	.		.
.	.		.
.	.		.
9900473.5780	9801468.8423		4998609.4034

Now let us take stock of what Napier had and had not achieved to this point. He had elaborated several tables and subtables based on the idea of using the arithmetic progressions of what we now call exponents to perform computations with the table numbers. These contain about 1600 entries, not quite a sufficient number; and, what is more, the various tables are based on different choices of r, so that the product or quotient of numbers from different tables is impossible. Furthermore, there is a third problem in that his table entries are laden with decimals, while the sines in the tables of his time were all whole numbers (for simplicity, we shall refer to a *sinus* as a sine from now on). Something else had to be done.

2.2 NAPIER'S LOGARITHMS

What Napier needed is the reverse of what he had. In his first three tables the exponents are integers but the sines are laden with decimals. What he needed was to start with sines that are whole numbers, available in published tables, and then compute the corresponding exponents whatever they may be.

His next idea was based on a graphical representation, and once polished he gave it to us as follows in the *Constrvctio*. First he chose a segment ST, of length 10000000, to represent the *Sinus Totus* [Art. 24], and let 1, 2, 3, and 4 be points located in the segment TS as shown. Assume that TS, $1S$, $2S$, $3S$, $4S, \ldots$ are sines in continued proportion

$$\frac{1S}{TS} = \frac{2S}{1S} = \frac{3S}{2S} = \frac{4S}{3S} = \cdots,$$

such as those of the First table, in which case the stated ratio is 0.9999999, $1S = 9999999$, and so on. Then he envisioned a point g that, starting at T with a given initial velocity, proceeds toward S with decreasing velocity, in such a manner that the moving point [*Descriptio*, p. 2]

> in equall times, cutteth off parts continually of the same proportion to the lines [segments] from which they are cut off.

This means, referring to the figure above and noting that subtracting each of the preceding fractions from 1 and simplifying gives

$$\frac{T1}{TS} = \frac{12}{1S} = \frac{23}{2S} = \frac{34}{3S} = \cdots,$$

that g traverses each of the segments $T1$, 12, 23, 34, ... in equal times. Napier referred to the described motion of g as *geometrical* (hence his use of the letter g). Differential equations were not known in Napier's time, and he could not explain this motion as clearly as we might wish.

Assume next that $1S$, $2S$, $3S$, $4S$, ... are the sines in the First table, and imagine an infinite half-line bi (drawn in the next figure), on which we choose coordinates $0, 1, 2, 3, 4, \ldots$ as follows. First, the origin is at b, and then let a new point a move on this line from b to the right with constant velocity, that of g when at T. We choose the coordinates $1, 2, 3, 4, \ldots$ (not shown in Napier's figure) to be the points on bi at which a arrives when g arrives at the points with labels $1, 2, 3, 4, \ldots$. Thus, as g passes through the points marking the sines of the First table in equal amounts of time, a reaches the exponents $1, 2, 3, 4, \ldots$ that generate these sines.

Napier chose to introduce this second line in a general situation, considering the case of an arbitrary sine as follows [Art. 26]:

> Let the line TS be radius, and dS a given sine in the same line: let g move geometrically from T to d in certain determinate moments of time. Again, let bi be another line, infinite towards i, along which, from b, let a move arithmetically [constant velocity] with the same velocity as g had at first when at T: and from the fixed point b in the direction of i let a advance in just the same moments of time [as it took g to advance to d] to the point c.

Then, in general, if the coordinates on the line bi are as we have just described, if dS is an arbitrary sine, if g and a start with the same initial velocity, and if a arrives at a point c in the same amount of time as g arrives at d, the distance bc represents the exponent—not necessarily an integer—corresponding to the sine dS.

In the *Constrvctio* [Article 26], Napier called the distance bc the "artificial number" of the "natural number" dS, but he later changed his terminology. Since the sines in Napier's basic tables are in continued proportion or ratio, he later coined the word logarithm, from the Greek words λόγων (*logon*) = ratio and ἀριθμός (*arithmos*) = number, to refer to each of these exponents. In the *Descriptio*, written a few years after the *Constrvctio*, he gave the definition as follows [pp. 4–5]:

> The Logarithme therefore of any sine is a number very neerely expressing the line, which increased equally in the meane time, while the line of the whole sine decreased proportionally into that sine, both motions being equal-timed, and the beginning equally swift.

By the time the translation of the *Constrvctio* was prepared, the word logarithm was already in common use, and that motivated the translator to use it there. We shall do so from now on.

At this point, Napier faced the task of computing the logarithms of sines given by whole numbers. The start is easy:

27. Whence nothing [zero] is the logarithm of radius.

Quite clearly: if g and a do not move they remain at T and b, respectively. But the rest, the exact computation of logarithms of whole numbers, was far from trivial with the tools available to Napier (the already mentioned lack of differential equations was responsible). On the face of that impossibility, he made a giant leap by realizing that it may be good enough to find lower and upper bounds for such logarithms. With a certain amount of ingenuity he found such bounds.

28. Whence also it follows that the logarithm of any given sine is greater than the difference between radius and the given sine, and less than the difference between radius and the quantity which exceeds it in the ratio of radius to the given sine. And these differences are therefore called the limits of the logarithm.

This can be explained, and then his proof abbreviated, as follows. Let dS be the given sine and let oS be the quantity that exceeds the radius in the ratio

$$\frac{oS}{TS} = \frac{TS}{dS}.$$

According to the geometric motion of g and the arithmetical motion of a, "oT, Td, and bc are distances traversed in equal times." Then, since the velocities

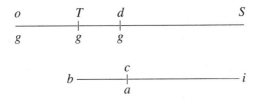

of g and a are equal when g is at T and a at b, and since the velocity of g is decreasing but that of a is constant,

$$oS - TS = oT > bc \ (= \log dS) > Td = TS - dS,$$

which was to be proved.

It goes without saying that a present-day reader would like a restatement of Napier's result in current notation.[7] That is not difficult if we denote the given sine dS by x. Then $oS = TS^2/x$, and we can write the inequalities of Article 28 in the form

$$TS - x < \log x < \frac{TS^2}{x} - TS = \frac{TS(TS - x)}{x},$$

to which we have appended an equality from Article 29. It was this result that enabled Napier to find the logarithms of many sines, starting with those in the

---

[7] Napier stated all of his results in narrative form. The *Constrvctio* does not contain a single modern-looking formula, nor is any notation used for "artificial number" or logarithm. I have, however, adopted log as a natural choice, since there is no possibility of confusing this with any current mathematical function at this point.

First table (they are not all whole numbers, but their logarithms will be useful nevertheless). Starting with $x = 9999999$ we have $TS - x = 1.0000000$ and

$$\frac{TS(TS - x)}{x} = \frac{10000000}{9999999} = 1.00000010000001.^{8}$$

Then, since these limits differ insensibly, Napier took [Art. 31]

$$\log 9999999 = 1.00000005.$$

Now, if we denote once more the sines in the First table—that is, the numbers on the right of the equal signs—by TS, $1S$, $2S$, $3S$, $\ldots$, we have already seen that as g goes through them in equal times, their logarithms follow the motion of a and we have [Art. 33]

$$\log 2S = 2 \log 1S, \quad \log 3S = 3 \log 1S,$$

and so on. In particular, for the last sine in the First table we have

$$\log 9999900.0004950 = 100 \log 1S = 100 \log 9999999,$$

and then, from the bounds on $\log 9999999$,

$$100 < \log 9999900.0004950 < 100.00001000.$$

We can pick $\log 9999900.0004950 = 100.000005$.

This method works well because the sines in the First table are very close to 10000000. But the theorem in Article 28 is not sufficiently accurate further down the line. For instance, if $x = 9000000$, which is still a lot closer to 10000000 than to zero, it would give

$$10000000 - 9000000 < \log 9000000 < \frac{10000000(10000000 - 9000000)}{9000000}.$$

That is, $1000000 < \log 9000000 < 1111111$, which is too wide a range to select $\log 9000000$ with any accuracy.

Thus Napier needed a better theorem, and actually found two. The first is:

36. The logarithms of similarly proportioned sines are equidifferent.

---

[8] A reader who is concerned that the application of this formula requires long division is invited to count the number of such divisions that will be necessary to elaborate Napier's complete table of logarithms. The total count will turn out to be extremely small.

This means that if v, x, y, and z are sines such that

$$\frac{x}{v} = \frac{z}{y},$$

then

$$\log x - \log v = \log z - \log y.$$

Napier stated [Art. 32]: "This necessarily follows from the definitions of a logarithm and of the two motions." This is because g traverses the distance $x - v$ in the same amount of time as $z - y$, and then a moves from $\log x$ to $\log v$ in the same amount of time as from $\log z$ to $\log y$.

His second theorem is a little more involved:

39. The difference of the logarithms of two sines lies between two limits; the greater limit being to radius as the difference of the sines to the less sine, and the less limit being to radius as the difference of the sines to the greater sine.

After providing a geometric proof similar to that in Article 28 but a little longer and requiring Article 36, he rephrased this result in a verbal form [Art. 40] that can be rewritten in today's terminology as follows. If x and y are the two sines and y is the larger, then

$$\frac{TS(y - x)}{y} < \log x - \log y < \frac{TS(y - x)}{x}.$$

Then he immediately put this result to good use by showing how to compute the logarithm of any number that is not a sine in the First table, but near one of them [Art. 41].

For example, let $x = 9999900$. In the First table the nearest sine is $y = 9999900.0004950 = 10000000 \times 0.9999999^{100}$ and, as shown in Article 33 (page 92),

$$100 < \log y < 100.0000100.$$

Then, with x and y as above, we have $TS(y - x) = 4950$, and the inequalities in Article 40 (stated here after quoting Article 39) give the approximate inequalities

$$0.00049500495002 < \log x - \log y < 0.00049500495005.$$

Napier did not state these inequalities, but only concluded that we can take $\log x - \log y = 0.0004950$, and then

$$\log x = \log y + 0.0004950.$$

Hence, from the bounds for log y,

$$100.0004950 < \log x < 100.0005050.$$

But $x = 9999900$ appears in the Second table as $10000000 \times 0.99999^1$, and the sines in this table are in continued proportion, just as TS, $1S$, $2S$, $3S, \ldots$ are in Napier's general formulation. As g goes through them in equal times their logarithms are found from the motion of a in the same equal times. They are double, triple, etc., the logarithm of x. Thus, from the limits found above for log x, the limits for the logarithms of the sines of the Second table are 200.0009900 and 200.0010100 for the second, 300.0014850 and 300.0015150 for the third, and so on until the limits for the logarithm of the last sine (the 50th after the first) are found to be

$$5000.0247500 < \log 9995001.222927 < 5000.0252500.$$

The actual logarithms of these sines can be chosen between the limits stated here [Art. 42]. Thus, the logarithms of the sines of the Second table have been quickly obtained by a cascading procedure identical to that giving those of the sines in the First table.

Next, in [Art. 43], "to find the logarithms of ... natural numbers ... not in the Second table, but near or between" the sines in this table, Napier used a new method. As above, start by naming a given sine and the one nearest to it in the Second table by x and y, with $y > x$. But now choose what Napier called a *fourth proportional*, that is, a number z such that (in our terminology)

$$\frac{z}{TS} = \frac{x}{y},$$

which is possible because the other three numbers are known. This places z within the bounds of the First table because $x \approx y$ implies that $z \approx TS$, and then the logarithm of z can be computed by the previous method. Then, the theorem in Article 36 and the fact that $\log TS = 0$ give

$$\log z = \log x - \log y.$$

In this equation, two of the logarithms are known and the third can then be found. Or, more cautiously, the limits for two of these logarithms can be determined, whence the limits for the third can be ascertained.

Napier applied this procedure to obtain the logarithm of $x = 9995000$, which is near the last entry, $y = 9995001.222927$, of the Second table. We have already found that

$$5000.0247500 < \log y < 5000.0252500,$$

and the fourth proportional is

$$z = TS \frac{9995000}{9995001.222927} = 9999998.7764614.$$

Then, replacing x with z and y with 9999999 in the limits of Articles 39 and 40, yields

$$\frac{TS(9999999 - z)}{9999999} = 0.2235386$$

and

$$\frac{TS(9999999 - z)}{z} = 0.2235386.$$

It follows that

$$\log z - \log 9999999 = 0.2235386.$$

Since we know that $1 < \log 9999999 < 1.0000001$, we obtain

$$1.2235386 < \log z < 1.2235387.$$

The equation $\log x = \log y + \log z$ and the limits found for the two logarithms on the right-hand side give (these are Napier's figures)

$$5001.2482886 < \log x < 5001.2487888.$$

From these limits Napier concluded as follows:

> Whence the number 5001.2485387, midway between them, is (by 31) taken most suitably, and with no sensible error, for the actual logarithm of the given sine 9995000.

Unfortunately, his result is a little off because the last entry of the Second table is in error, as mentioned above. On the good side of things, this got him into the Third table, whose second entry is precisely 9995000. Then the logarithms of the first two entries in the first column of the Third table are known, and the result in Article 36 allowed him to cascade down, computing the logarithms of all the entries in this column [Art. 44]. Then the logarithms of numbers that are near or between the entries of this column can be computed by the method of the fourth proportional illustrated above. Using this method, he computed the logarithm of 9900000, which is near the last entry of the first column of the Third table [Art. 45]. But this is the first entry in the second column. Now denote the first two sines in the first column by x and y and the first two sines in the second column by u and v. We know that

$$\frac{u}{x} = \frac{v}{y},$$

and from this that

$$\frac{v}{u} = \frac{y}{x}.$$

By Article 36, $\log v - \log u = \log y - \log x$, and, since the last difference is known, $\log v$ can be computed. Cascading down, the logarithms of all the sines in this second column can be found. Finally, once the logarithms of the numbers in the first two columns were obtained, the fact that the entries in each row of the Third table are in continued proportion means that their logarithms are equidifferent. This allowed Napier to compute the logarithms of all the entries in the Third table [Art. 46].

Having completed this second stage of his work, Napier now discarded the First and Second tables, keeping only the Third table with the logarithms of all its entries. This—the framework for the construction of the complete logarithmic table—he called *The Radical Table* [Art. 47]. A portion of it is shown on the next page. He considered it sufficient to leave just one decimal digit in the logarithms (*Artificiales*), while he had used seven in all preliminary computations to avoid error accumulation. It is not important to us that these logarithms are slightly erroneous due to the faulty last entry in the Second table (the last one of the 69th column should be 6934253.4), because we are interested in Napier's invention and how he constructed the table, but not in using it.

Then Napier embarked on the third stage of his work: the elaboration of a table of logarithms, which he called the *principal table*, for the sines of angles from 45° to 90° in steps of 1 minute. These can be obtained from any table of common sines. For each sine he found first the entry in the Radical table that is closest to it. But now that seven-decimal-digit accuracy was not needed, the method of the fourth proportional becomes irrelevant, and even the use of the inequalities in Article 38. What counted at this stage was speed, so that the principal table can actually be constructed. So, given a sine and its closest entry in the Radical table, x and y with $y > x$, compute $TS(y - x)$, and then, instead of dividing this product first by y and then by x to find the limits of $\log x - \log y$, just divide it by the easiest possible number between x and y. Then the result is an approximation of the true value of $\log x - \log y$, and the unknown logarithm can be approximately found from it.

RADICALIS TABVLÆ.

Columna prima.		Columna secunda.	
Naturales.	Artificiales	Naturales.	Artificiales
10000000.0000	0	9900000.0000	100503.3
9995000.0000	5001.2	9895050.0000	105504.6
9990002.5006	10001.5	9890102.4750	110505.8
9985007.4987	15003.7	9885157.4237	115507.1.
9980014.9950	20005.0	9880214.8451	120508.3
Reciminoau ad	*usque ad*	*usque ad*	*usque ad*
9900473.5780	100025.0	9801468.8423	200528.2

Columna 69.

	Naturales.	Artificiales.
& cæteri vsque ad	5048858.8900	6834225.8
	5046333.4605	6839227.1
	5043811.2932	6844228.3
	5041289.3879	6849229.6
	5038768.7435	6854230.8
	&tandem	*usque ad*
	4998609.4034	6934250.8

Reproduced from page 25 of the 1620 edition of the *Constrvctio*.

For example [Art. 50], if $y = 7071068$, the nearest sine in the Radical table is $x = 7070084.4434$, and then $TS(y - x) = 9835566000$. Dividing this result by the easiest number between x and y, which Napier took to be 7071000, "there comes out" 1390.9. Thus, taking $\log x$ from the Radical table, gives us

$$\log y \approx \log x - 1390.9 = 3467125.4 - 1390.9 = 3465734.5.$$

"Wherefore 3465735 is assigned for the required logarithm of the given sine 7071068."

All it took from this point on was time and effort, and Napier was eventually able to compute the logarithms of the sines of all angles between 45° and 90° in steps of 1 minute. All of this by dipping his quill in ink and partly by candlelight, long division after long division.

The final stage would be to deal with the angles between 0° and 45°, since we know that what really matters is what happens in the first quadrant. To do this, Napier proposed two methods, both of which are based on the theorem in Article 36.

First, if we use the equations after the quotation of this theorem with $x = 2v$, $y = 5000000$, and $z = TS$, we have

$$\log 2v - \log v = \log 10000000 - \log 5000000,$$

and taking the last logarithm from the previously computed part of the principal table, but before rounding it as an integer, yields

$$\log v = \log 2v + \log 5000000 = \log 2v + 6931469.22.$$

Or as Napier put it [Art. 51],

> Therefore, also, 6931469.22 will be the difference of all logarithms whose sines are in the proportion of two to one. Consequently the double of it, namely 13862938.44, will be the difference of all logarithms whose sines are in the ratio of four to one; and the triple of it, namely 20794407.66, will be the difference of all logarithms whose sines are in the ratio of eight to one.

Not content with this, he showed, in a similar manner, that

> 52. All sines in the proportion of ten to one have 23025842.34 for the difference of their logarithms.

That this figure should have been 23025850.93 is neither here nor there. What matters is that he could now find the logarithms of the sines of angles below 45°. To quote his own statement [Art. 54]:

> This is easily done by multiplying the given sine by 2, 4, 8, 10, 20, 40, 80, 100, 200, or any other proportional number you please ... until you obtain a number within the limits of the Radical table.

The second method is based on a trigonometric identity that he expressed as follows:

> 55. As half radius is to the sine of half a given arc, so is the sine of the complement of the half arc to the sine of the whole arc.

He provided an ingenious geometric proof of this result, which, if we keep in mind that his sine is the product of the radius R and our sine, can be rewritten as follows:

$$\frac{R/2}{R \sin \frac{1}{2}\alpha} = \frac{R \sin\left(90° - \frac{1}{2}\alpha\right)}{R \sin \alpha}.$$

Today we would write this result in the familiar form $\sin \alpha = 2 \sin \frac{1}{2}\alpha \cos \frac{1}{2}\alpha$. Then he used the theorem in Article 36 again to obtain in verbal form a result that can be rewritten as

$$\log \frac{R}{2} - \log\left(R \sin \frac{1}{2}\alpha\right) = \log\left[R \sin\left(90° - \frac{1}{2}\alpha\right)\right] - \log\left(R \sin \alpha\right).$$

Then, if $22.5° < \frac{1}{2}\alpha \leq 45°$, the two logarithms on the right-hand side as well as that of $R/2$ are in the principal table, and the remaining logarithm can be determined. Napier concluded [Art. 58]:

> From these, again, may be had in like manner the logarithms of arcs down to 11 degrees 15 minutes. And from these the logarithms of arcs down to 5 degrees 38 minutes. And so on, successively, down to 1 minute.

And now there is nothing left but the actual construction of the logarithmic table. Napier gave precise instructions about how to do it step by step, starting with the following advice [Art. 59]:

> Prepare forty-five pages, somewhat long in shape, so that besides margins at the top and bottom, they may hold sixty lines of figures.

A page of the resulting table is reproduced on the next page. The publisher did not seem to have pages that could accommodate 60 lines of figures, so he put two columns of 30 lines on each page. The first column of this sample page contains the angles from 28° 0′ to 28° 30′, the second column lists their sines (probably taken from Erasmus Rheinhold's table of sines, which Napier mentioned in particular), and the third their logarithms, computed as explained in the preceding articles. The rest of the angles corresponding to 28° would be on the next page, not shown. On arrival at the page containing the sines and logarithms for angles from 44° 30′ to 44° 60′ no new pages were used. Instead, the table continued on the right of that last page with angles increasing from 45° 0′ at the bottom of the seventh column to 45° 30′ at the top. The angles located on both ends of each horizontal line are complementary. Their sines and logarithms are listed in the sixth and fifth columns, respectively. Then the table continued in the same manner on the preceding page and then on those

Deg. 28 +—

m.	Sines.	Logarith	Differen.	Logarit.	Sines.	
0	469472	756147	631658	124489	882948	60
1	469728	755600	630956	124644	882811	59
2	469985	755054	630255	124799	882674	58
3	470242	754508	629553	124954	882537	57
4	470499	653962	628853	125109	882401	56
5	470755	753416	628152	125264	882264	55
6	471012	752871	627452	125420	882127	54
7	471268	752327	626752	125575	881990	53
8	471525	751783	626052	125730	881853	52
9	471781	751239	625353	125886	881715	51
10	472038	750695	624654	126041	881578	50
11	472294	750152	623955	126197	881441	49
12	472551	749610	623256	126353	881303	48
13	472807	749067	622558	126509	881166	47
14	473063	748525	621860	126665	881028	46
15	473320	747984	621162	126822	880891	45
16	473576	747443	620465	126978	880753	44
17	473832	746902	619768	127134	880615	43
18	474088	746362	619071	127291	880477	42
19	474344	745822	618374	127448	880339	41
20	474600	745282	017677	127604	880101	40
21	474856	744743	616981	127761	880063	39
22	475112	744204	616285	127918	879925	38
23	475368	743665	615590	128075	879787	37
24	475624	743127	614894	128233	879649	36
25	475880	742589	614199	128390	879510	35
26	476136	742052	613506	128547	879372	34
27	476392	741515	112810	128705	879233	33
28	476647	740978	611115	128863	879095	32
29	476907	740443	611421	129020	878956	31
30	477259	739906	610727	129178	878817	30

Min.

Deg. 61

Reproduced from Wright's translation of the *Descriptio*.

previously used, as shown on the sample page. The central column contains the differences between the logarithm on its left and that on its right. These are the logarithms of the sine of a given angle and of its cosine (which is equal to the sine of the complement), and it is useful in evaluating the logarithms of tangents when viewed as quotients of sine and cosine.

It must be said for the sake of fairness that the earliest discoverer of logarithms was Joost, or Jobst, Bürgi (1552–1632), a Swiss clockmaker, about 1588. He seems to have been inspired by Stifel's table, but he chose 1.0001 as his starting number close to 1. Then, if $N = 100000000 \times 1.0001^{L}$, Bürgi called $10L$, not L, the "red number" (*Rote Zahl*) corresponding to the "black number" (*Schwarze Zahl*) N. These are the colors in which these numbers were printed in his 1620 tables *Aritmetische vnd geometrische Progress Tabulen*. Thus, he lost publication priority to Napier.

Bürgi was in Prague as an assistant to the astronomer Johannes Kepler (1571–1630). Kepler had become an assistant to Tycho Brahe in 1598, and in 1599 Brahe moved to Prague, attracted by an offer from the emperor Rudolf II (who reigned over the *Sacrum Romanum Imperium* in the period 1576–1612) and Kepler followed the year after. On Brahe's death in 1601, Kepler became the official astronomer and went on to discover the three laws of planetary motion. His work, based on Brahe's astronomical observations, was assisted by the use of logarithms, and this helped spread their use in Europe. Bürgi remained an unknown not just because of questions of priority, but because most copies of his tables were lost during the Thirty Years' War. The crucial Battle of the White Mountain, in which 7,000 men lost their lives, was fought just outside Prague, in November 1620.

2.3 BRIGGS' LOGARITHMS

The first mathematics professorship in England, a chair in geometry, was endowed by Sir Thomas Gresham in 1596 at Gresham College in London, and Henry Briggs (1561–1631) was its first occupant. He was quite impressed by the invention of logarithms, and on March 10, 1615, he wrote to his friend James Usher:

> *Napper*, Lord of *Markinston*, hath set my Head and Hands a Work, with his new and admirable Logarithms. I hope to see him this Summer if it please God, for I never saw Book that pleased me better, and made me more wonder.[9]

The astrologer William Lilly (1602–1681) wrote his autobiography in 1667–1668 in the form of a letter to a friend. It contains an interesting—but rather fantastic—account of Briggs' arrival at Merchiston, stating that on meeting Napier, "almost one quarter of an hour was spent, each beholding the other almost with admiration, before one word was spoke."[10]

---

[9] Quoted from Parr, *The life of the Most Reverend Father in God, James Usher*, 1686, p. 36 of the collection of three hundred letters appended at the end of the volume.

[10] Lilly finished his book with an account of the first encounter between Napier and Briggs, pp. 235–238. This quotation is from p. 236. The entire account of Briggs' arrival is reprinted (almost faithfully) in Bell, *Men of mathematics*, 1937, p. 526.

Having overcome this mutual admiration hiatus, they set down to work. For this there was ample time, since Briggs remained at Merchiston for about a month. Both men had become aware of some serious shortcomings in the logarithm scheme published by Napier. For instance, if $x = yz$, then

$$\frac{x}{y} = \frac{z}{1},$$

and, according to Article 36, $\log x - \log y = \log z - \log 1$, or

$$\log yz = \log y + \log z - \log 1.$$

This is as true today as it was then, but today $\log 1 = 0$. In Napier's scheme, it was the logarithm of the *sinus totus* that was zero. The logarithm of 1, a number very close to S in the top segment TS on page 90, is an enormous number on the bottom line. It was computed by Napier to be 161180896.38 (while this is not entirely correct, because of the errors already mentioned, the true value in his own system is very close to it). Thus, the Neperian logarithm of a product is not the sum of the logarithms of the factors, and the Neperian logarithm of 1 must be remembered every time the logarithm of a product is to be computed. Similarly for a quotient: if $x = y/z$, then $\log x - \log 1 = \log y - \log z$ and

$$\log \frac{y}{z} = \log y - \log z + \log 1.$$

The second difficulty in using Napier's logarithms is in finding the logarithm of ten times a number whose logarithm is known. If that number is called x, the obvious equation

$$\frac{10x}{x} = \frac{10^7}{10^6}$$

leads to

$$\log 10x - \log x = \log 10^7 - \log 10^6,$$

and, since the logarithm of the *sinus totus* is zero, we obtain

$$\log 10x = \log x - \log 10^6.$$

What makes the computation of $\log 10x$ in this manner inconvenient is the fact that $\log 10^6 = 23025842$.

Both Napier and Briggs had been separately thinking of ways to solve these deficiencies, and each of them made some proposals to accomplish that. These involved, in either case, the construction of a new set of logarithms from scratch. But they differed about the key properties on which the new logarithms should be based. Here is Briggs' proposal (see below for reference):

When I explained their [the logarithms'] doctrine publicly in London to my
auditors at Gresham College; I noticed that it would be very convenient in the
future, if 0 were kept for the Logarithm of the whole Sine (as in the *Canone
Mirifico*) while the Logarithm of the tenth part of the same whole Sine . . .
were 1,00000,00000 . . .

In short, Briggs proposed and started the construction of a new set of logarithms
based on the values $\log 10^7 = 0$ and $\log 10^6 = 10^{10}$. This certainly simplifies
the computation of $\log 10x$ from $\log x$.

Briggs wrote to Napier with this proposal, and later he journeyed to Edinburgh, where they met for the first time. It turns out that Napier had also been
thinking about the change in the logarithms, and Briggs described Napier's
counterproposal as follows:

But [he] recommended to therefore make this change, that 0 should be the
Logarithm of Unity, & 1,00000,00000 [that of] the whole Sine: which I could
not but acknowledge was most convenient by far.

Indeed, with $\log 1 = 0$, it follows from the preceding discussion that the
logarithm of a product or quotient becomes the sum or difference of the logarithms. So Briggs put aside the logarithms that he had already computed, and
in the following summer he again journeyed to Edinburgh and showed Napier
the new recomputed logarithms. They appeared first in a small pamphlet
of 16 pages, *Logarithmorum chilias prima* (The first thousand logarithms),
whose preface is dated 1617, the year of Napier's death. In 1624 he published
Arithmetica logarithmica sive logarithmorvm chiliades triginta (Logarithmic
arithmetic or thirty thousands of logarithms), containing the new logarithms
of the integers 1 to 20,000 and 90,000 to 100,000 (the reason for the large gap
in the table will be given later). The quotations translated above are from the
Preface to the Reader in this book.

However, we should not create the impression that he elaborated the new
logarithms on the basis of Napier's proposal of the summer of 1615: that
$\log 1 = 0$ and $\log 10^7 = 10^{10}$. The first of these equations he fully accepted
[p. 2; 2–1],[11] for it makes the logarithm of a product or quotient the sum or
difference of their logarithms [pp. 2,3; 2–2, 2–3].[12] Then it is easy to see

[11] References in square brackets are to the *Arithmetica logarithmica*. The first page
reference is to the original in Latin; the second, after the semicolon, to Bruce's translation
(see the bibliography) by chapter page (he does not have global page numbering).

[12] Briggs referred to these lemmas as axioms. These same properties, as well as the next
one about the logarithm of a rational power, had been more clearly stated by Napier in the
Appendix to the *Constrvctio*, pp. 50–51 of Macdonald's translation.

by repeated application of the product rule that if n is a positive integer, then $\log x^n = n \log x$. Also, if m is a positive integer and $y = x^{m/n}$, then $x^m = y^n$ and $m \log x = n \log y$, so that

$$\log x^{\frac{m}{n}} = \frac{m}{n} \log x,$$

which means that for a rational exponent $\log x^r = r \log x$. Thus, accepting the equation $\log 1 = 0$, we get these very convenient properties that Napier's own logarithms lack. However, Briggs rejected the second equation in favor of a new proposal made in the Appendix to the *Constrvctio*. This Appendix starts with the words [Macdonald translation, p. 48]:

Among the various improvements of Logarithms, the more important is that which adopts a cypher as the Logarithm of unity, and 10,000,000,000 as the

Logarithm of either one tenth of unity or ten times unity.

This is the suggestion that Briggs actually quasi-accepted, in the form $\log 10 = 10^{14}$ [p. 3; 3–1]. The large number of zeros was intended to provide sufficient accuracy without resorting to numbers containing the still too new decimal point and digits. Soon thereafter, when decimal fractions were fully accepted, the best choice was $\log 10 = 1$. This results in the same logarithms that Briggs computed but with a decimal point in their midst.

But how are all these logarithms computed? At the time that Briggs was about to embark on the elaboration of his tables (Napier's health was failing in his 65th year, so it was up to Briggs to start a new series of computations)

JOHN NAPIER IN 1616
Engraving by Samuel Freeman from the portrait
in the possession of the University of Edinburgh.
From Smith, *Portraits of Eminent Mathematicians*, II.

there were two main methods to calculate logarithms, and both of them had been published in the Appendix to Napier's *Constrvctio*, as Briggs himself stated [p. 5; 5–2]. We do not know whether these are due to Napier or are a product of his collaboration with Briggs at Merchiston. The first method was based on the extraction of fifth roots and the second on the extraction of square roots. Napier felt that "though this method [of the fifth root] is considerably more difficult, it is correspondingly more exact" [Appendix, p. 50]. Briggs chose the method of the square root, and started evaluating (by hand, of course, which may take several hours) the square root of 10 [p 10; 6–2]:[13]

$$10^{1/2} = \sqrt{10} = 3.16227766016837933199889354,$$

so that $\log 3.16227766016837933199889354 = 0.5$. Then, by successive extraction of square roots, he evaluated $10^{1/4}$, $10^{1/8}$, and so on, down to

$$10^{1/2^{53}} = 1.0000000000000000255638298640064707$$
and [p. 10; 6–3]

$$10^{1/2^{54}} = 1.00000000000000000127819149320032345.$$

Briggs' values can be seen on the next page, reproduced from Chapter 6, page 10, of the second edition of *Arithmetica logarithmica*. Note that he did not use the decimal point, but inserted commas to help with counting spaces. On the logarithm side he omitted many zeros, which can be confusing. We know he chose the logarithm of 10 to be what we express as 10^{14}, but in this table it appears as 1,000. Down at the bottom, these logarithms are given with forty-one digits, making it difficult to know what an actual logarithm in this range is. To avoid possible misinterpretations, we follow the modern practice of using a decimal point for the numbers, but shall retain instead the first of Briggs' commas for their logarithms. If this comma is read as a decimal point, these are today's decimal logarithms, while the Briggsian logarithms are 10^{14} times larger. It must be pointed out that Briggs made a mistake in his computation of $10^{1/4}$. His digits are wrong from the twentieth on. This mistake trickles down through his entire table, but because the wrong digits are so far on the right, the error becomes smaller as it propagates, and his last two entries are almost in complete agreement with values obtained today using a computer.

The end result of this stage of Briggs' work is a table of logarithms, the table shown on page 107, displaying in the second column the logarithms of

---

[13] Briggs denoted the square root of 10 by $l.10$, meaning the *latus* (side) of 10; that is, the side of a square of area 10. The square root of the square root of 10 would then be $ll.10$, the eighth root $l.(8)10$, and so on. I have adopted our usual exponential notation, unknown at that time, as most convenient for our purposes.

10	D *ARITHMTICA* E	
	Numeri continue Medij inter Denarium & Vnitatem.	*Logarithmi Rationales.*
10		1,000
1	31622,77660,16837,93319,98893,54	0,50
2	17782,79410,03892,28011,97304,13	0,25
3	13335,21432,16332,40256,65389,308	0,125
4	11547,81984,68945,81796,61918,213	0,0625
5	10746,07828,32131,74972,13817,6538	0,03125
6	10366,32928,43769,79972,90627,3131	0,01562,5
7	10181,51721,71818,18414,73723,8144	0,00781,25
8	10090,35044,84144,74377,59005,1391	0,00390,625
9	10045,07364,25446,25156,64670,6113	0,00195,3125
10	10022,51148,29291,29154,65611,7367	0,00097,65625
11	10011,24941,39987,98758,85395,51805	0,00048,82812,5
12	10005,62312,60220,86366,18495,91839	0,00024,41406,25
13	10002,81116,78778,01323,99249,64325	0,00012,20703,125
14	10001,40548,51694,72581,62767,32715	0,00006,10351,5625
15	10000,70271,78941,14355,38811,70845	0,00003,05175,78125
16	10000,35135,27746,18566,08581,37077	0,00001,52587,89062,5
17	10000,17567,48442,26738,33846,78274	0,00000,76293,94531,25
18	10000,08783,70363,46121,46574,07431	0,00000,38146,97265,625
19	10000,04391,84217,31672,36281,88083	0,00000,19073,48632,8125
20	10000,02195,91867,55542,03317,07719	0,00000,09536,74316,40625
21	10000,01097,95873,50204,09754,72940	0,00000,04768,37158,20312,5
22	10000,00548,97921,68211,14626,60250,4	0,00000,02384,18579,10156,25
23	10000,00274,48957,07382,95091,25449,9	0,00000,01192,09289,55078,125
24	10000,00137,24477,59510,83282,69572,5	0,00000,00596,04544,77539,0625
25	10000,00068,62238,56210,25737,18748,2	0,00000,00298,02322,38769,53125
26	10000,00034,31119,22218,83912,75020,8	0,00000,00149,01161,19384,76562,5
27	10000,00017,15559,59637,84719,93879,1	0,00000,00074,50580,59692,38281,25
28	10000,00008,57779,79451,03051,17588,8	0,00000,00037,25290,29846,19140,625
29	10000,00004,28889,89633,54198,42901,3	0,00000,00018,62645,14923,09570,3125
30	10000,00002,14444,94793,77767,42970,4	0,00000,00009,31322,57461,54785,15625
31	10000,00001,07222,47391,14050,76926,8	0,00000,00004,65661,28730,77392,57812,5
32	10000,00000,53611,23694,13317,14831,4	0,00000,00002,32830,64365,38696,28906,25
33	10000,00000,26805,61846,70731,51508,7	0,00000,00001,16415,32182,69348,14453,125
34	10000,00000,13402,80923,26383,99277,7	0,00000,00000,58207,66091,34674,07226,5625
35	10000,00000,06701,40461,60946,55519,6	0,00000,00000,29103,83045,67337,03613,28125
36	10000,00000,03350,70230,79911,91730,0	0,00000,00000,14551,91522,83668,51806,64062,5
37	10000,00000,01675,35115,39815,61857,6	0,00000,00000,07275,95761,41834,25903,32031,25
38	10000,00000,00837,67557,69872,72426,9	0,00000,00000,03637,97880,70917,12951,66015,625-
39	10000,00000,00418,83778,84927,59087,9	0,00000,00000,01818,98940,35458,56475,83007,8125
40	10000,00000,00209,41889,42461,60262,5	0,00000,00000,00909,49470,17729,28237,91503,90625
41	10000,00000,00104,70944,71230,25311,0	0,00000,00000,00454,74735,08864,64118,95751,95312
42	10000,00000,00052,35472,35614,98950,4	0,00000,00000,00227,37367,54432,32059,47875,97656
43	10000,00000,00026,17736,17807,46048,9	0,00000,00000,00113,68683,77216,16029,73937,98828
44	10000,00000,00013,08868,08903,72167,8	0,00000,00000,00056,84341,88608,08014,86968,99414
45	10000,00000,00006,54434,04451,85869,75	0,00000,00000,00028,42170,94304,04007,43484,49707
46	10000,00000,00003,27217,02225,92881,337	0,00000,00000,00014,21085,47152,02003,71742,24853
47	10000,00000,00001,63608,51112,96427,283	0,00000,00000,00007,10542,73576,01001,85871,12426
48	10000,00000,00000,81804,25556,48210,295	0,00000,00000,00003,55271,36788,00500,92935,56213
49	10000,00000,00000,40902,12778,24104,311	0,00000,00000,00001,77635,68394,00250,46467,78106
50	10000,00000,00000,20451,06389,12051,946	0,00000,00000,00000,88817,84197,00125,23233,89053
51	10000,00000,00000,10225,53194,56025,921 *L*	0,00000,00000,00000,44408,92098,50062,61616,94526
52	10000,00000,00000,05112,76597,28012,947 *M*	0,00000,00000,00000,22204,46049,25031,30808,47263
53	10000,00000,00000,02556,38298,64006,470 *N*	0,00000,00000,00000,11102,23024,62515,65404,23631
54	10000,00000,00000,01278,19149,32003,235 *P*	0,00000,00000,00000,05551,11512,31257,82702,11815

the 54 weird numbers in the first column. But we would prefer to have a table of the logarithms of more ordinary numbers such as 2, 5, 17, or—to be systematic—what about the logarithms of the first 1000 positive integers? This was, precisely, Briggs' original goal, and the logarithms of such numbers can also be obtained by the successive extraction of square roots.

To see how this is done we can examine the last four rows of the reproduced table, those containing the entries labeled L to P. If we write P in the form $1 + p$ with $p = 0.0000000000000000127819149320003235$, a short calculation shows that $N = 1+2p$, $M \approx 1+4p$, and $L \approx 1+8p$. The approximation loses accuracy as we progress through these numbers, in reverse alphabetical order, so we shall not go any further. Now, log P is the last number in the right column of the table, namely 0,00000000000000005551115123125782702118815, and using the fact that $p \ll 1$ it is easy to see that

$$\log N = \log(1 + 2p) \approx \log(1 + p)^2 = 2 \log P,$$

$$\log M \approx \log(1 + 4p) \approx \log(1 + p)^4 = 4 \log P,$$

and

$$\log L \approx \log(1 + 8p) \approx \log(1 + p)^8 = 8 \log P.$$

This may be generously interpreted to mean that if a number of the form $1 + r$ is "near the numbers L M N & P" then

$$\log(1 + r) = \log\left(1 + \frac{r}{p}p\right) \approx \frac{r}{p} \log P.$$

Briggs did not state any formulas and simply asserted that "the Logarithm of this number [a number in the stated range] is easily found, from the rule of proportion" [p. 11; 6–4], which he called the "golden rule" (*auream regulam*). He illustrated it with the number 1.0000000000000001, in which case

$$\frac{r}{p} = \frac{10000000000000000}{12781914932003235},$$

and asserted that its logarithm is 0,0000000000000000434294481903251804 (the last three digits are in error).

It may be more interesting to find log 2, although Briggs did not do it at this point in his book for reasons that will become apparent only too soon. We would start by computing 53 consecutive square roots of 2, ending with

$$2^{1/2^{53}} = 1.0000000000000000076954795931116620.$$

This number is in the L to P range in the first column of Briggs' table, and then the golden rule (in which we have replaced $\approx$ with $=$ for convenience) and the stated value of $\log P$ give

$$\log 2^{1/2^{53}} = \frac{07695479593116620}{12781914932003235} \log P$$

$$= 0,00000000000000003342104322889629327919 31.$$

Therefore,

$$\log 2 = (2^{53})(0,00000000000000003342104322889629327919 31)$$

$$= 0,30102999566398116973197.$$

This method has two drawbacks. First, the accuracy is somewhat limited (the last six decimal digits are incorrect), although it may be considered acceptable for a table of logarithms. But the method is too labor-intensive.[14]

In Chapter VII, Briggs started exploring how to shorten his methods, and actually computed $\log 2$ as follows. First he noticed that it is sufficient to compute the logarithm of $1\frac{024}{1000}$ (since he did not use a decimal point, this is how he wrote 1.024) by the square root method, because, since 1024 is the tenth power of 2,

$$\log 2 = \frac{\log 1024}{10}$$

and

$$\log 1024 = \log 1000 + \log 1\tfrac{024}{1000} = 3 \log 10 + \log 1\tfrac{024}{1000}.$$

Since 1.024 is closer to unity than 2, then 47 (as opposed to 53) square root extractions of 1.024 give the number 1.00000000000000016851605705394977, which is in the proportionality region L to P, and "the Logarithm of that number is found by the golden rule 0,000000000000000007318559369062 39336" [p. 13; 7–3]. Thus,

$$\log 1.024 = (2^{47})(0,0000000000000000007318559369062 39336)$$

$$= 0,01029995663981195.$$

Adding $3 \log 10$ and dividing by 10 yields $\log 2 = 0,301029995663981195$, as given in the *Arithmetica logarithmica* [p. 14; 7–4].

---

[14] Bruce has estimated that the computation of about fifty successive square roots is a process that may require between 100 and 200 hours of hard work [p. 7–6]. This is equivalent to several weeks if you have other commitments.

Then Briggs used the equation $5/10 = 1/2$ and the properties of logarithms to obtain $\log 5 - \log 10 = \log 1 - \log 2$, from which [p. 14; 7–5]

$$\log 5 = \log 10 - \log 2$$
$$= 1{,}000000000000000000 - 0{,}301029995663981195$$
$$= 0{,}698970004336018805.$$

Next, he gave a list of numbers whose logarithms can be easily computed from those previously computed using the properties of logarithms:

> From the multiplication of Two alone, by itself & into its factors, 4. 8. 16. 32. 64. &c. Likewise of Five by itself & into its factors, 25. 125. 625. 3125. &c, Two in factors of Five, 250. 1250. 6250. &c. Two into Ten 20. 200. 2000. 40. 400. 80. 800. &c.

The next prime after two is three, "whose Logarithm is most conveniently found from the Logarithm of Six," since $\log 3 = \log 6 - \log 2$. But six is very far from one, and it would take too many roots to bring it down to the $L - P$ proportionality region. Thus, Briggs decided to find the logarithm of

$$\frac{6^9}{10^7} = 1.0077696$$

instead, which he did again by the golden rule, and then obtained

$$\log 6 = \frac{7 + \log 1.0077696}{9}.$$

It should be evident that the elaboration of a complete table of logarithms by the square root method would be impossibly time-consuming. However, the logarithm of 6 was the last one that Briggs computed in this manner, for he made an additional discovery. He announced it at the start of Chapter 8 in the following terms [p. 15; 8–1]:

> We can find the Logarithm of any proposed number according to this method, by continued Means, which supplies our need by finding the square root laboriously enough, but the vexation of this enormous labor is considerably lessened through differences.

To illustrate this idea he chose the example of finding consecutive square roots of 1.0077696, the number involved in finding the logarithm of six. After computing some roots, as shown in the table reproduced on page 112 [p. 16;

8–2], he noticed that if each of them is written in the form $1 + A$, then each value of A is approximately twice the next value. For instance, the A part of the root in line 43 of this table (0.0004838402688466298549 2535) is about twice the A part in the top line of box 42 (0.000241890787 82468563808727). Then Briggs decided to investigate the difference between half of each A and the next A, which he called B. The first of his B differences is seen in the third line of box 42, and it is a very small number in comparison with the A values under consideration. Using the next two A values, Briggs computed a second B, which is in the third line of box 41, and observed that it is about one quarter of the previous B. So he denoted by C the difference between this quarter and the second B, and noticed that C is very small in comparison to the B values.

The fact that Briggs—or any of his contemporaries for that matter—did not use subscripts, added to the fact that his table contains only one column, may make things difficult for the present-day reader, so another table has been included on page 113 to show things in current notation. In its first column, n indicates the order of the root; in the second these roots are given in the form $1 + A_n$, using the decimal point that Briggs omitted; and the remaining columns give the differences of several orders as computed by Briggs (some of his digits are incorrect, if we redo his work using a computer) and labeled by the headers. Leading zeros have been omitted, as originally done by Briggs, but without his vertical arrangement this may be confusing if not noted.

What is clear is that these successive differences become smaller and smaller, to the point that

$$F_9 = \frac{1}{32} E_8 - E_9 \approx 0$$

with the 26-decimal-digit accuracy used in these computations. At this point Briggs got the idea of working backward through the differences. If we take $F_{10} = 0$, since F_{10} is even smaller than F_9, then

$$0 = F_{10} = \frac{1}{32} E_9 - E_{10}$$

implies that $E_{10} = \frac{1}{32} E_9 = 0.0000,00000,00000,00000,00000,065$. In the same manner, $D_{10} = \frac{1}{16} D_9 - E_{10} = 0.0000,00000,00000,00000,02855,524$, and so on, as shown in the lower table on page 113, until we obtain $A_{10} = 0000,07558,20443,63012,14290,760$, and then $1 + A_{10} = \sqrt{1 + A_9}$. It is evident that this method of evaluation of differences is much faster than directly finding the square root. Briggs illustrated this procedure again to compute A_{11}

	10077,696	
46	10038,77283,33696,24566,38465,51	
45	10019,36766,13694,66167,58702,29	
44	10009,67914,63909,90172,88907,20	
43	10004,83840,26884,66298,54925,35	A
42	10002,41890,87882,46856,38087,27	A
	2,41920,13442,33149,27462,67	$\frac{1}{2}A$
	29,25559,86292,89375,40	B
41	10001,20938,12639,71345,94391,94	A
	1,20945,43941,23428,19043,63	$\frac{1}{2}A$
	7,31301,52082,24651,69	B
	7,31389,96573,22343,85	$\frac{1}{4}B$
	88,44490,97692,15	C
40	10000,60467,23505,53096,80160,05	A
	60469,06319,85672,97195,97	$\frac{1}{2}A$
	1,82814,32576,17035,92	B
	1,82825,38026,56162,92	$\frac{1}{4}B$
	11,05444,39127,00	C
	11,05561,37211,52	$\frac{1}{8}C$
	116,98084,52	D
39	10000,30233,16050,56577,59647,94	A
	30233,61752,76548,40080,02	$\frac{1}{2}A$
	45702,19970,80432,08	B
	45703,58144,04258,98	$\frac{1}{4}B$
	1,38173,23826,90	C
	1,38180,54890,87	$\frac{1}{8}C$
	7,31063,97	D
	7,31130,28	$\frac{1}{16}D$
	66,31	E
38	10000,15116,46599,90567,29504,88	A
	15116,58025,28288,79823,97	$\frac{1}{2}A$
	11425,37721,50319,09	B
	‡ 11425,54992,70108,02	$\frac{1}{4}B$
	17271,19788,93	C
	17271,65478,36	$\frac{1}{8}C$
	45689,43	D
	45691,50	$\frac{1}{16}D$
	2,07	E
	2,07	$\frac{1}{32}E$

Hucusque Differentiæ minores sunt inventæ per subductionem majorum è partibus homogenearum præcedentium.

n	$1 + A_n$	$B_n = \frac{1}{2}A_{n-1} - A_n$	$C_n = \frac{1}{4}B_{n-1} - B_n$	$D_n = \frac{1}{8}C_{n-1} - C_n$	$E_n = \frac{1}{16}D_{n-1} - D_n$
0	1.0077,696				
1	1.0038,77283,33696,24566,38465,51				
2	1.0019,36766,13694,66167,58702,29				
3	1.0009,67914,63909,90172,88907,20				
4	1.0004,83840,26884,66298,54925,35				
5	1.0002,41890,87882,46856,38087,27	29,25559,86292,89375,40			
6	1.0001,20938,12639,71345,94391,94	7,31301,52082,24651,69	88,44490,97692,15		
7	1.0000,60467,23505,53096,80160,05	1,82814,32576,17035,92	11,05444,39127,00	116,98084,52	
8	1.0000,30233,16050,56577,59647,94	45702,19970,80432,08	1,38173,23826,90	7,31063,97	66,31
9	1.0000,15116,46599,90567,29504,88	11425,37721,50319,09	17271,19788,93	45689,43	2,07

n	$1 + A_n = 1 + \frac{1}{2}A_{n-1} - B_n$	$B_n = \frac{1}{4}B_{n-1} - C_n$	$C_n = \frac{1}{8}C_{n-1} - D_n$	$D_n = \frac{1}{16}D_{n-1} - E_n$	$E_n = \frac{1}{32}E_{n-1}$
10	1.0000,07558,20443,63012,14290,760	2856,32271,50461,680	2158,87118,092	2855,524	65
11	1.0000,03779,09507,73708,05241,254	714,07798,01904,126	269,85711,294	178,468	2

(his own table of backward differences, which was printed right below that of forward differences, shows a similar vertical arrangement of the A to D differences).[15]

---

[15] Briggs gave a second method for computing the differences, which is based on evaluating successive powers. Fix a value of n and note that if $1 + A_n = \sqrt{1 + A_{n-1}}$, then

$$A_{n-1} = (1 + A_n)^2 - 1, \quad A_{n-2} = (1 + A_{n-1})^2 - 1 = (1 + A_n)^4 - 1,$$

and so on. Therefore,

$$B_n = \frac{1}{2} A_{n-1} - A_n = \frac{1}{2}\left[(1 + A_n)^2 - 1\right] - A_n = \frac{1}{2} A_n^2$$

and

$$
\begin{aligned}
C_n &= \frac{1}{4} B_{n-1} - B_n = \frac{1}{4}\left[\frac{1}{2} A_{n-2} - A_{n-1}\right] - \frac{1}{2} A_n^2 \\
&= \frac{1}{8}\left[(1 + A_n)^4 - 1\right] - \frac{1}{4}\left[(1 + A_n)^2 - 1\right] - \frac{1}{2} A_n^2 \\
&= \frac{1}{2} A_n^3 + \frac{1}{8} A_n^4.
\end{aligned}
$$

Similarly, we would obtain

$$D_n = \frac{7}{8} A_n^4 + \frac{7}{8} A_n^5 + \frac{7}{16} A_n^6 + \frac{1}{8} A_n^7 + \frac{1}{64} A_n^8,$$

$$E_n = \frac{21}{8} A_n^5 + 7 A_n^6 + \frac{175}{16} A_n^7 + \frac{1605}{128} A_n^8 + \frac{715}{64} A_n^9 + \frac{301}{28} A_n^{10} + \cdots,$$

and higher-order differences. We know that Briggs did not use subscripts, but in writing these equations he did not even use the letter A. The stated equations were printed in the *Arithmetica logarithmica* as

Secunda :②
Tertia :③+:④
Quarta :④+:⑤+:⑥+:⑦:⑧
Quinta 2:⑤+7⑥+10:⑦+12:⑧+11:⑨+7:⑩

He went down to "Decima" but he never wrote beyond the tenth power of A_n.

This method is not faster than the previous one, but it is of some theoretical interest, if only because of what Briggs could have done with it but didn't. Using the equations in the header of the lower table on page 113, we see that

$$
\begin{aligned}
(1 + A_n)^{1/2} &= 1 + A_{n+1} = 1 + \tfrac{1}{2} A_n - B_{n+1} = 1 + \tfrac{1}{2} A_n - \left(\tfrac{1}{4} B_n - C_{n+1}\right) \\
&= 1 + \tfrac{1}{2} A_n - \tfrac{1}{4} B_n + \left(\tfrac{1}{8} C_n - D_{n+1}\right) \\
&= 1 + \tfrac{1}{2} A_n - \tfrac{1}{4} B_n + \tfrac{1}{8} C_n - \left(\tfrac{1}{16} D_n - E_{n+1}\right) \\
&= 1 + \tfrac{1}{2} A_n - \tfrac{1}{4} B_n + \tfrac{1}{8} C_n - \tfrac{1}{16} D_n + \tfrac{1}{32} E_n - \cdots.
\end{aligned}
$$

Giving a full account of the rest of Briggs' methods to save time in the computation of logarithms would be a very long story. In Chapter 9 he used the methods already explained and some clever ways of writing certain numbers as sums and products to evaluate the logarithms of all the primes from 2 to 97. Then the rules of logarithms provide those of whole composite numbers, and in Chapter 10 he dealt with fractions. In the next three chapters, Briggs presented several methods of interpolation, starting with simple proportion in Chapter 11. In Chapter 12 he considered a region of the table in which second-order differences for successive numbers whose logarithms are known remain nearly constant, and developed a method to find nine logarithms of equally spaced numbers between every two of those successive numbers. This is known today as *Newton's forward difference method* because it was later rediscovered by Newton (it is briefly described in Section 4.3).

With the aid of the method of finite differences of Chapter 8, Briggs computed a first table of the new logarithms of the numbers 1 to 1000 in *Logarithmorum chilias prima*. In 1620, between this publication and that of the *Arithmetica logarithmica*, he had become the first Savilian professor of geometry at Oxford—a chair endowed by Sir Henry Savile. As we have already mentioned, the 1624 edition of the *Arithmetica logarithmica* contains no logarithms of the numbers between 20,000 and 90,000. The reason for this is that the methods of Chapter 12, used to compute the printed logarithms, are not applicable to this range. Briggs proposed a new method in a longer Chapter 13, bearing the following very long title:

Bringing now to the right-hand side the values obtained for B_n to E_n by Briggs' second method, we obtain

$$(1 + A_n)^{1/2} = 1 + \tfrac{1}{2}A_n - \tfrac{1}{8}A_n^2 + \tfrac{1}{8}\left(\tfrac{1}{2}A_n^3 + \tfrac{1}{8}A_n^4\right)$$

$$-\tfrac{1}{16}\left(\tfrac{7}{8}A_n^4 + \tfrac{7}{8}A_n^5 + \tfrac{7}{16}A_n^6 + \tfrac{1}{8}A_n^7 + \tfrac{1}{64}A_n^8\right)$$

$$+\tfrac{1}{32}\left(\tfrac{21}{8}A_n^5 + 7A_n^6 + \tfrac{175}{16}A_n^7 + \tfrac{1605}{128}A_n^8 + \cdots\right)$$

$$= 1 + \tfrac{1}{2}A_n - \tfrac{1}{8}A_n^2 + \tfrac{1}{16}A_n^3 - \tfrac{5}{128}A_n^4 + \tfrac{7}{256}A_n^5 - \cdots$$

$$= 1 + \tfrac{1}{2}A_n + \frac{\tfrac{1}{2}\left(\tfrac{1}{2} - 1\right)}{1 \cdot 2}A_n^2 + \frac{\tfrac{1}{2}\left(\tfrac{1}{2} - 1\right)\left(\tfrac{1}{2} - 2\right)}{1 \cdot 2 \cdot 3}A_n^3 + \cdots.$$

Today we recognize this as the binomial series expansion of $(1 + A_n)^{1/2}$. Although we may like to think that this particular case of the binomial theorem was implicitly contained in Briggs' work (this was first pointed out by Whiteside in "Henry Briggs: the binomial theorem anticipated," 1961), the fact is that Briggs, who had bigger fish to fry at that point, was not aware of it.

It is desired to find the Logarithms of any Chiliad [that is, one thousand consecutive numbers]. Or given any set of equidistant numbers, together with the Logarithms of these four numbers, to find the Logarithms of intermediate numbers for each interval [between the given numbers].

The second paragraph begins as follows:[16]

Take the first, second, third, fourth, &c. differences of the given Logarithms; & divide the first by 5, the second by 25, the third by 125, &c; ... the quotients are called the first, second, third, &c. mean differences, ...

He hoped to use it to compute the logarithms of the missing chiliads. He expressed this wish, four years later, in a letter to John Pell of 25 October 1628:[17]

My desire was to have these Chiliades that are wanting betwixt 20 and 90 calculated and printed, and I had done them all almost by myselfe, and by some frends whom my rules had sufficiently informed, and by agreement the business was conveniently parted amongst us : but I am eased of that charge and care by one Adrian Vlacque ... But he hathe cut off 4 of my figures throughout : and hathe left out my dedication, and to the reader, and two chapters the 12 and 13, in the rest he hath not varied fromme at all.

What happened is that the large gap in Briggs' *Arithmetica logarithmica* of 1624 was filled by the Dutch publisher Adriaan Vlacq (c. 1600–1667) (or Ezechiel de Decker, a Dutch surveyor, assisted by Vlacq), but reducing the accuracy of the logarithms from 14 to 10 digits and eliminating Chapters 12 and 13, the summit of Briggs' work on finite and mean differences. Their work was published, without any notice to Briggs, in a second edition of the *Arithmetica logarithmica*.

Briggs turned his attention to a new project, the *Trigonometria britannica*. After expressing his displeasure about the omission of Chapter 13 in the Dutch edition of the *Arithmetica logarithmica*, he included his method of mean differences in Chapter 12 of this project. Briggs would work on it for the rest of his life but could not finish it. He left it in the hands of his friend Henry Gellibrand (1597–1636), professor of astronomy at Gresham College, who was able to complete it before his premature death and published it in 1633.

---

[16] This and the preceding interpolation methods of Chapters 11 and 12 can be seen in modern notation in Goldstine, *A history of numerical analysis from the 16th through the 19th century*, 1977, pp. 23–32. See also Bruce's translation and his notes to these chapters.

[17] From Bruce, "Biographical Notes on Henry Briggs," p. 7.

2.4 HYPERBOLIC LOGARITHMS

The loss of Bürgi's tables was not the only effect of the Thirty Years' War on
the development of logarithms. A Belgian Jesuit residing in Prague, Grégoire
de Saint-Vincent (1584–1667), had previously made a discovery while inves-

GRÉGOIRE DE SAINT-VINCENT IN 1653
Engraving by Richard Collin.
From *Opus geometricum posthumum ad mesolabium*, Ghent, 1688.

tigating the area under the hyperbola that would eventually greatly facilitate
the computation of logarithms. But its publication was delayed because Saint-
Vincent fled from Prague (the Swedes are coming!) in 1631, at the start of the
third phase of the Thirty Years' War.

King Gustavus Adolphus II had landed with his troops on the coast of
Pomerania in 1630 and set about to the invasion of central Europe. He won a
brilliant victory at the Battle of Leipzig on September 17, 1631, but later died
at the battle of Lützen on November 16, 1632. The Swedes were victorious,
but Gustavus Adolphus was mortally wounded. In his haste to depart, Saint-
Vincent left all his papers behind, but they were returned to him about ten

years later, and his research on the squaring of the circle and the hyperbola was published in 1647 in a book of over 1250 pages: *Opvs geometricvm quadratvræ circvli et sectionvm coni.*

The result on the hyperbola mentioned above is contained in several propositions in Book VI, in particular in Proposition 109, on page 586, which is reproduced here. The text of the proposition can be translated as follows:

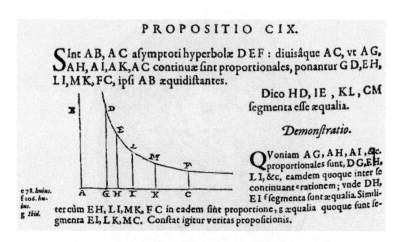

Let AB, AC be the asymptotes of the hyperbola DEF : break up AC, as AG, AH, AI, AK, AC [so that they] are in continued proportion, place GD, EH, LI, MK, FC, equidistant from [meaning parallel to] AB.

I say that HD, IE, KL, CM are equal [area] patches.

In our terms, this means that if

$$\frac{AH}{AG} = \frac{AI}{AH} = \frac{AK}{AI} = \frac{AC}{AK} = \cdots,$$

then the hyperbolic areas *DGHE, EHIL, LIKM, MKCF,* ... are equal.

In other words, if we denote the ratio AH/AG by r, the abscissas AG, $AH = AGr$,

$$AI = AHr = AGr^2, \quad AK = AIr = AGr^3, \quad AC = AKr = AGr^4,$$

and so on, form a geometric progression, and then the hyperbolic areas *DGHE, DGIL, DGKM, DGCF,* ... form an arithmetic progression. If $AG = 1$, this is the same type of relationship that Michael Stifel had pointed out between successive powers of 2 and the corresponding exponents (page 83). We can conclude, in current terminology, that the relationship between the area under the hyperbola from $x = 1$ to an arbitrary abscissa $x > 1$ and the value of x is logarithmic. Saint-Vincent did not explicitly note this, but one of his students, Alfonso Antonio de Sarasa (1618–1667), did in his solution to a problem proposed by Marin Mersenne, a Minimite friar, in 1648. The next year, in his *Solvtio problematis a R P Marino Mersenno Minimo propositi*, de Sarasa stated the problem in this way:[18]

> Given three arbitrary magnitudes, rational or irrational, and given the logarithms of two, to find the logarithm of the third geometrically.

De Sarasa solved Mersenne's problem in developing his Proposition 10, which simply restates a particular case of the problem as follows:

> Given three magnitudes, A, B, and C, which can be shown in one and the same geometric progression, and given the logarithms of two of these magnitudes, say those of A and B, to determine the logarithm of the third, C, geometrically.

That is, de Sarasa assumed that A, B, and C are terms of a geometric progression, because this is the only case he thought he could solve. They are shown in the next figure as ordinates $GH = A$, $IK = B$, and $LF = C$ of a hyper-

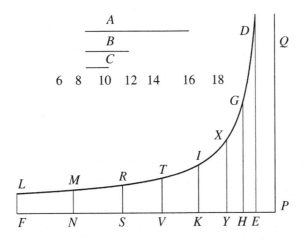

bola, and we can think of the remaining shown ordinates as equally spaced

---

[18] This and the next two quotations are from Burn, "Alphonse Antonio de Sarasa and logarithms," 2001. This paper contains a detailed analysis of de Sarasa's *Solutio*. I follow a small part of Burn's presentation here since I have been unable to see the original work.

terms of the same geometric progression. If some ordinates of the hyperbola $y = 1/x$ (de Sarasa did not give an equation) are in geometric progression, then the corresponding abscissas are also in geometric progression. Then, by Saint-Vincent's Proposition 109, the area KL is four times the area NL, and the area GK is twice the area NL, at which point de Sarasa observed:

> Whence these areas can fill the place of the given logarithms (*Unde hae superficies supplere possunt locum logarithmorum datorum*).

This observation allowed de Sarasa to solve the modified Mersenne problem, which can be done quickly in current notation. Assume that there are real numbers $a > 0$ and $0 < r < 1$, and positive integers m and n such that

$$GH = ar^m, \quad XY = ar^{m+n}, \quad IK = ar^{m+2n}, \quad \text{and} \quad LF = ar^{m+6n}.$$

Let S denote the area under the hyperbola between any two of these consecutive ordinates. For example, $NL = S$. If we refer to the exponents m, $m + 2n$, and $m + 6n$ as the logarithms of $A = GH$, $B = IK$, and $C = LK$, respectively, then it is clear that the differences of these logarithms, $2n$ and $4n$, are equal to n times the area ratios GK/S and KL/S. We can write this as follows:

$$\log B - \log A = n\frac{GK}{S}$$

and

$$\log C - \log B = n\frac{KL}{S}.$$

As an example, de Sarasa considered the case in which the numbers 6 to 18 shown in his figure are the logarithms of the ordinates GH to LF. Then $n = 2$, and the shown ordinates represent every other term of their progression. We have $\log A = 6$, $\log B = 10$, and $KL/S = 4$. It follows that $\log C - 10 = 8$ and $\log C = 18$.

What is important to us is de Sarasa's realization that areas between the hyperbola and the horizontal axis are like logarithms. While he kept things general, we need to specify. Thus, consider the hyperbola $y = 1/x$ and let $A(x)$ denote the area under it from 1 to some $x \geq 1$. It can be called, at least for now, the *hyperbolic logarithm* of x. If we denote this logarithm by Hlog, we have $A(x) = \text{Hlog}\, x$. Since we have $A(1) = \text{Hlog}\, 1 = 0$, then the hyperbolic logarithm shares this property with the Briggsian logarithm. Is it the Briggsian logarithm? We shall have the answer shortly.

The Scottish mathematician James Gregory (1638–1675) elaborated on the work of Saint-Vincent and de Sarasa and computed a number of hyperbolic logarithms in a 1667 short book published in Padua, where he resided at the time: *Vera circvli et hyperbolæ qvadratvra, in propria sua proportionis specie, inuenta, & demonstrata* (page references are to this original). In Proposition XXXII [p. 46], Gregory posed the problem of finding areas under the hyperbola *DIL* with asymptotes *AO* and *AK*. To do that he chose

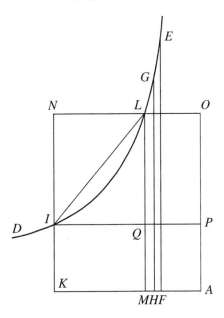

the lengths $IK = 1000000000000$, $LM = 10000000000000 = 10IK$, and $AM = 1000000000000 = IK$ (Gregory's figure is clearly not to scale). Then he employed a series of polygons with an increasing number of sides, inscribed and circumscribed to the hyperbolic space *LIKM* [pp. 47–48], and was able to approximate its area to be 23025850929940456240178700 [p. 49].

Gregory was, at the time, using very large numbers to avoid the still unfamiliar decimal point. If we choose a positive x-axis in the direction AO and a positive y-axis in the direction AK (a choice that is as good as the opposite, given that both are asymptotes of the hyperbola), we would write the equation of Gregory's hyperbola as $y = 10^{25}/x$. If instead we divide all the y-values by 10^{12} and all the x-values by 10^{13} (which is equivalent to the choice $IK = 1$, $LM = 10$, and $AM = 0.1$), then the equation of the hyperbola becomes the familiar $y = 1/x$, and the area stated above must be divided by 10^{25}.[19] Thus,

---

[19] Gregory himself made the choice $IK = 1$ and $LM = 10$ in Proposition XXXIII, entitled: *It is proposed to find the logarithm of any number whatever* [p. 49].

$LIKM = 2.3025850929940456240178700$. Notice that for $y = 1/x$ we have $AK = AP = 1$, and therefore each of the rectangles $IQMK$ and $LQPO$ has area 0.9. Then the hyperbolic areas $LIKM$ and $ILOP$ are equal. The figure clearly shows that $ILOP = \text{Hlog}\,AO$, that is, that $LIKM = \text{Hlog}\,LM$. Therefore, since $LM = 10$,

$$\text{Hlog}\,10 = 2.3025850929940456240178700.$$

It is clear from a comparison of this value with the Briggsian logarithm of 10 that Hlog is a new logarithm. One, we must say in a spoiler mood, that would go on to a position of prominence in the future.

In the few remaining pages of this book Gregory showed how to evaluate additional hyperbolic logarithms [pp. 53–55] and considered the problem of finding a number from its logarithm [p. 56].

At the end of his stay in Italy, in 1668, Gregory published a second book: *Geometriæ pars vniversalis* (The universal part of geometry). In the preface, on pages six to eight, he gave the first graph ever of the hyperbolic logarithm (shown below, with the positive x-axis in the direction OL and $OC = 1$) and stated for the first time in print the properties of this curve.

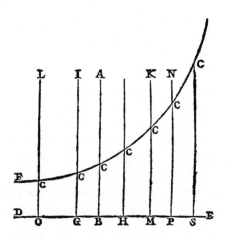

2.5 NEWTON'S BINOMIAL SERIES

Sometime between the publication of de Sarasa's and Gregory's works, a young Isaac Newton (1642–1727) was fooling around with Pascal's triangle

ISAAC NEWTON IN 1689
Portrait by Sir Godfrey Kneller.
Farleigh House, Farleigh Wallop, Hampshire.

(which was known way before Pascal), giving the coefficients of the expansion
of $(a + b)^n$ for $n = 0, 1, 2, \ldots$ This triangle can be rewritten in a rectangular
arrangement, with all its entries moved to the left and zero-filled on the right,
as shown in the next table:

$n = 0$	1	0	0	0	0	0
$n = 1$	1	1	0	0	0	0
$n = 2$	1	2	1	0	0	0
$n = 3$	1	3	3	1	0	0
$n = 4$	1	4	6	4	1	0

This had already been done by Stifel in his *Arithmetica integra*, and Newton
wrote it also as a rectangular array, but turning the rows into columns, in some
manuscripts now preserved at the University Library, Cambridge, but never
published in his lifetime. The second of these manuscripts, which started as

a redraft of the first and was possibly written in the autumn of 1665,[20] is reproduced on page 126, and we shall present first the result obtained on the top third of the page: the computation of the area under the hyperbola

$$be = \frac{1}{1+x}$$

from the origin at d (it looks like ∂ in Newton's hand) to an arbitrary point e at some $x > 0$. Newton made the following observation about his vertical arrangement of Pascal's triangle:

> The composition of wch table may be deduced from hence, viz: The sume of any figure & y^{e} figure above it is equall to y^{e} figure following it.

Referring instead to the horizontal arrangement by Stifel, which is more familiar to us, this can be translated as follows: any entry m, other than the first, in a given row is the sum of two entries in the previous row: the one above m and the one to its left. Viewing things in this way is important because it allowed Newton to construct a new column to the left of the one for $n = 0$ with the same property. If we insist on a horizontal arrangement, this gives us a new row above the one for $n = 0$, as follows:

$n = -1$	1	-1	1	-1	1	-1
$n = 0$	1	0	0	0	0	0
$n = 1$	1	1	0	0	0	0
$n = 2$	1	2	1	0	0	0
$n = 3$	1	3	3	1	0	0
$n = 4$	1	4	6	4	1	0

Notice that now there is no zero-fill on the right in the new first row. It continues indefinitely as an unending string of alternating 1's and -1's. If the new row is valid, it gives the following expansion:

$$(a+b)^{-1} = a^{-1} - a^{-2}b + a^{-3}b^2 - a^{-4}b^3 + a^{-5}b^4 - a^{-6}b^5 + \cdots$$

and, in the particular case $a = 1$ and $b = x$, we obtain the infinite series

$$(1+x)^{-1} = 1 - x + x^2 - x^3 + x^4 - x^5 + \cdots.$$

[20] It appears under the modern title "Further development of the binomial expansion" in Whiteside, *The mathematical papers of Isaac Newton*, I, 1967, pp. 122–134. The table in question is on p. 122.

This is the first instance of what we call today *Newton's binomial series*, and it is known to be a valid expansion for small values of x. Newton did not write this expansion explicitly in the manuscript under discussion, but it is clear that this is what he had in mind when he computed the area under the hyperbola $be = (1 + x)^{-1}$.

To do that, he considered first the graphs of the polynomials

$$1, \quad 1 + x, \quad 1 + 2x + x^2, \quad 1 + 3x + 3xx + x^3, \quad \&c.$$

(in the style of his time, Newton frequently wrote xx instead of x^2), whose coefficients are given by the other rows of the table above, and stated that the areas under these graphs from 0 to $x > 0$ are given by

$$x, \quad x + \frac{xx}{2}, \quad x + \frac{2xx}{2} + \frac{x^3}{3}, \quad x + \frac{3xx}{2} + \frac{3x^3}{3} + \frac{x^4}{4}, \quad \&c.^{21}$$

And then, making a leap of faith, he assumed that the same procedure would apply to finding the area under $(1 + x)^{-1}$, represented by the infinite series whose coefficients are given by the entries in the top row of the extended table. Newton concluded with the following statement:

> By wch table [this is where he implicitly assumes the series expansion for the hyperbola] it may appear y^t y^e area of the hyperbola *abed* [meaning the area under $y = (1 + x)^{-1}$ from 0 to $x > 0$] is
>
> $$x - \frac{xx}{2} + \frac{x^3}{3} - \frac{x^4}{4} + \frac{x^5}{5} - \frac{x^6}{6} + \frac{x^7}{7} - \frac{x^8}{8} + \frac{x^9}{9} - \frac{x^{10}}{10} \quad \&c.$$

If we note that Newton's hyperbola, $y = 1/(1+x)$, is a translation of $y = 1/x$ to the left by one unit, then the area that Newton obtained is, in the notation introduced in Section 2.4, the same as the area $A(1 + x)$ under $y = 1/x$ from $x = 1$ to $1 + x > 1$. Thus, we can rewrite Newton's discovery as

$$H \log (1 + x) = A(1 + x) = x - \frac{x^2}{2} + \frac{x^3}{3} - \frac{x^4}{4} + \frac{x^5}{5} - \frac{x^6}{6} + \&c.$$

Newton obtained this unpublished result

between the years 1664 & 1665. At wch time I found the method of Infinite

---

[21] Newton had already figured out (although he was not the first to do so) that for $n = 0, 1, 2, \ldots$ the area under $y = x^n$ between 0 and $x > 0$ is $x^{n+1}/(n + 1)$.

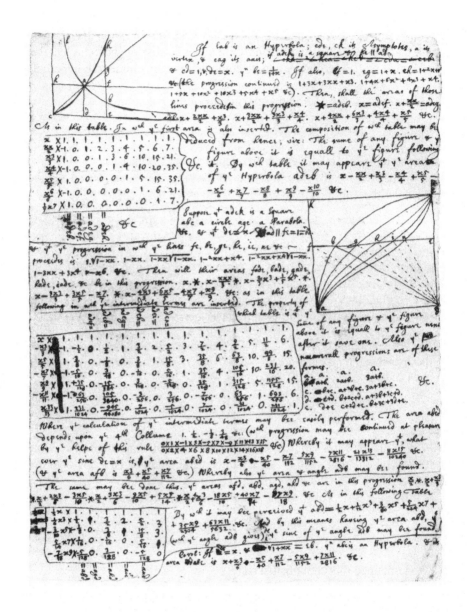

From Whiteside, *The mathematical papers of Isaac Newton*, I.
Facing the title page.

series. And in summer 1665 being forced from Cambridge by the Plague [22]
I computed y^e area of y^e Hyperbola at Boothby in Lincolnshire to two & fifty
figures by the same method.[23]

Then, in a manuscript probably written in 1667,[24] Newton considered
again the area under the hyperbola $y = (1 + x)^{-1}$, represented in the next
figure with the origin at b and in which $ab = bc = 1$. After restating the area

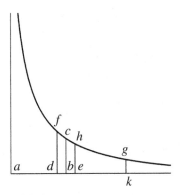

under this hyperbola as an infinite series and making some calculations, he
made the following statement [p. 186]:

> Now since the lines ad, ae, &c: beare such respect to y^e superficies [areas]
> $bcfd$, $bche$, &c: as numbers to their logarithmes; (viz: as y^e lines ad, ae,
> &c: increase in Geometricall Progression, so y^e superficies $bcfd$, $bche$, &c:
> increase in Arithmeticall Progression): Therefore if any two or more of those
> lines multiplying or dividing one another doe produce some other like ak,
> their correspondent superficies, added or subtracted one to or from another
> shall produce y^e superficies $bcgk$ correspondent to y^t line ak.

To interpret this statement in today's notation let x and y be the abscissas of the
points d and e, respectively, so that the lengths ad and ae are $1 + x$ and $1 + y$.

---

[22] This was the great plague that took nearly 70,000 lives in London alone, and Cambridge
University was closed.

[23] Newton made this statement on July 4, 1699, in one of his notebooks containing old
annotations on John Wallis' work. It is quoted from Whiteside, *The mathematical papers
of Isaac Newton*, I, p. 8.

[24] Reproduced in Whiteside, *The mathematical papers of Isaac Newton*, II, 1968, pp. 184–
189. Page references are to this printing.

In view of the previous interpretation of areas under this hyperbola as values of H log, the areas $bcfd$ and $bche$ are $H\log(1+x)$ and $H\log(1+y)$. Then, if ak is the product of the lengths ad and ae, the area $bcgk$ is $H\log(1+x)(1+y)$, and Newton's statement is that

$$H\log(1+x)(1+y) = H\log(1+x) + H\log(1+y).$$

Similarly,

$$H\log\frac{1+x}{1+y} = H\log(1+x) - H\log(1+y).$$

Thus, what Newton gave us in this manuscript is a statement (possibly the first ever) of the properties of hyperbolic logarithms. He illustrated these rules by the computation of a number of hyperbolic logarithms to 57 decimal figures [pp. 187–188]. Finally, to show that he could trust these rules, he computed the hyperbolic logarithm of 0.9984 in two ways: first by repeatedly using the stated rules with

$$0.9984 = \frac{2\times2\times2\times2\times2\times2\times2\times2\times3\times13}{10000},$$

and then writing it as $0.9984 = 1 + (-0.0016)$ and using the infinite series. These two results (the first contains a minor error) agreed "in more y^n 50 figures" [p. 189].

Newton, who did not publish the preceding work, may have been disappointed in 1668, when other mathematicians published their work on the quadrature of the hyperbola using infinite series. The first, in April, was William, Viscount Brouncker (1620–1684), at the time the first president of the Royal Society.[25] The hyperbola $y = 1/x$ is represented by the curve EC in the next figure, in which AB is a segment of the x-axis from $x = 1$ to $x = 2$, and the positive y-axis, not shown, is directed down. The area $ABCdEA$ is bounded between the sum of the areas of the parallelograms with diagonals $CA, dF, bn, fk, ap, cm, el, gh$, &c. and the area of the parallelogram $ABDE$ minus the sum of the areas of the triangles (not explicitly drawn) $EDC, EdC,$ $Ebd, dfC, Eab, bcd, def, fgC$, &c. In this manner Brouncker was able to conclude that

$$ABCdEA = \frac{1}{1\times2} + \frac{1}{1\times2} + \frac{1}{3\times4} + \frac{1}{5\times6} + \frac{1}{7\times8} + \frac{1}{9\times10} \text{ \&c.}$$

---

[25] "The squaring of the hyperbola, by an infinite series of rational numbers, together with its demonstration, by that eminent mathematician, the Right Honourable the Lord Viscount Brouncker," 1668.

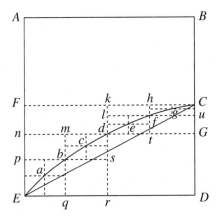

in infinitum [p. 646]. This is what we now call the logarithm of two. Brouncker's method can be generalized, but it is sufficiently unappealing to discourage such a course of action.

Next, Nicolaus Mercator (1620–1687)—born in the province of Holstein, Denmark (now Germany), with the last name Kaufmann but working in England—published the quadrature of the hyperbola in his 1668 book *Log-arithmotechnia: sive methodus construendi logarithmos nova, accurata, & facilis*. Like Newton, but independently, Mercator based his quadrature on the infinite series for the quotient $1/(1 + a)$, which he gave on page 30 using a as the variable. Using this series and referring to the next figure, in which

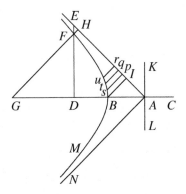

$AI = 1$ and IE is "divisa in partes æquales innumeras" of length a, he found the start of a series for the sum of the altitudes $ps + qt + ru + \cdots$, which multiplied by a gives the area $BIru$ [Proposition XVII, pp. 31–32]. Mercator

did not explicitly give the logarithmic series, but it can be readily obtained from this, and two men showed how to do it. The first was John Wallis,[26] who showed that, defining $A = Ir$, the hyperbolic space $BIru$ is equal to the sum of the series

$$A - \tfrac{1}{2}A^2 + \tfrac{1}{3}A^3 - \tfrac{1}{4}A^6 + \tfrac{1}{4}A^5, \quad \&c.$$

[p. 754].[27] The second commentary on Mercator's work was published later in the same year by James Gregory in his *Exercitationes geometricæ*, 1668. He devoted the second *Exercitatio*, entitled *N. Mercatoris quadratura hyperbolæ geometricè demonstrata*, to prove Mercator's quadrature of the hyperbola using term-by-term integration. Here, in Consectario 2 (what we call corollaries he called *consectaria*, meaning conclusions or inferences) to Proposition IIII [p. 11] he gave the logarithmic series as follows. Referring to x as the "primus terminus" of his expansion, he stated it as: "primus terminus $- \tfrac{1}{2}$ secundi [the second power of x] $+ \tfrac{1}{3}$ tertii $- \tfrac{1}{4}$ quartii $+$ &c. in infinitum."

The next advance in the computation of logarithms and in determining the exact nature of this elusive "Hlog" was also based on work of Newton: his generalization of the binomial series to fractional exponents. He stated his discovery in a letter of June 13, 1676, to Henry Oldenburg, secretary of the Royal Society, in response to an indirect request from Leibniz, who wished to have information on Newton's work on infinite series. It was in this letter that Newton introduced the notation $a^{\frac{m}{n}}$ for a number raised to a fractional power. In a subsequent letter of October 24,[28] he endeavored to explain how he arrived at the statement of his theorem, for which he never provided a proof, recalling his original work with the aid of some old manuscript.

He explained [p. 130] how at the beginning of his study of mathematics, he "happened on the works of our most Celebrated Wallis," in particular, on his *Arithmetica infinitorvm* of 1655.[29] In this work Wallis had managed to

---

[26] "Logarithmotechnia Nicolai Mercatoris," 1668.

[27] An abbreviated explanation in English was given by Edwards in *The historical development of the calculus*, 1979, pp. 162–163.

[28] Page references given below for this letter refer to the English translation in Turnbull, *The correspondence of Isaac Newton*, II. See the bibliography for additional sources.

[29] Wallis was the first to use of the symbol ∞ for infinity. It is in Proposition 1 of the First Part of *De sectionibus conicis* of 1655, p. 4, where he stated *esto enim ∞ nota numeri infiniti* (let ∞ be the symbol for an infinite number); reproduced in *Opera mathematica*, **1**, 1695, p. 29. It also appeared in Proposition XCI of *Arithmetica infinitorvm* of 1655, p. 70, in which he stated ... *erit ∞ vel infinitus* (... will be ∞ or infinity); reproduced in *Opera mathematica*, **1**, p. 405. Also in Stedall, *The arithmetic of infinitesimals. John Wallis 1656*, 2004, p. 71.

JOHN WALLIS IN 1658
Savilian Professor of Geometry at Oxford.
Portrait by Ferdinand Bol.
Photograph by the author from the original
at the *Musée de Louvre*, Paris.

find the areas under the curves that we now express by the equations

$$y = (1 - x^2)^0, \quad y = (1 - x^2)^1, \quad y = (1 - x^2)^2, \quad y = (1 - x^2)^3,$$

and so on, from the origin to $x > 0$, which are, in current notation,

$$x, \quad x - \tfrac{1}{3}x^3, \quad x - \tfrac{2}{3}x^3 + \tfrac{1}{5}x^5, \quad x - \tfrac{3}{3}x^3 + \tfrac{3}{5}x^5 - \tfrac{1}{7}x^7,$$

and so on.[30] To find the area under the circle $y = (1 - x^2)^{\frac{1}{2}}$, he noticed that the exponent is the mean between 0 and 1, and attempted to use his new method

---

[30] In Wallis' notation, these can be seen in Proposition CXVIII of the *Arithmetica infinitorvm*. In *Opera mathematica*, **1**, p. 415. In Stedall, *The arithmetic of infinitesimals. John Wallis 1656*, p. 88.

of intercalation to find it between the first two areas stated above. This led him to a most interesting formula for π,[31] but in the end he was unable to do the intercalation.

Newton's power of observation, when he happened on Wallis' work, made him notice that in all the expressions for the areas given above [p. 130],[32]

the first term was x, and that the second terms $\frac{0}{3}x^3 \cdot \frac{1}{3}x^3 \cdot \frac{2}{3}x^3 \cdot \frac{3}{3}x^3$ &c were in Arithmetic progression, and hence

the first two terms of the areas to be intercalated, those under the graphs of $y = (1 - x^2)^{\frac{1}{2}}$, $y = (1 - x^2)^{\frac{3}{2}}$, $y = (1 - x^2)^{\frac{5}{2}}$, and so on (why stop with the circle), "should be

$$x - \frac{\frac{1}{2}x^3}{3} \cdot \quad x - \frac{\frac{3}{2}x^3}{3} \cdot \quad x - \frac{\frac{5}{2}x^3}{3} \cdot \quad \&c."$$

From his study of the area under the hyperbola, Newton was already sure that these areas would be given by infinite series, so he had to find the rest of the terms. Wallis' denominators from the third on are 5, 7, etc., which Newton kept for his intercalated series. Next he looked at the numerators of Wallis' coefficients, including that of the x term, and discovered that

these were the figures of the powers of the number 11, namely of these 11^0. 11^1. 11^2. 11^3. 11^4. that is first 1. then 1,1. third 1,2,1. fourth 1,3,3,1. fifth 1,4,6,4,1, &c.[33]

The first figure, to use Newton's own word, in a power of 11 is always 1, and what he sought was a method to determine all the remaining figures if given the first two. Here is his own description of the discovery that he made:

---

[31] Proposition CXCI of the *Arithmetica infinitorvm*. In *Opera mathematica*, **1**, p. 469.

[32] I shall closely follow Newton's own recollections in his October 24 letter but will replace his expression $\overline{1 - xx}$ with $(1 - xx)$. In his original work of 1665, already quoted at the start of this section, Newton expressed himself less clearly than in his recollections.

[33] Newton used the notation $\overline{11}\big|^0$, rather than 11^0, and so on for the other powers. Note that the word "figure" cannot be interpreted to mean "digit" starting with $11^5 = 161051$; but it means each of the coefficients of the powers of 10 in the expansion $11^5 = 1 \times 10^5 + 5 \times 10^4 + 10 \times 10^3 + 10 \times 10^2 + 5 \times 10^1 + 1 \times 10^0$, that is, the Pascal triangle coefficients in the expansion of $(10 + 1)^5$.

. . . and I found that on putting m for the second figure [in a power of 11], the rest would be produced by continual multiplication of the terms of this series.

$$\frac{m-0}{1} \times \frac{m-1}{2} \times \frac{m-2}{3} \times \frac{m-3}{4} \times \frac{m-4}{5} \quad \&c.$$

E. g. let $m = 4$, and $4 \times \dfrac{m-1}{2}$ that is 6 will be the third term [figure], & $6 \times \dfrac{m-2}{3}$ that is 4 the fourth, and $4 \times \dfrac{m-3}{4}$ that is 1 the fifth, & $1 \times \dfrac{m-4}{5}$ that is 0 the sixth, at which place in this case the series ends.

This procedure gives the figures of 11^4.

Assuming that the same rule applies to fractional exponents, which is a stretch to say the least, Newton considered the case $m = \frac{1}{2}$, and stated [p. 131]:

. . . since for a circle the second term was $\dfrac{\frac{1}{2}x^3}{3}$, I put $m = \frac{1}{2}$, and the terms appearing were

$$\frac{1}{2} \times \frac{\frac{1}{2}-1}{2} \text{ or } -\frac{1}{8}, \quad -\frac{1}{8} \times \frac{\frac{1}{2}-2}{3} \text{ or } +\frac{1}{16}, \quad +\frac{1}{16} \times \frac{\frac{1}{2}-3}{4} \text{ or } -\frac{5}{128},$$

& so to infinity. From which I learned that the desired area of a segment of a circle is

$$x - \frac{\frac{1}{2}x^3}{3} - \frac{\frac{1}{8}x^5}{5} - \frac{\frac{1}{16}x^7}{7} - \frac{\frac{5}{128}x^9}{9} \quad \&c.$$

And by the same reasoning the areas of the remaining curves to be inserted came forth . . .

The area under the hyperbola was relevant to our discussion of logarithms, but would the remaining areas be equally relevant? Not really; it is Newton's next insight that matters:

But when I had learnt this I soon considered that the terms

$$(1 - xx)^{\frac{0}{2}}. \quad (1 - xx)^{\frac{2}{2}}. \quad (1 - xx)^{\frac{4}{2}}. \quad (1 - xx)^{\frac{6}{2}}. \quad \&c$$

that is 1. $1 - xx$. $1 - 2xx + x^4$. $1 - 3xx + 3x^4 - x^6$ &c could be interpolated in the same way as the areas generated by them [the ones found by Wallis] : and that nothing else was required [for this purpose] but to omit the denominators 1, 3, 5, 7, &c . . .

He forgot to say that we must also divide by x. This is what we now call term-by-term differentiation. Newton applied this procedure to his newly found areas to obtain the curves themselves.

A portion of the first page of Newton's October 1676 letter to Oldenburg.
From Turnbull, *The correspondence of Isaac Newton*, II.

Thus as *e.g.* $(1 - xx)^{\frac{1}{2}}$ would have the value

$$1 - \tfrac{1}{2}x^2 - \tfrac{1}{8}x^4 - \tfrac{1}{16}x^6 \quad \&c$$

and $(1 - xx)^{\frac{3}{2}}$ would have the value

$$1 - \tfrac{3}{2}xx + \tfrac{3}{8}x^4 + \tfrac{1}{16}x^6 \quad \&c$$

and $(1 - xx)^{\frac{1}{3}}$ would have the value

$$1 - \tfrac{1}{3}xx - \tfrac{1}{9}x^4 - \tfrac{5}{81}x^6 \quad \&c.$$

But are these valid expansions or is it all pie in the sky? By way of demonstration, Newton just offered the following:

To prove these operations I multiplied

$$1 - \tfrac{1}{2}x^2 - \tfrac{1}{8}x^4 - \tfrac{1}{16}x^6 \quad \&c$$

into itself, & it became $1 - xx$, the remaining terms vanishing into infinity by the continuation of the series. On the other hand,

$$1 - \tfrac{1}{3}xx - \tfrac{1}{9}x^4 - \tfrac{5}{81}x^6 \quad \&c$$

twice multiplied into itself also produced $1 - xx$.

From these examples Newton was able to infer and state (in the June 13, 1676, letter) a general theorem. If m/n "is integral, or (so to speak) fractional" (what we now call a rational number), Newton's theorem stated that [34]

$$(1+x)^{m/n} = 1 + \frac{m}{n}x + \frac{\frac{m}{n}\left(\frac{m}{n}-1\right)}{1\cdot 2}x^2 + \frac{\frac{m}{n}\left(\frac{m}{n}-1\right)\left(\frac{m}{n}-2\right)}{1\cdot 2\cdot 3}x^3 + \&c.$$

This is now called the *binomial theorem*. Newton used it, as well as other series expansions, in the development of what has become known as the calculus. The matter of which logarithm is the one previously referred to as "Hlog" will be taken up in the next section.

---

[34] It is stated here in modern notation. Newton wrote his equation as

$$\overline{P + PQ}\,|^{\frac{m}{n}} = P^{\frac{m}{n}} + \frac{m}{n}AQ + \frac{m-n}{2n}BQ + \frac{m-2n}{3n}CQ + \frac{m-3n}{4n}DQ + \&c,$$

in which "I employ ... A for the first term, $P^{\frac{m}{n}}$; B for the second, $\frac{m}{n}AQ$, & so on." Putting $P = 1$, $Q = x$, and rearranging yields the stated result.

Newton did not state the complete theorem until he wrote this letter of 1676, but he had already given the form of the general coefficient (which amounts to the same thing) at the end of the 1665 manuscript already quoted at the beginning of this section. Using x/y instead of m/n, he wrote this coefficient as

$$\frac{1 \times x \times \overline{x-y} \times \overline{x-2y} \times \overline{x-3y} \times \overline{x-4y} \times \overline{x-5y} \times \overline{x-6y}}{1 \times y \times \quad 2y \times \quad 3y \times \quad 4y \times \quad 5y \times \quad 6y \times \quad 7y} \quad \&c.$$

2.6 THE LOGARITHM ACCORDING TO EULER

The identification of the logarithm temporarily labeled "Hlog" in the preceding discussion was made by Leonhard Euler (1707–1783) of Basel. The son

LEONHARD EULER IN 1756
Portrait by Jakob Emanuel Handmann.

of a preacher and destined to enter the ministry, his ability in mathematics soon convinced his father to let him switch careers, and he went on to became the most prolific mathematics writer of all time. In 1727, the year of Newton's death, he was invited to join the newly founded Academy of Saint Petersburg, in Russia, and soon began producing first-rate research. It was the next year, in a manuscript on the firing of cannon, that he introduced a soon to become famous number as follows: "Write for the number whose logarithm is unity, *e*," but he did not give a reason for this choice of letter.[35] By that time, he

---

[35] In "Meditatio in Experimenta explosione tormentorum nuper instituta," published posthumously in 1862, p. 800. The earliest printed appearance of the number *e* is in Euler's

had already defined the exponential and logarithmic functions, but the mathematical community at large had to wait until Euler was ready to publish. In 1741 he accepted a position at the Academy of Berlin, where he would remain for twenty-five years, and in 1744 he wrote his enormously influential treatise *Introductio in analysin infinitorum*. Published in Lausanne in 1748, it became

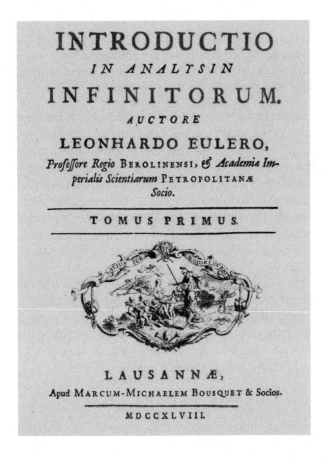

INTRODUCTIO

IN ANALYSIN

INFINITORUM.

AUCTORE

LEONHARDO EULERO,

Profeſſore Regio BEROLINENSI, *&* *Academiæ Imperialis Scientiarum* PETROPOLITANÆ *Socio.*

TOMUS PRIMUS.

LAUSANNÆ,

Apud MARCUM-MICHAELEM BOUSQUET & Socios.

MDCCXLVIII.

the standard work on analysis during the second half of the eighteenth century.

In the first volume of this treatise Euler considered the definition of a^z (at this point he chose z as the symbol for a real variable) a trivial matter "easy

---

Mechanica sive motvs scientia analytice exposita, I, 1736 = *Opera omnia*, Ser. 2, **1**, p. 68. For English translations of the relevant passages of these works, see Smith, *A source book in mathematics*, pp. 95–96.

to understand from the nature of Exponents" [Art. 101].[36] It is not; while the meaning of $a^{\frac{m}{n}}$ was clear after Newton, what does $a^{\sqrt{7}}$ mean? Euler simply said "a certain value comprised between the limits a^2 and a^3" [Art. 97]. In short, he was not too clear about it but seemed to have in mind the existence of what we now call a continuous extension $y = a^z$ of $y = a^{\frac{m}{n}}$. Accepting this as a fact, then he defined its inverse function [Art. 102]:

> In the same manner as, given the number a, it is possible to find the value of y from any value of z, conversely, given any affirmative [positive] value of y, there is a convenient value of z, such that $a^z = y$; this value of z, regarded as a Function of y, is usually called LOGARITHM of y.

There is, of course, no such thing as *the* logarithm of y. What there is, instead, is one logarithm of y for each choice of the number a, "which, for this reason is called *base* of the logarithms" [Art. 102]. Then, in our present notation, $a^z = y$ if and only if $\log_a y = z$. However, logarithms as exponents were not new when the *Introductio* was published. They had been introduced by Euler in the unpublished manuscript cited at the beginning of this section, but the first systematic exposition of logarithms as exponents had already been printed—without knowledge of Euler's work—in the introduction to William Gardiner's *Tables of Logarithms* of 1742, "collected wholly from the papers" of William Jones.

The success of the *Introductio* rests on the amount and importance of the mathematical discoveries that Euler included in it, making it one of the most significant mathematics books of all times. Its readers might have been bewildered about the fact that there is one logarithm for each base a [Art. 107], but Euler easily showed that all logarithms of y are multiples of each other [Art. 108]. Indeed, if $z = \log_a y$ then $y = a^z$ and

$$\log_b y = \log_b a^z = z \log_b a = \log_a y \log_b a, \text{ }^{37}$$

so that any two logarithmic functions, as we would say today, are constant multiples of each other. Thus it appears that we need retain only one logarithm, and the question is: what should be its base?

---

[36] References to the *Introductio* are by Article number rather than by page. In this way the reader can refer to any of the available editions cited in the bibliography.

[37] This is a modernized version. What Euler actually did is to show that if M and N are two numbers whose logarithms in base a are m and n, and whose logarithms in base b are μ and ν, then $m/n = \mu/\nu$. Taking $M = a$, so that $m = 1$, gives $\nu = n\mu$, and with $N = y$ this is the equation stated above.

The answer came from Euler's work in expanding both the exponential function $y = a^z$, $a > 1$, and the corresponding logarithm in infinite series. To do that, "let ω be an infinitely small number" [Art. 114]. Then, $a^0 = 1$ means that $a^\omega = 1 + \psi$, where ψ is also infinitely small. Write $\psi = k\omega$, where, as Euler remarked, "k is a finite number that depends on the value of the base a," and then for any number i [Art. 115],

$$a^{i\omega} = (1 + k\omega)^i = 1 + \frac{i}{1} k\omega + \frac{i(i-1)}{1 \cdot 2} k^2\omega^2 + \frac{i(i-1)(i-2)}{1 \cdot 2 \cdot 3} k^3\omega^3 + \&c.$$

Euler did not give a reason, but we know this to be true by Newton's binomial theorem if i is an integer or a quotient of integers. Now, for any number z let $i = z/\omega$. For Euler i was "infinitely large," but we may prefer to think of ω as a very small number chosen so that i is a very large quotient of integers. Putting $\omega = z/i$, the previous equations become

$$a^z = \left(1 + \frac{kz}{i}\right)^i = 1 + \frac{1}{1} kz + \frac{1(i-1)}{1 \cdot 2i} k^2z^2 + \frac{1(i-1)(i-2)}{1 \cdot 2i \cdot 3i} k^3z^3 + \&c.$$

Insisting on the fact that i is infinitely large, Euler stated that [Art. 116]

$$\frac{i-1}{i} = 1, \qquad \frac{i-2}{i} = 1,$$

and so on, which is approximately true if i is very large. Therefore,

$$a^z = 1 + \frac{kz}{1} + \frac{k^2z^2}{1 \cdot 2} + \frac{k^3z^3}{1 \cdot 2 \cdot 3} + \frac{k^4z^4}{1 \cdot 2 \cdot 3 \cdot 4} + \&c.$$

For $z = 1$, he obtained the following relationship between a and k:

$$a = 1 + \frac{k}{1} + \frac{k^2}{1 \cdot 2} + \frac{k^3}{1 \cdot 2 \cdot 3} + \frac{k^4}{1 \cdot 2 \cdot 3 \cdot 4} + \&c.,$$

and then we can ask the question: for which particular base a is $k = 1$? Clearly, this is true for [Art. 122]

$$a = 1 + \frac{1}{1} + \frac{1}{1 \cdot 2} + \frac{1}{1 \cdot 2 \cdot 3} + \frac{1}{1 \cdot 2 \cdot 3 \cdot 4} + \&c.$$

Euler found the sum of this series to be 2.71828182845904523536028 &c. Later [Art. 123] he denoted this number by e—the first letter of the word exponential—and gave the series

$$e^z = 1 + \frac{z}{1} + \frac{z^2}{1 \cdot 2} + \frac{z^3}{1 \cdot 2 \cdot 3} + \frac{z^4}{1 \cdot 2 \cdot 3 \cdot 4} + \&c.$$

Having expanded the general exponential function in an infinite series, Euler turned to the logarithmic function, which he denoted by l regardless of its base. To obtain its series expansion, he noted [Art. 118] that the equation $a^\omega = 1 + k\omega$ yields $\omega = l(1 + k\omega)$ and, consequently,

$$i\omega = il(1 + k\omega) = l(1 + k\omega)^i.$$

"It is clear, that the larger the number chosen for i, the more the Power $(1 + k\omega)^i$ will exceed unity;" that is, for any $x > 0$ (Euler switched from z to x at this point) we can choose i so that $x = (1 + k\omega)^i - 1$. Thus,

$$1 + x = (1 + k\omega)^i$$

and

$$l(1 + x) = l(1 + k\omega)^i = i\omega,$$

and i must be infinitely large because ω is infinitely small. From the definition of x it follows that [Art. 119]

$$i\omega = \frac{i}{k}(1 + x)^{1/i} - \frac{i}{k},$$

and then we can use Newton's binomial theorem to obtain

$$(1+x)^{1/i} = 1 + \frac{1}{i}x + \frac{\frac{1}{i}\left(\frac{1}{i}-1\right)}{1\cdot2}x^2 + \frac{\frac{1}{i}\left(\frac{1}{i}-1\right)\left(\frac{1}{i}-2\right)}{1\cdot2\cdot3}x^3$$

$$+ \frac{\frac{1}{i}\left(\frac{1}{i}-1\right)\left(\frac{1}{i}-2\right)\left(\frac{1}{i}-3\right)}{1\cdot2\cdot3\cdot4}x^4 + \&c.$$

$$= 1 + \frac{1}{i}x - \frac{1(i-1)}{i\cdot2i}x^2 + \frac{1(i-1)(2i-1)}{i\cdot2i\cdot3i}x^3$$

$$- \frac{1(i-1)(2i-1)(3i-1)}{i\cdot2i\cdot3i\cdot4i}x^4 + \&c.$$

(Euler did not write the first of these two series, but gave only the simplified form.) Since i is infinitely large,

$$\frac{i-1}{2i} = \frac{1}{2}; \quad \frac{2i-1}{3i} = \frac{2}{3}; \quad \frac{3i-1}{4i} = \frac{3}{4}, \quad \&c.;$$

from which

$$i(1 + x)^{1/i} = i + \frac{x}{1} - \frac{xx}{2} + \frac{x^3}{3} - \frac{x^4}{4} + \&c.$$

According to the last equation in Article 19, $i\omega$ is obtained by dividing this result by k and subtracting i/k. Then the equation $l(1 + x) = i\omega$, established at the end of Article 18, shows that

$$l(1 + x) = \frac{1}{k}\left(\frac{x}{1} - \frac{xx}{2} + \frac{x^3}{3} - \frac{x^4}{4} + \&c.\right).$$

In the particular case in which $a = e$ and $k = 1$ this becomes [Art. 123]

$$l(1 + x) = \frac{x}{1} - \frac{xx}{2} + \frac{x^3}{3} - \frac{x^4}{4} + \&c.$$

This is the same series obtained by Newton and Gregory, which shows that the elusive "Hlog" is actually l with base e. Euler called the values of this l "Logarithmi *naturales* seu *hyperbolici*" (natural or hyperbolic logarithms) [Art. 122]. From now on we shall reserve the letter l for natural logarithms.

Since every logarithm is a constant times l, the values of any logarithm can be computed from this series. So, what about $\log_{10} 2$ and $\log_{10} 5$? Is the use of this series faster than evaluating 54 square roots of 10 and then 52 square roots of 2, as we did when discussing the work of Briggs? In Chapter VI of the *Introductio* [Art. 106], Euler had used a variant of the square root method to find $\log_{10} 5$, obtaining the value 0.6989700, but then, at the very end of Article 106, he mentioned the discovery of "extraordinary inventions, from which logarithms can be computed more expeditiously." This refers to the use of the series derived above, which he would obtain in Chapter VII. However, it is not possible just to plug in, for instance, $x = 4$, because then, as Euler observed, the terms of this series "continually get larger" [Art. 120]. Mathematicians knew at that time—or they felt in their bones, in the absence of a theory of convergence—that if an infinite series is to have a finite sum, then its terms must decrease to zero, so that x cannot exceed 1 in this case. Euler found his way around this obstacle as follows. Replacing x with $-x$ in the equation giving $l(1 + x)$ and subtracting the result from that equation [Art. 121] yields

$$l(1 + x) - l(1 - x) = \frac{x}{1} - \frac{x^2}{2} + \frac{x^3}{3} - \frac{x^4}{4} + \&c.$$

$$-\left(-\frac{x}{1} - \frac{x^2}{2} - \frac{x^3}{3} - \frac{x^4}{4} - \&c. \right)$$

$$= 2\left(\frac{x}{1} + \frac{x^3}{3} + \frac{x^5}{5} + \&c. \right)$$

Using the properties of logarithms, this is equivalent to [Art. 123]

$$l\frac{1 + x}{1 - x} = \frac{2x}{1} + \frac{2x^3}{3} + \frac{2x^5}{5} + \frac{2x^7}{7} + \frac{2x^9}{9} + \&c.,\text{ }^{38}$$

"which Series converges strongly, if x is replaced by an extremely small fraction." This device allows us to compute the logarithms of numbers larger than 1 from values of x smaller than one. Thus, for $x = 1/5$, one has $(1 + x)/(1 - x) = 3/2$, and Euler obtained (except for the numbers in parentheses)

$$l\frac{3}{2} = \frac{2}{1 \cdot 5} + \frac{2}{3 \cdot 5^3} + \frac{2}{5 \cdot 5^5} + \frac{2}{7 \cdot 5^7} + \frac{2}{9 \cdot 5^9} + \&c.$$

$$(= 0.40546510810816438197801310).$$

Next, substituting first $x = 1/7$ and then $x = 1/9$,

$$l\frac{4}{3} = \frac{2}{1 \cdot 7} + \frac{2}{3 \cdot 7^3} + \frac{2}{5 \cdot 7^5} + \frac{2}{7 \cdot 7^7} + \frac{2}{9 \cdot 7^9} + \&c.$$

$$(= 0.28768207245178092743921901),$$

and

$$l\frac{5}{4} = \frac{2}{1 \cdot 9} + \frac{2}{3 \cdot 9^3} + \frac{2}{5 \cdot 9^5} + \frac{2}{7 \cdot 9^7} + \frac{2}{9 \cdot 9^9} + \&c.$$

$$(= 0.22314355131420975576629509).$$

---

[38] This formula was found first by James Gregory in his *Exercitationes geometricæ*, as Consectario 4 to Proposition IIII [p. 12], in which Gregory denoted by H and 4 the abscissas that we denote by $1 - x$ and $1 + x$ and by S and 3 the corresponding ordinates. Once again referring to x (which he viewed as the area of a certain parallelogram) as "primus terminus," he stated his result as follows: "spatium Hyperbolicum $SH43 =$ duplo primi termini $+ \frac{2}{3}$ tertii $+ \frac{2}{5}$ quintii $+ \frac{2}{7}$ septimi $+ \frac{2}{9}$ nonii $+$ &c. in infinitum." Variants of the same formula were also used in logarithmic computation by Newton in his work on fluxions and infinite series of 1670–1671 (see Whiteside, *The mathematical papers of Isaac Newton*, III, 1969, p. 227) and by Edmund Halley in "A most compendious and facile Method for Constructing the Logarithms," 1695, pub. 1697.

Then, by the properties of logarithms,

$$l2 = l\frac{3}{2} + l\frac{4}{3} = 0.693147180559945309417232l,$$

$$l5 = l\frac{5}{4} + 2l2 = l\frac{5}{4} + 2\left(l\frac{3}{2} + l\frac{4}{3}\right) = 1.6094379124341003746007593,$$

and

$$l10 = l5 + l2 = 2.3025850929940456840179914.$$

This value, which Euler gave as stated here, may be compared to that obtained by Gregory at the end of Section 2.4. Finally [Art. 124], Euler returned to the computation of common logarithms and showed that if the hyperbolic logarithms are divided by the hyperbolic logarithm of 10, the common logarithms will be obtained. He did not give an example, but we shall show one, while simplifying his explanation (and reducing his accuracy) in the process. Using the already established equation $\log_b y = \log_a y \, \log_b a$ with $b = e$, $y = 2$, and $a = 10$, we obtain

$$\log_{10} 2 = \frac{l(2)}{l(10)} = \frac{0.693147180559945309}{2.302585092994045684} = 0.301029995663981195.$$

This is approximately the same value provided by the square root method, and it was in this manner that the natural logarithm became indispensable.

Now that the logarithm is a function, we can ask whether it has a derivative and how to find it. The series found above allowed Euler to find it as follows in his book on differential calculus [Article 180]:[39]

> We put $x + dx$ in place of x, so that y is transformed into $y + dy$; whereby we have
>
> $$y + dy = l(x + dx) \quad \& \quad dy = l(x + dx) - l(x) = l\left(1 + \frac{dx}{x}\right).$$
>
> As above the hyperbolic logarithm of this kind of expression $1 + z$ can be expressed by an infinite series, as
>
> $$l(1 + z) = \frac{z}{1} - \frac{z^2}{2} + \frac{z^3}{3} - \frac{z^4}{4} + \&c.$$

---

[39] *Institutiones calculi differentialis*, 1755. This quotation is from Blanton's translation.

Therefore if we substitute $\dfrac{dx}{x}$ for z, we obtain:

$$dy = \frac{dx}{x} - \frac{dx^2}{2x^2} + \frac{dx^3}{3x^3} - \&c.$$

Since all the terms of the series vanish in front of [meaning: compared to] the first term, it will be

$$d.lx = dy = \frac{dx}{x}.$$

In short, the derivative of lx is $1/x$. Of all the logarithms this is the only one with such a neat derivative. For if

$$\log_b x = \log_e x \log_b e = lx \log_b e,$$

then it is clear that

$$\log_b' x = \frac{1}{x} \log_b e$$

is not quite as good-looking as the derivative of lx. We conclude that the logarithm to keep is the natural logarithm, which is a very appropriate name for it.

So, what about the logarithms of negative numbers? In 1712 and 1713 a dispute over this had flared up in the correspondence between Gottfried Wilhelm Leibniz (1646–1716) and Johann Bernoulli (1667–1748). Leibniz had published an article [40] expressing his opinion on the logarithm of -1 [p. 167]:

> Indeed it is not positive, for such numbers are the Logarithms of positive numbers larger than unity. And yet it is not negative; because such numbers are the Logarithms of positive numbers smaller than unity. Therefore the Logarithm of -1 itself, which is not positive, nor negative, is left out as not true but imaginary.

An unfortunate choice of word (*imaginarius*), because it has a clear meaning in today's mathematics. What Leibniz meant is that $l(-1)$ does not exist. Bernoulli disagreed, stating his opinion that

$$lx = l(-x)^{[41]}$$

---

[40] "Observatio, quod rationes sive proportiones non habeant locum circa quantitates nihilo minores, & de vero sensu methodi infinitesimalis," 1712.

[41] I have inserted parentheses where he had none, and will continue this practice to the end of this chapter for the benefit of the modern reader. However, the original notation of Euler, who did not use parentheses around expressions such as $-x$ and $+x$, is kept in the quotations.

in a letter to Leibniz of May 25, 1712. He gave four reasons for it, the first being that for $x > 0$ the differentials of both lx and $l(-x)$ are identical, and then the stated equation follows [pp. 886–887]. This was not accepted by Leibniz in his reply of June 30 because he believed that differentiating lx is legitimate for $x > 0$ only [p. 888]. Euler had also exchanged some correspondence with Bernoulli on this subject from 1727 to 1731, but, although he remained unconvinced by Bernoulli's arguments, he had no alternative theory of his own to propose at that time.

The controversy between Leibniz and Bernoulli resurfaced in 1745 when their correspondence was first published. By this time Euler had found the solution: $l(-a) = la + \pi(1 \pm 2n)\sqrt{-1}$, where n is any positive integer, and presented it in a letter of December 29, 1746, to the French mathematician Jean le Rond d'Alembert (1717–1783).[42] But d'Alembert responded:

> However, although your reasons are very formidable and very learned, I admit, Sir, that I am not yet completely convinced, because ...

at which point he stated three of his own reasons for this lack of conviction, but we know now that they are invalid.[43] In fact, d'Alembert sided with Bernoulli, convinced by Bernoulli's fourth stated reason: that $(-x)^2 = x^2$ implies that $2l(-x) = 2lx$ and, once more,

$$l(-x) = l(+x).$$

Euler replied on April 15, 1747, but d'Alembert remained unconvinced and Euler eventually gave up the argument. Instead, on August of that year he sent an article, *Sur les logarithmes des nombres négatifs et imaginaires* (On the logarithms of negative and imaginary numbers), to the Academy of Berlin to give his solution; but he may have withdrawn it later because it was not published until 1862!

However, he presented another complete solution in 1749.[44] With both correspondents now dead, Euler felt free to express his own opinion. First, he explained how Bernoulli's first reason was wrong because it was [p. 144; 200]

---

[42] In a previous letter of September 24, Euler had communicated the value of $l(-1)$ to Gabriel Cramer as $(\pi \pm 2m\pi) \cdot \sqrt{-1}$. See Euler, *Opera omnia*, Ser. 4a, **1**, Birkhäuser Verlag, Basel, 1975, R. 469, pp. 93–94.

[43] Euler's solution and d'Alembert's reply can be seen in Euler, *Opera omnia*, Ser. 4a, **5**, Birkhäuser Verlag, Basel, 1980, pp. 251–253 and 256–259.

[44] "De la controverse entre Messrs. Leibnitz et Bernoulli sur les logarithmes négatifs et imaginaires," 1749, pub. 1751. Page references are to the original paper first and, after a semicolon, to the *Opera omnia*. Quotations are from the original paper.

clear, that since the differential of $l-x$ & of $l+x$ is the same $\frac{dx}{x}$, the quantities $l-x$ and $l+x$ differ from one another by a constant quantity, which is equally evident, in view [of the fact] that $l-x = l-1 + l+x$.

Then he pronounced himself against Leibniz' belief that the logarithm of -1 does not exist [p. 154; 208]:

> Because, if $l-1$ were imaginary, its double, i.e. the logarithm of $(-1)^2 = +1$, would be too, which does not agree with the first principle of the doctrine of logarithms, according to which it is assumed that $l+1 = 0$.

However, Euler had a greater difficulty disposing of Bernoulli's fourth reason, because it rests on the belief that for any power p^n it is true that $l(p^n) = nlp$. But accepting this leads to contradiction [p. 147; 202]:[45]

> Because it is certain that $(a\sqrt{-1})^4 = a^4$, thus we'll also have $l(a\sqrt{-1})^4 = la^4$, & furthermore $4\,l(a\sqrt{-1}) = 4\,la$, consequently $l(a\sqrt{-1}) = la$.

Putting $a = 1$ leads to $l\sqrt{-1} = 0$, and this is impossible because Euler was already aware of the fact that (this will be explained in Section 3.5)

$$\frac{1}{2}\pi = \frac{l\sqrt{-1}}{\sqrt{-1}}.$$

But rejecting Bernoulli's belief that $l(-1) = l(+1) = 0$ also leads to contradiction [p. 148; 203]:

> To make this more evident, let $l-1 = \omega$, & if isn't $\omega = 0$, its double 2ω will not be $= 0$ either, but 2ω is the logarithm of the square of -1, and this being $+1$, the logarithm of $+1$ will no longer be $= 0$, which is a new contradiction.

These contradictions cannot be allowed to stand or the "enemies of Mathematics," as Euler called them [p. 154; 209], would have a field day. Not to fear. He found the source of these contradictions in a very insidious assumption that mathematicians, including himself, implicitly made [pp. 155–156; 210]:

> it is that one ordinarily assumes, almost without noticing, that each number has a unique logarithm ... Therefore I say, to make all these difficulties & contradictions disappear, that just as a consequence of the given definition each number has an infinitude of logarithms; which I will prove in the following theorem.

---

[45] The square root of -1 is shown here in Euler's original notation.

The theorem just restated that each number has an infinitude of logarithms. And how would he find the additional logarithms? By widening the search. Now that $\sqrt{-1}$ has become involved, it is not just the logarithms of negative numbers that we must seek, but those of complex numbers as well.

3

COMPLEX NUMBERS

3.1 THE DEPRESSED CUBIC

In his *Artis magnæ, sive de regvlis algebraicis* (a title usually shortened to *Ars magna*) of 1545, Girolamo Cardano (1501–1576) set out to split the number 10 into the sum of two numbers whose product is 40. Thus, $x(10 - x) = 40$, which has the solutions $5 + \sqrt{-15}$ and $5 - \sqrt{-15}$. And indeed, Cardano invited his readers to consider that the product

$5 + \sqrt{-15}$ times $5 - \sqrt{-15}$, disregarding the cross products, makes $25 - (-15)$, which is $+15$, therefore this product is 40.[1]

This was the first time that the square root of a negative number was used in computation. However, this was not a motivation for the introduction of complex numbers. The given equation is a quadratic, and the fact that it has no real solutions is, and was then, a commonplace. Mathematicians of the

---

[1] The original passage in Cardano's own notation reads: "5 p: ℞ m: 15 in 5 m: ℞ m: 15, dismissis incruciationibus, fit 25 m:m: 15, quod est p: 15, igitur hoc productum est 40," and the Latin words translated above as "disregarding the cross products" are "dismissis incruciationibus." The translation by Witmer has a different interpretation of these words. Since the Latin *cruciare* means to torture, this passage is rendered as: "Putting aside the mental tortures involved, multiply $5 + \sqrt{-15}$ by $5 - \sqrt{-15}$ making $25 - (-15)$ which is $+15$. Hence this product is 40." See p. 219 of the Dover Publications edition. Witmer's translation of this problem and solution is reproduced in Calinger, *Classics of Mathematics*, 1995, p. 265. For another English translation plus a photographic reproduction of the Latin original, see Struik, *A source book in mathematics, 1200–1800*, 1969, pp. 67–69.

HIERONYMI CAR

DANI, PRÆSTANTISSIMI MATHE-
MATICI, PHILOSOPHI, AC MEDICI,

ARTIS MAGNÆ,

SIVE DE REGVLIS ALGEBRAICIS,
Lib.unus. Qui & totius operis de Arithmetica, quod
OPVS PERFECTVM
inscripsit,est in ordine Decimus.

HAbes in hoc libro,studiose Lector,Regulas Algebraicas (Itali, de la Cos-
sa uocant) nouis adinuentionibus,ac demonstrationibus ab Authore ita
locupletatas,ut pro pauculis antea uulgo tritis.iam septuaginta euaserint,Ne-
q solum , ubi unus numerus alteri,aut duo uni,uerum etiam,ubi duo duobus,
aut tres uni æquales fuerint,nodum explicant.   Hunc aũt librum ideo seor-
sim edere placuit,ut hoc abstrusissimo, & plane inexhausto totius Arithmeticæ thesauro in lucem eruto, & quasi in theatro quodam omnibus ad spectan
dum exposito, Lectores incitarétur,ut reliquos Operis Perfecti libros, qui per
Tomos edentur,tanto auidius amplectantur,ac minore fastidio perdiscant.

Title page of the Nuremberg edition of Cardano's *Ars magna*
From Smith, *Portraits of Eminent Mathematicians*, II, 1938.

sixteenth century would have been content with branding the new equation "impossible" or with simply saying that it has no solution.

In the sixteenth century quadratic equations were old hat. Abraham bar Hiyya Ha-Nasi (1070–1136), of Barcelona, one of the twelfth-century translators, better known as Savasorda, made the complete solution known in his book *Hibbur ha-Meshihah ve-ha-Tishboret* (Treatise on Measurement and Calculation), the earliest algebra written in Europe, later translated into Latin by Plato of Tivoli as *Liber enbadorum* in 1145. But the cubic equation was a hot topic and everybody was trying to find a method of solution. Umar ibn Ibrahim al-Khayyami (1048–1131) had already been able to solve cubics geometrically by the method of intersecting conics, but the Europeans were interested now in a purely algebraic solution.

By the end of the fifteenth century no one had succeeded, but sometime in the second decade of the new century, a professor of arithmetic and geometry at the University of Bologna, Scipione di Floriano di Geri Dal Ferro (1465–1526), learned how to solve the particular case $x^3 + px = q$, where p and q are positive numbers. This equation is called the "depressed cubic" because it lacks the term in x^2.

Dal Ferro never published his solution and kept his knowledge to himself. Mathematics was then as magic is today: you kept your secrets to yourself and impressed others with the results, in this case with your ability to solve equations. Furthermore, mathematicians could make some extra money—and enhance their necessary prestige to keep their nontenured positions—by challenging competitors to public contests in which they posed each other problems. However, shortly before his death Dal Ferro confided his secret to one of his students, Antonio Maria Fior of Venice. Much later, Fior—a poor mathematician himself—had the audacity to challenge none other than Nicolò Fontana (*c.* 1499–1577) of Brescia to a cubic-solving contest, to take place on the 22nd of February, 1535. But Fior was not as good at keeping his secret as Dal Ferro had been, and there was a circulating rumor that he knew how to solve $x^3 + px = q$.

Fontana, usually known as Tartaglia, "the stammerer"—as a consequence of an injury inflicted by a French soldier when he was a child—was one of the great intellects of his time. Born in poverty, he was self-educated in mathematics, and in this very same year, 1535, he had discovered how to solve another type of cubic, $x^3 + px^2 = q$. Now he applied himself to the task of rediscovering the solution of the depressed cubic, and succeeded eight days before the contest. Each contestant then submitted thirty cubics to the other, to be solved within the next fifty days. Actually, it was no contest. Tartaglia solved all of Fior's depressed cubics in just two hours, but Fior could never solve any of Tartaglia's, who had chosen the type $x^3 + px^2 = q$.

3.2 CARDANO'S CONTRIBUTION

As a youngster Cardano persuaded his father to let him enter the University of Pavia—his hometown—to study medicine, earning his doctorate in this discipline in 1525 at the University of Padua, but when denied membership in

NICCOLÒ FONTANA
Frontispiece of the princeps edition of the *Qvesiti*, Venice, 1546.
Reproduced from the virtual exhibition *El legado de las matemáticas:
de Euclides a Newton, los genios a través de sus libros*, Sevilla, 2000.

the College of Physicians in Milan—due to the fact that he was an illegitimate
son—he had difficulty making ends meet, and by 1533 things had gotten so
bad that he had to pawn his wife's jewelry. Happily enough, he was also
proficient in mathematics, astrology, and gambling. In this last discipline he
wrote the book *Liber de lvdo aleae* (Book on games of chance), probably
about 1563, and with this expertise plus no aversion to cheating he won more
than he lost. He also came out ahead in astrology, for, although imprisoned
in 1570 for casting the horoscope of Jesus, he was later hired as astrologer
to the Pope. Eventually, he became a successful physician all over Europe,
with a reputation for achieving miraculous cures. Earlier in his life he had
become lecturer in mathematics at the Piatti Foundation in Milan, a post that
he resigned in 1540.

GIROLAMO CARDANO
From Smith, *Portraits of Eminent Mathematicians*, II.

It was during his tenure at the Piatti that Cardano heard of the cubic-solving contest, and he wasted no time in trying to convince Tartaglia to reveal his secret. He succeeded in 1539, but only under the promise of secrecy, and he received the solution without proof. But Cardano was able to supply his own, and later realized that Tartaglia's solution was the same as the one he eventually found in Dal Ferro's surviving papers. Somehow this finding made Cardano feel free to publish it in the *Ars magna*, while giving proper credit to both Tartaglia and Dal Ferro. He started Chapter XI with these words:[2]

> Scipio dal Ferro of Bologna about thirty years ago invented the rule in this chapter, and passed it on to Antonio Maria Florido of Venice, who when engaged in a contest with Niccolò Tartaglia of Brescia, gave Niccolò occasion

---

[2] Alternative translations can be found in Cardano, *Ars magna*; Witmer, *Ars magna or the rules of algebra*, 1993, p. 96; reproduced in Calinger, *Classics of Mathematics*, p. 263; Smith, *A source book in mathematics*, 1929; reprinted by Dover Publications, 1959, p. 204; and Struik, *A source book in mathematics, 1200–1800*, p. 63.

to discover it; and he gave it to us because of our requests, but suppressed the demonstration. Armed with this help, we sought its demonstration in various forms, which was very difficult, and write out as follows.

Of course that stole the thunder from Tartaglia, who wanted to be the first to eventually publish it. He claimed later, in his *Qvesiti et inventioni diverse* of 1546, that he had given his proof to Cardano. In any event, let us proceed to Cardano's demonstration, shown as first published in the next two pages.

But before we embark on the details of the proof it is necessary to explain the difference between Cardano's writing, which would be very difficult to follow today, and our current style. We have written the depressed cubic as $x^3 + px = q$ instead of Cardano's *De cubo & rebus æqualibus numero* (on the cube and the thing equal to the number). In his time there was no habit of denoting either the unknown or the coefficients by letters or of obtaining general formulas. Instead, they taught and learned by example and their descriptions were mostly verbal.

Cardano started his proof by example under the heading DEMONSTRATIO. Referring to his own figure, shown on the next page, he stated the equation to be solved as follows: "Then let for example the cube of *GH* & six times the side *GH* equal 20," which can be written as $x^3 + 6x = 20$ if we use the letter x instead of *GH*.[3] Then, without interruption, he stated a proposition that is the most important step in solving the equation. It can be translated as follows and will be explained immediately following the translation:

> & set two cubes *AE* & *CL* [meaning cubes built on square bases with diagonals *AE* and *CL*], whose difference be 20, such that the product of side *AC*, and side *CK*, be 2, the third part of the number of things ["thing" meaning the unknown *GH*], & cutting off *CB*, equal to *CK*, I say that, if this were to be, the remaining line *AB*, is equal to *GH*, & is therefore the value of the thing, for *GH* has already been assumed, that it was [the thing].

Since the second term in the equation is $6x$, there are six "things" and then "the number of things" is 6. The rest of the proposition simply states that if two segments *AC* and *CK* are chosen such that $AC^3 - CK^3 = 20$ and

---

[3] Semiverbal ways of writing equations were already in use, although there were no universally accepted norms. For instance, in folio 30 *r.* (see the reproduction on page 155), Cardano wrote this equation as "cub⁹ p: 6 reb⁹ æqlis 20," which reads "*cubus* plus six *rebus* equals 20." The Latin word *res*, meaning "thing," or a variation of it was frequently used for the unknown. It was then translated into Italian as *la cosa*, and later algebraists became known as "cossists."

Hieronymi Cardani

relinquitur prima 6 m: ℞ 30⅖,hæ autem quantitates proportionales
sunt,& quadratum secundæ est æquale duplo producti secundæ in
primam,cum quadruplo primæ,ut proponebatur.

De cubo & rebus æqualibus numero.   Cap. XI.

Cipio Ferreus Bononiensis iam annis ab hinc triginta fer
mé capitulum hoc inuenit , tradidit uero Anthonio Ma
riæ Florido Veneto,qui cũ in certamen cũ Nicolao Tar
talea Brixellense aliquando uemssēt, occasionem dedit, ut
Nicolaus inuenerit & ipse,qui cum nobis rogantibus tradidissēt, sup
pressa demonstratione,freti hoc auxilio, demonstrationem quærluui
mus,eamcg in modos, quod difficillimum fuit, redactam sic subieci
mus.   Demonstratio.

Sit igitur exempli causa cubus g h & sexcuplum lateris g h æqua
le 20,& ponam duos cubos a e & c l,quorum differentia sit 20 , ita
quod productum a c lateris, in c k latus,
sit 2, tertia scilicet numeri rerum pars , &
abscindam c b,æqualem c k,dico, quod si
ita fuerit,lineam a b residuum , esse æqua
lem g h,& ideo rei æstimationem, nam de
g h iam supponebatur,quod ita essēt, per
ficiam igitur per modum primi suppositi
6' capituli huius libri, corpora d a,d c,d e
d f,ut per d c intelligamus cubum b c,per

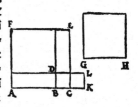

d f cubum a b,per d a triplum c b in quadratum a b,per d e triplum
a b in quadratũ b c. quia igitur ex a c in c k sit 2,ex a c in c k ter fiet
6 numerus rerum, igitur ex a b in triplum a c in c k fiunt 6 res a b,
seu sexcuplum a b,quare triplum producti ex a b, b c,a c, est sexcu
plum a b,at uero differentia cubi a c , à cubo c k , & existenti à cubo
b c ei æqle ex supposito,est 20,& ex supposito primo 6' capituli , est
aggregatum corporum d a,d e,d f,tria igitur hæc corpora sunt 20,
posita uero b c m:cubus a b,æqualis est cubo a c,& triplo a c in qua
dratum c b,& cubo b c m:& triplo b c in quadratum a c m: per de
monstrata illic,differentia autem tripli b c in quadratum a c, à triplo
a c in quadratum b c est productum a b,b c,a c,quare cum hoc,ut de
monstratum est,æquale sit sexcuplo a b, igitur addito sexcuplo a b,
ad id quod fit ex a c in quadratum b c ter,fiet triplum b c in quadra
tum a c,cum igitur b c sit m:iam ostensum est,quod productum c b

DE ARITHMETICA LIB. X.   30

in quadratum A C ter,eſt m:& reliquum quod ei æquatur eſt p:igitur
triplum C B in q̄dratum A B,& triplum A C in q̄dratū C B, & ſexcuplū
A B nihil faciunt. Tanta igitur eſt differentia,ex cōmuni animi ſententia,ipſius cubi A C,à cubo B C,quantum eſt quod cōflatur ex cubo A C,
& triplo A C in quadratum C B,& triplo C B in quadratum A C m:& cu
bo B C m:& ſexcuplo A B,hoc igitur eſt 20,quia differentia cubi A C,à
cubo C B,fuit 20,quare per ſecundum ſuppoſitum 6' capituli , poſita
B C m:cubus A B æquabitur cubo A C , & triplo A C in quadratum B C,
& cubo B C m:& triplo B C in quadratum A ç m:cubus igitur A B,cum
ſexcuplo A B,per communem animi ſententiam , cum æquetur cubo
A C & triplo A C in quadratum C B, & triplo C B in quadratum A B m:
& cubo C B m:& ſexcuplo A B , quæ iam æquatur 20 , ut probatum
eſt,æquabuntur etiam 20,cum igitur cubus A B & ſexcuplum A B æ-
quentur 20,& cubus G H,cum ſexcuplo G H æquentur 20,erit ex com
muni animi ſententia,& ex dictis,in 35² p' & 31² undecimi elemento-
rum,G H æqualis A B,igitur G H eſt differentia A C & C B , ſunt autem
A C & C B,uel A C & C K,numeri ſeu liniæ continentes ſuperficiem , æ-
qualem tertiæ parti numeri rerum,quarum cubi differunt in numero
æquationis,quare habebimus regulam.

REGVLA.

Deducito tertiam partem numeri rerum ad cubum , cui addes
quadratum dimidij numeri æquationis,& totius accipe radicem, ſcili
cet quadratam,quam ſeminabis,uniǫ dimidium numeri quod iam
in ſe duxeras,adijcies,ab altera dimidium idem minues,habebisǫ Bi
nomium cum ſua Apotome, inde detracta ℞ cubica Apotomæ ex ℞
cubica ſui Binomij,reſiduū quod ex hoc relinquitur,eſt rei eſtimatio.

Exemplum.cubus & 6 poſitiones, æquan- | cub⁹ p:6 reb⁹ æq̄lis 20
tur 20,ducito 2 , tertiam partem 6 , ad cu- | 2   20
bum,fit 8,duc 10 dimidium numeri in ſe, | 8 ————— 10
fit 100,iunge 100 & 8,fit 108,accipe radi- | 108
cem quæ eſt ℞ 108, & eam geminabis,alte | ℞ 108 p:10
ri addes 10,dimidium numeri,ab altero mi | ℞ 108 m:10
nues tantundem,habebis Binomiū ℞ 108 |
p:10,& Apotomen ℞ 108 m:10 , horum | ℞ v: cu.℞ 108 p:10
accipe ℞ᵘ cub" & minue illam quę eſt Apo | m:℞ v:cu.℞ 108 m:10
tomæ,ab ea quæ eſt Binomij, habebis rei æſtimationem, ℞ v: cub: ℞
108 p:10 m:℞ v: cubica ℞ 108 m:10.

Aliud,cubus p: 3 rebus æquetur 10,duc 1,tertiam partem 3 , ad
cubum,fit 1,duc 5,dimidium 10,ad quadratum,fit 25,iunge 25 & 1,
H 2   fiunt

$AC \times CK = 2$, then $GH = AB$. Thus $x = AB$, which will later make solving the equation a simpler matter.

For the sake of generality and simplicity, we shall make some changes in Cardano's presentation from this point on. First, we shall keep p and q instead of 6 and 20. Then if, as shown in the next figure, we write $u = AC$

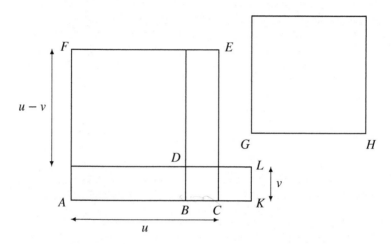

and $v = BC = CK$, it is clear that $AB = u - v$, and the proposition can be restated as follows: if u and v are positive numbers such that

$$u^3 - v^3 = q \qquad \text{and} \qquad uv = \frac{p}{3},$$

then $u - v = x$. Thus we see that if the two preceding equations can be solved for u and v in terms of p and q, then x can found as the difference of u and v.

Returning for a moment to Cardano's original statement, note its geometric flavor. The unknown GH is the side of a square, and the demonstration will be purely geometric. Algebra as we know it was in its infancy, but the accomplishments of Greek geometry were fresh in the minds of mathematicians of the Renaissance, and it is to geometric methods that they turned for inspiration and geometric results that they tried to emulate.

However, while keeping the geometric character of the demonstration, we shall rearrange Cardano's own argument to shorten it and to avoid using a proposition from his earlier Chapter VI. Then, using a bit of three-dimensional imagination, think of a cube of height u erected on the square with diagonal AE shown in the previous figure (we can think of the cube's height in a direction perpendicular to the page). Its volume, u^3, is equal to the sum of the volumes of the eight solid bodies listed in the next table.

Solid	Base diagonal	Height	Volume
Cube	FD	$u - v$	$(u - v)^3$
Cube	DC	v	v^3
Parallelepiped	AD	$u - v$	$(u - v)^2 v$
Parallelepiped	DE	$u - v$	$(u - v)^2 v$
Parallelepiped	FD	v	$(u - v)^2 v$
Parallelepiped	AD	v	$(u - v) v^2$
Parallelepiped	DE	v	$(u - v) v^2$
Parallelepiped	DC	$u - v$	$v^2 (u - v)$

Then

$$\begin{aligned} u^3 &= (u - v)^3 + v^3 + 3(u - v)^2 v + 3(u - v)v^2 \\ &= v^3 + (u - v)^3 + (u - v)[3(u - v)v + 3v^2] \\ &= v^3 + (u - v)^3 + 3uv(u - v). \end{aligned}$$

Therefore, $(u - v)^3 + 3uv(u - v) = u^3 - v^3$ or, recalling the definition of u and v,

$$(u - v)^3 + p(u - v) = q.$$

Comparing this with the given cubic $x^3 + px = q$ shows that $u - v$ is a solution. This is the end of the demonstration proper, which Cardano expressed with the words: "therefore *GH* is the difference of *AC* and *CB*" (*igitur GH est differentia AC & CB*).[4]

What remains now is to solve the equations

$$u^3 - v^3 = q \qquad \text{and} \qquad uv = \frac{p}{3}$$

for u and v. But Cardano skipped this step, simply saying "whence we have the rule" (*quare habebimus regulam*) and proceeding to give the solution in

---

[4] Algebra may have been in its infancy in Cardano's time, but it has been grown up for some time now. A nongeometric demonstration is very simple: expanding $(u - v)^3$ as $u^3 - 3u^2 v + 3uv^2 - v^3$ and rewriting the right-hand side as $u^3 - 3uv(u - v) - v^3$ provides at once the equation that took a number of solid bodies to assemble in the sixteenth century.

the general case. Today we can see that adding the square of the first equation above to four times the cube of the second and simplifying gives

$$(u^3 + v^3)^2 = q^2 + \frac{4p^3}{27}.$$

Finding now the square root of both sides and solving the resulting equation simultaneously with $u^3 - v^3 = q$ yields

$$u^3 = \sqrt{\frac{q^2}{4} + \frac{p^3}{27}} + \frac{q}{2} \quad \text{and} \quad v^3 = \sqrt{\frac{q^2}{4} + \frac{p^3}{27}} - \frac{q}{2}.$$

Finally, extracting cube roots and putting $x = u - v$ gives Dal Ferro's formula:

$$x = \sqrt[3]{\sqrt{\frac{q^2}{4} + \frac{p^3}{27}} + \frac{q}{2}} - \sqrt[3]{\sqrt{\frac{q^2}{4} + \frac{p^3}{27}} - \frac{q}{2}},$$

usually known today as Cardano's formula.[5]   Clearly, this formula gives a real solution of the equation, in fact the only real solution, because, under Cardano's assumption that p is positive, the derivative of $x^3 + px - q$ is positive everywhere. Application of this formula to Cardano's example $x^3 + 6x = 20$ yields the solution

$$x = \sqrt[3]{\sqrt{108} + 10} - \sqrt[3]{\sqrt{108} - 10},$$

although in an inconvenient form, since it would take a little time to figure out that this simplifies to $x = 2$.[6]

---

[5] Cardano's actual rule, stated under the word REGULA. on page 155, is: "Cube the third part of the number of things, to which you add the square of half the number of the equation, & take the root of the whole, of course the square [root], which you will duplicate, to one of them add half of the number which you just squared, from the other subtract the same half, you will have a *Binomium* and its *Apotome* [these terms are from Book X of Euclid's *Elements*], then subtract the cube root of the *Apotome* from the cube root of its *Binomium*, the remainder that is left from this, is the value of the thing."

[6] This formula can be seen in Cardano's notation in the last two lines inside the box on page 155. Cardano used the letter R with its tail crossed for root (*radix*), and the V: (the initial of *vniversalis*) next to it means that it is to be applied to the entire expression that follows, while "cu." indicates that it is a cube root. This cube root applies to the sum of the square root of 108 (only of 108, since there is no V:) plus 10, using the symbol p: for plus. This explains the first of these two lines, and the second has a similar explanation but replacing p: with m: for minus. In this explanation we have followed Smith, *History of Mathematics*, II, p. 416. Actually, the location of the V: in Cardano's writing is confusing, at least for English-language speakers, since it is easy to interpret the two symbols before the number 108 as "cube root," while in Latin the proper order is *radix vniversalis cubica*.

Cardano's work goes beyond that of his predecessors in that he was able to solve the general cubic, not just the depressed case. But Cardano is rather long-winded on this subject. The fact that negative numbers were not acceptable motivated him to divide this case into many subcases, all with positive coefficients, which he studied in Chapters XVII to XXIII, stating a number of rules to follow in each subcase and providing individual geometric demonstrations [*Ars magna or the rules of algebra*, pp. 121–154]. However, from our point of view all this work boils down to making the substitution $x = y - b/3$ in the general cubic $x^3 + bx^2 + cx + d = 0$, which reduces it to the depressed cubic $y^3 + py = q$ with

$$p = c - \frac{b^2}{3} \quad \text{and} \quad q = \frac{bc}{3} - \frac{2b^3}{27} - d.$$

This provides the complete solution of the cubic.

Cardano must also be credited with having noticed that cubics have three solutions and that their sum is $-b$, the opposite of the coefficient of x^2. After giving three examples numbered 5 to 7 in Chapter XVIII, he made the following statement:[7]

> From this it is evident that the number of x^2's,[8] in the three examples in which there are three solutions for x, is always the sum of the three solutions: as in the fifth example, where $2 + \sqrt{2}$, 2, and $2 - \sqrt{2}$ make up 6, the number of x^2's[9] ... Hence, knowing two such solutions, the third always emerges.

But before we can pronounce the cubic solved let us admit that there is a hole in the bucket. We have not considered the case in which $p < 0$, and we do not know whether the stated formula is valid in that case. In Chapter XII Cardano studied the cubic $x^3 = px + q$ with $p, q > 0$, and found the solution to be much like that for $x^3 + px = q$ except for replacing the $+$ sign under each square root with a $-$ sign. This is equivalent to saying that the solution stated above is also valid for $p < 0$. But then, what happens if

$$\frac{q^2}{4} + \frac{p^3}{27} < 0?$$

---

[7] With minor modifications, this is Witmer's translation in *Ars magna or the rules of algebra*, p. 134.

[8] In Cardano's arrangement the x^2 term is on the right-hand side of the equal sign, so there is no need for a negative sign. He would have been horrified by such a need.

[9] In Cardano's own words and notation, " ... *uelut in quinto exemplo*, 2 p: ℞2, & 2, & 2 m: ℞2, *componunt* 6, *numerum quadratorum*," f. 39ᵛ.

In such a case, Cardano's formula was meaningless in his time, and there was no way to proceed. Tartaglia limited himself to calling such cubics "irreducible," but it is too bad to think of them in these terms because it is easily shown that all depressed cubics with $p < 0$ that have three real solutions are irreducible. Indeed, a straightforward application of calculus shows that if $p < 0$, then the expression $x^3 + px - q$ has the maximum value

$$M = 2\left(-\frac{p}{3}\right)^{3/2} - q \quad \text{for} \quad x = -\sqrt{-\frac{p}{3}}.$$

But it should be clear from the figure below that if $x^3 + px - q$ has three real

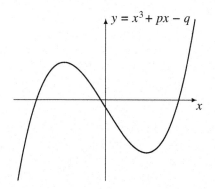

roots, then $M > 0$. Thus,

$$q < 2\left(-\frac{p}{3}\right)^{3/2} \quad \text{or} \quad \frac{q^2}{4} + \frac{p^3}{27} < 0.$$

Is Cardano's formula powerless to compute one real solution of a cubic that has three?

3.3 THE BIRTH OF COMPLEX NUMBERS

Rafael Bombelli (c. 1526–1572), a hydrologist from Bologna who made his fame in mathematics, pursued this matter further, demonstrating that the formula of Scipione Dal Ferro is valid in every case. He did this in his *Opera su l'algebra*, written about 1550 in two parts. Only the purely algebraic part (the second part is based on geometric constructions), in three books, was printed

First page of Book II of the 1579 edition of Bombelli's *L'Algebra*.
Except for the title page and the dedication, it is identical to the 1572 edition.
Reproduced from the virtual exhibition *El legado de las matemáticas*:
de Euclides a Newton, los genios a través de sus libros, Sevilla.

for the first time in 1572, and then again in 1579, under the title *L'Algebra.
Opera di Rafael Bombelli da Bologna, divisa in tre libri.*

As an example of the *casus irreducibilis*, Bombelli considered the cubic $x^3 = 15x + 4$, which has the positive solution $x = 4$, found by inspection. Bombelli used Cardano's rule to solve this cubic in the 1550 manuscript, as shown here. The equation itself is written in the center of the second line as

$$\overset{3}{1} \, a \, \overset{1}{15} \, p. \, \overset{0}{4} \, .$$

Here the symbol $\smile$ represents the variable, the superscript on it is the power

Reproduced from the preface to the 1966 edition of *L'Algebra*.

to which it must be raised, and the number under ⌣ is the coefficient of that power (this is the origin of the exponent notation that we presently use). The letter *a* stands for the equal sign (either as the initial of the Latin *æqualis* or as a contraction of the Italian *eguale a*) and *p.* is for plus. He gave Cardano's solution of the cubic on the eighth line, to the right of the fancy summary box and the word *farà*. He followed Cardano in using the letter *R* with its tail crossed to indicate a root, and there are four of these in the formula. Those with a 3 on top are cube roots and the other two are square roots. The underboxing indicates the range of each root. With *p* meaning plus and *m* meaning minus, this solution can be rewritten using current symbols as

$$\sqrt[3]{2 + \sqrt{0 - 121}} + \sqrt[3]{2 - \sqrt{0 - 121}}.$$

This is, indeed, the solution provided by Cardano's rule, because rewriting the cubic as $x^3 - 15x = 4$, we have $p = -15$ (adopting modern usage of negative numbers, inconceivable to Cardano) and $q = 4$. Then, $q^2/4 = 4$ and

$p^3/27 = -125$, and the solution provided by Cardano's rule is

$$x = \sqrt[3]{\sqrt{4-125}+2} - \sqrt[3]{\sqrt{4-125}-2},$$

which is equivalent to that stated by Bombelli.

Cardano might have stopped right at this point, but Bombelli stated that the sum of his two cube roots is 4 and "that is the value of the Thing,"[10] the value found by inspection. This may be a surprising statement in view of those square roots of negative numbers, so it would be of interest to reproduce Bombelli's reasoning in arriving at such a conclusion. But before we do that it is convenient to prepare the ground, as he did well before this point. On page 169 of the 1572 edition he explained how to deal with square roots of negative numbers, such as those that arise in the irreducible case. To avoid using negative coefficients, such a cubic and Cardano's formula for its solution (the one from his Chapter XII) can be rewritten as $x^3 = px + q$ with $p > 0$ and

$$x = \sqrt[3]{\frac{q}{2} + \sqrt{\frac{q^2}{4} - \frac{p^3}{27}}} + \sqrt[3]{\frac{q}{2} - \sqrt{\frac{q^2}{4} - \frac{p^3}{27}}}.$$

Of course, Bombelli wrote this solution in narrative form at the start of the *Capitolo di Cubo eguale a Tanti e numero* (Chapter of the Cube equal to the Unknowns and number) [p. 222]:[11]

> Wanting to equate the cube to the Unknowns and number [that is, considering the equation $x^3 = px + q$] take one-third of the Unknowns and cube it and the result is subtracted from the square of half the number, and of what remains the square root is taken which is added to and subtracted from half the number, and of the sum and difference the cube root of each of these is taken, and these two roots together are the value of the Unknown (as will be seen in the examples below).

The following is a translation of Bombelli's groundbreaking proposal [p. 133], but some words remain in Italian because they need additional explanation. It is provided below.

---

[10] At the bottom of the summary box we read *Somma 4 : et tanto uale la Cosa*. In 1550 Bombelli still used the word *cosa* for the unknown, but in preparing the manuscript for publication he introduced a number of changes, and one of them was to discard this term. In print he used *Tanto* (so much) instead of *Cosa* for the unknown, giving the reason for it at the start of Book II, as shown on page 161: "*Tanto* is a word appropriate to the quantity of numbers," while *Cosa* "is common to every substance whether known or unknown."

[11] Page references are to the 1966 edition.

I have found some other kind of extended cube root [cube root of a sum] that is very different from all the others, which has its origin in the chapter of the cube equal to the unknowns and number, when the cube of one-third of the unknowns [the cube of $p/3$] is greater than the square of half the number [when $p^3/27 > q^2/4$], ... the excess cannot be called either plus or minus, but I shall call it *più di meno* when it must be added, and when it must be subtracted I shall call it *men di meno* ...

Actually, he did not mean the excess. He meant the square root of the excess, that is, any of the two square roots in the previous equation. In fact, the terms *più di meno* and *men di meno* are contractions of *più radice di meno* 1 and *men radice di meno* 1; that is, "plus the root of minus 1" and "minus the root of minus 1." In our symbols, *più di meno* means $+\sqrt{-1}$ and *men di meno* means $-\sqrt{-1}$. Later, he further abbreviated *più di meno* to *p. di m.* and *men di meno* to *m. di m.* Thus, Bombelli accepted and introduced imaginary numbers in this passage and created a notation for them. Translating his into ours, if a is a real number, *p. di m. a* means $+ai$ and *m. di m. a* means $-ai$.

He devoted the rest of Book I to developing the arithmetic of complex numbers, starting with the most elementary rules of multiplication for "real" and "imaginary" numbers, which he called *la regola del più et meno*. If we agree that *via* means "times" and *fa* means "makes" or "equals," it is simpler to quote them in Italian with a symbolic representation on the right [pp. 133–134].

Più via più di meno, fa più di meno	$+(+i) = +i$
Meno via più di meno, fa meno di meno	$-(+i) = -i$
Più via meno di meno, fa meno di meno	$+(-i) = -i$
Meno via meno di meno, fa più di meno	$-(-i) = +i$
Più di meno via più di meno, fa meno	$(+i)(+i) = -1$
Più di meno via men di meno, fa più	$(+i)(-i) = +1$
Meno di meno via più di meno, fa più	$(-i)(+i) = +1$
Meno di meno via men di meno, fa meno	$(-i)(-i) = -1$

Next he gave a definition that includes that of what we now call the conjugate of a complex number [p. 134]:

Notice that when we say the Residue of a Binomial [what Cardano called *Apotome*], what is called *più di meno* in the Binomial, will be called *meno di meno* in the Residue.

Thus, if the Binomial is a complex number, then its Residue is what we now call its conjugate.

Bombelli then went on to explain, by a wealth of examples rather than by stating a rule, how to multiply complex numbers. Omitting, as trivial, how to multiply a complex number by a real number, we give one of his many examples of complex multiplication. We have changed the original notation as follows: we write $\sqrt[3]{}$ instead of R.c. and $+\sqrt{-}$ instead of *più di meno* R.q. Except for these changes in notation, his example is as follows [p. 135]:

> Multiply $\sqrt[3]{3+\sqrt{-5}}$ by $\sqrt[3]{6+\sqrt{-20}}$, to do it start similarly [referring to the previous, simpler example] by multiplying $+\sqrt{-5}$ by $+\sqrt{-20}$, which will be -10, then we multiply 3 times 6 which makes 18, which together with -10 makes $+8$, and then we multiply 3 times $+\sqrt{-20}$, makes $+\sqrt{-180}$, and then we multiply 6 times $+\sqrt{-5}$, makes $+\sqrt{-180}$, which together with $+\sqrt{-180}$ makes $+\sqrt{-720}$, and this together with $+8$ and with the removed cube root makes $\sqrt[3]{8+\sqrt{-720}}$ which is the result of the multiplication.[12]

It is clear that the cube roots are spurious here. Other than this cube root fixation of Bombelli's, this is a pure example of complex multiplication. After many more examples [pp. 135–140], only the dimmest of Bombelli's readers could fail to get the hang of complex multiplication. Addition is, of course, also used in the preceding computations in the obvious way: add the "real" and "imaginary" parts separately.

Perhaps surprisingly, the next operation that Bombelli considered is the extraction of cube roots. In a chapter entitled *Modo di trovare il lato Cubico di simil qualità di Radici* (A way to find the Cube side of a similar kind of Root), he gave a general rule (to be illustrated by examples, surely, but a general rule) to extract the cube root of a composite number that has a *più di meno* part. To be precise, he gave a general procedure to find the cube root of an expression of the form $a+\sqrt{-b}$. Referring to a as "the number" and to $\sqrt{b}$ as "the square root," his method, entirely in narrative form, is as follows [pp. 140–141]:

> Add the square of the number to the square of the root and of this sum extract the Cube root, then search by trial-and-error to find a number and a square root such that when their squares are added together they amount to as much as the cube root mentioned above and such that when we subtract from the cube of the number three times the product of the number times the square of the square root, what remains is the number of the cube root that we seek,

---

[12] As an example of Bombelli's equation writing, the last equation in this quotation appears in the original as R.c. $\lfloor 8\ p.\ di\ m.\ \text{R.q.}\ 720 \rfloor$.

and then proceeded to give his first example without even writing a period. But before we present it we shall interpret this statement in current terminology. Basically, we set

$$\sqrt[3]{a + \sqrt{-b}} = x + \sqrt{-y}$$

and then find the unknown x and y by trial and error but subject to the following constraints:

$$x^2 + y = \sqrt[3]{a^2 + b} \quad \text{and} \quad x^3 - 3xy = a.$$

Bombelli did not give any reasons for doing this. His readers would have been more interested in the fact that the rule works in example after example than in how he found it. But we know today that the first constraint results from equating the absolute values of both sides of the previous equation, and the second constraint from equating the real parts of the cubes of both sides.

Bombelli's first example was to find

$$\sqrt[3]{2 + \sqrt{-121}},$$

in which case the constraints become

$$x^2 + y = \sqrt[3]{2^2 + 121} = 5 \quad \text{and} \quad x^3 - 3xy = 2.$$

Then he started his trial-and-error method as follows [p 141]:

> Now it is necessary to find a number [x] whose square is smaller than 5 and whose cube is larger than 2, which if we assume that it is 1 the square root [$\sqrt{y}$] will by necessity be the square root of 4, which squares added together make 5 and the cube of the number is 1 and the product of this number with the square [y] of the square root [$\sqrt{y}$] makes 4 which tripled makes 12, which cannot be subtracted from the cube of the number that is only 1, therefore 1 is no good [as the value of x], nor can 3 be good because its square surpasses 5, ...

The next logical choice was $x = 2$, and then both constraints are satisfied by choosing $y = 1$. Therefore, the desired cube root is $2 + \sqrt{-1}$.

Bombelli was working only with integers in his first and second examples, but his third was a little more adventurous. He proposed to find the cube root of [p. 142]

$$8 + \sqrt{-232 \tfrac{8}{27}},$$

in which case the stated constraints become

$$x^2 + y = \sqrt[3]{8^2 + 232\tfrac{8}{27}} = 6\tfrac{2}{3} \qquad \text{and} \qquad x^3 - 3xy = 8.$$

The trial-and-error method must yield a number x whose square is smaller than $6\tfrac{2}{3}$ and whose cube is larger than 8. Then he said that $x = 2$ "is no good" because its cube is not larger than 8 and that $x = 3$ "is equally no good" because its square is not smaller than $6\tfrac{2}{3}$. Bombelli continued:

> therefore it is necessary to find a quantity that is larger than 2 and smaller than 3, that $\sqrt{2} + 1$ has this property, that its square, which is $3 + \sqrt{8}$ is smaller than $6\tfrac{2}{3}$, and its cube is $\sqrt{50} + 7$, which is larger than 8. Now let us see whether it satisfies the rest: square $\sqrt{2} + 1$ makes $3 + \sqrt{8}$ and subtract it from $6\tfrac{2}{3}$, it remains $3\tfrac{2}{3} - \sqrt{8}$ and this must be

the value of y. Routine computation shows that this choice of x and y satisfies the second constraint.[13]

The next chapter deals with division, and includes the following example (written as a quotient in today's manner) [p. 145]:

$$\frac{10}{\sqrt[3]{2 + (\sqrt{-1})11}}.$$

Bombelli's solution consists in cubing both "parts," numerator and denominator. Then

> we multiply the denominator by $2 - (\sqrt{-1})11$, its residue, makes 125, which divided into 1000 it becomes 8 and this is multiplied by $2 - (\sqrt{-1})11$ makes $16 - (\sqrt{-1})\,88$,

---

[13] It can be argued that Bombelli did not guess that $x = \sqrt{2} + 1$, that he must have posited this value and that of y and then obtained the original number from them. But this does not invalidate his trial-and-error method if a sufficiently good approximation is wanted instead of an exact root. To show this using decimal notation, not yet available in Bombelli's time, start by trying $x = 2.5$. Then $x^2 = 6.25$, $y = 0.41\overline{6}$, and $x^3 - 3xy = 12.5 > 8$. Since the cube is dominant in the last equation, this suggests that 2.5 is too large a value for x, and we try $x = 2.4$ next. This gives $x^2 = 5.76$, $y = 0.90\overline{6}$, and $x^3 - 3xy = 7.296 < 8$, suggesting that 2.4 is too small a value for x; but it seems to be closer to the true value than 2.5. Three additional attempts will produce the required approximation: $x = 2.42$ yields $x^3 - 3xy = 8.289952 > 8$, $x = 2.414$ yields $x^3 - 3xy = 7.989335776 < 8$, and $x = 2.4142$ yields $x^3 - 3xy = 7.999322685 \approx 8$. Thus, $x \approx 2.4142 \approx \sqrt{2} + 1$.

and restoring the cube root, we have $\sqrt[3]{16 - (\sqrt{-1})\,88}$. Disregarding once again the intrusive cube root, this is a perfect example of the procedure: to divide complex numbers, multiply both numerator and denominator by the conjugate of the denominator to make the operation trivial.

The next four short chapters [pp. 147–150] are devoted to addition and subtraction, and contain the unavoidable sprinkling of cube roots. It is not necessary to give examples, since these operations have already been used in the preceding ones. The remaining four chapters of Book I [pp. 151–154] do not involve complex numbers.

Book II of *L'Algebra* is devoted to the study of algebraic polynomials and to the solution of equations of the first to the fourth degree. It is here that, as we have already seen, Bombelli named the variable *il tanto* and that he introduced a half-circle [14] surmounted by an integer for the power to which it is raised. A remarkable contribution by Bombelli is the complete treatment of all possible 42 cases of fourth-degree equations, with which he concluded Book II [pp. 268–314]. In this aim at complete generality, he can be considered a precursor of Viète.

In this second book, Bombelli freely used complex numbers to solve equations, starting with second-order equations. In the *Capitolo di potenze e numero equali a Tanti* (Of the square and the number equal to the Unknowns), he considered the equation $x^2 + 20 = 8x$, solving it by stating that (the reader should keep in mind the formula for solving a quadratic equation) [p. 201]

> the square of half of the Unknowns is 16, which is smaller than 20 and this equation cannot hold except in this sophistic manner. Subtract 20 from 16 it remains -4, its root is $+(\sqrt{-1})\,2$, and this is subtracted and added to one-half of the Unknowns, which will be $4 + (\sqrt{-1})\,2$ and the other $4 - (\sqrt{-1})\,2$, and any of these quantities by itself will be the value of the Unknown.

But Bombelli's greatest achievement was the solution of the cubic in the irreducible case.[15] Once in possession of his method of cube root extraction, he could easily solve the cubic $x^3 = 15x + 4$. He gave the details of Cardano's rule, but no longer had to explain how to find the cube roots, since that had been done in Book I. The following translation is not from the manuscript page shown above but from the printed edition, which was intended to be the final version [p. 225].

---

[14] For typographical reasons, it was rendered as $\smile$ in the printed version of the book.

[15] The study of all kinds of cubic equations is contained in pages 214 to 268 of Book II.

Equate x^3 to $15x + 4$; take one-third of the Unknowns, which is 5, cubed it makes 125 and this is subtracted from the square of half the number, which is 4, it remains $- 121$ (which will be called *più di meno*) which taking the square root of this it will be $(\sqrt{-1})11$, which added to half the number makes $2 + (\sqrt{-1})11$, of which having taken the cube root and added to its residue makes $2 + (\sqrt{-1})1$ and $2 - (\sqrt{-1})1$, which joined together make 4, and 4 is the value of the Unknown.

Bombelli admitted that this solution may seem extravagant to many, and that he himself was of the same opinion at one time, but that he had a demonstration of the existence of a real root. In fact, on page 298 of the 1572 edition he gave the following geometric proof on a plane surface [pp. 228–229].

In Bombelli's time, letters were not used as coefficients, and he used the cubic $x^3 = 6x + 4$ in his proof, but we shall reproduce his argument very closely using $x^3 = px + q$ instead. He chose first a segment q (not the q in the equation), shown in the next figure, of length one, for which we (and he)

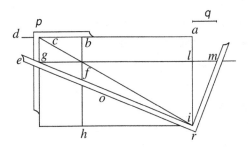

will have no further use. Then draw the half-line me (he denoted it by *.m.e.*, using dots around the letters in all his segment notations). On it, choose l such that $ml = 1$, and then f such that $lf = p$ (Bombelli used p to denote one of the square rulers shown in the figure). Over this lf draw a rectangle abf whose area is q. Now extend ab to a half-line ad and al to a half-line ar. Then take two square rulers (just think of them as right angles), and place the first so that its angle is at a point i on the half line ar and one of its arms goes through m. Lower or raise it in such a way that if the segment if is extended until it touches bd at c, and if the second square ruler is placed with its angle at c and one arm on da, then the intersection of the remaining arms at g is on me,

and this done I say that the segment from the point *.l.* to the angle of the [first] square ruler is the value of the Unknown and prove it in this manner.

Using fractions instead of narrative, the similarity of the triangles ilm and gli yields

$$\frac{il}{lm} = \frac{gl}{li},$$

and denoting il by x and with $ml = 1$, we obtain $gl = x^2$. Then the area of the rectangle ilg will be x^3 and the area of the rectangle ilf will be px because $il = x$ and $lf = p$. The rectangle hfg has the same area as the rectangle alf, which is q (because—but Bombelli did not explain this, thinking it was obvious—the triangle iac has the same area as the other right triangle with hypotenuse ic, the triangles ilf and ihf also have equal areas, and so do the triangles fbc and fgc). Thus, the area of the rectangle ilg, which is x^3, is the sum of the area of the rectangle ilf, which is px, and the area of the rectangle hfg, which is q. We conclude that $x^3 = px + q$ and that $il = x$ is the solution of the proposed cubic.

Bombelli found only one real solution in his examples of the form $x^3 = px + q$, but some of them have three real solutions. The remaining two are easily found by factoring. For example, knowing the solution $x = 4$ of $x^3 = 15x + 4$, we find the other two by writing $x^3 - 15x - 4 = (x-4)(x^2+4x+1)$, which Bombelli could do just as easily in his own notation, and then solving the quadratic $x^2 + 4x + 1 = 0$ to obtain the remaining roots: $-2 \pm \sqrt{3}$.

The cubic was a hot topic, as has already been said, and the fact that a real solution can be obtained from numbers with an imaginary part is what motivated the mathematical community, albeit rather reluctantly, to start using complex numbers.

However, a different solution of the cubic, which does not use square roots of negative numbers, was provided in 1591 by François Viète on the basis of his formula for $\cos 3\theta$. It is in his first treatise in *De æqvationvm recognitione et emendatione tractatvs dvo*,[16] as an alternative Theorem III. We reproduce the part of Viète's statement that is of interest to us [p. 174]:

ANOTHER,
THIRD THEOREM.

If A cubed − B squared 3 in A is equal to B squared in D, while B is greater than half of D [that is, if $A^3 - 3B^2A = B^2D$ and $D/2 < B$] ...

---

[16] Page references are to the English translation by Witmer in *The analytic art by François Viète*, 1983.

*And there are two triangles of equal hypotenuse B, such that the acute
angle subtended by the perpendicular of the first is triple the angle subtended
by the perpendicular of the second; and twice the base of the first is D, and A
is twice the base of the second.*

Admittedly, Viète seems to be talking in riddles. But this is only a temporary impression, since this result bears interpretation. In the given cubic A is
the unknown, so we can call it x. Then, if we write $p = 3B^2$ and $q = B^2D$,
the cubic becomes $x^3 = px + q$. Next, the hypothesis $D/2 < B$ becomes

$$\frac{q}{2B^2} = \frac{3q}{2p} < \sqrt{\frac{p}{3}},$$

which, squaring and simplifying, reduces to

$$\frac{q^2}{4} < \frac{p^3}{27}.$$

In short, we are dealing with the irreducible depressed cubic.

Then, to obtain the apparently cryptic solution, start by dividing the original
cubic, $A^3 - 3B^2A = B^2D$, by $2B^3$. The result can be written as

$$4\left(\frac{A}{2B}\right)^3 - 3\frac{A}{2B} = \frac{D}{2B}.$$

But this brings to mind the equation

$$4\cos^3\theta - 3\cos\theta = \cos 3\theta,$$

which is just a rearrangement of $\cos 3\theta = \cos^3\theta - 3\cos\theta\sin^2\theta$ after replacing
$\sin^2\theta$ with $1 - \cos^2\theta$. This was easy for Viète, who had already discovered this
identity, as we saw in Chapter 1. Now a comparison of the stated trigonometric
identity and the cubic equation of the theorem becomes inescapable. The two
equations are one and the same if

$$\cos\theta = \frac{A}{2B} \quad \text{and} \quad \cos 3\theta = \frac{D}{2B}.$$

This is what Viète expressed by means of his two triangles, which will be of
more help if they are actually drawn, as shown below.

The solution of the cubic is now very clear. The condition $D/2 < B$
means that $D/2B$ is a valid cosine, and then 3θ can be obtained from a table

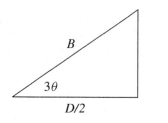
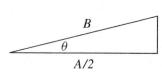

of cosines. Once θ is known, the solution is $A = 2B\cos\theta$. Or, if we prefer to write the cubic as $x^3 = px + q$ and if we note that

$$\frac{D}{2B} = \frac{q}{2B^3} = \frac{q/2}{(p/3)^{3/2}},$$

then the solution is

$$x = 2\sqrt{\frac{p}{3}}\cos\left(\frac{1}{3}\arccos\frac{q/2}{\sqrt{p^3/27}}\right).$$

The last quotient is a valid cosine because $q^2/4 < p^3/27$, that is, precisely because the cubic is irreducible.

Using an electronic device rather than a table of cosines, we can re-obtain Bombelli's solution of $x^3 = 15x + 4$ as follows:

$$x = 2\sqrt{5}\cos\left(\frac{1}{3}\arccos\frac{2}{\sqrt{3375/27}}\right) = 2\sqrt{5}\cos\left(\frac{79.69515353°}{3}\right) = 4.$$

Note that the angles $360° \pm 79.69515353°$ have the same cosine, and can be used in place of $79.69515353°$ in the stated formula to obtain the remaining real roots: $-3.732050808 = -2 - \sqrt{3}$ and $-0.267949192 = -2 + \sqrt{3}$.

It is fortunate that Viète's solution without imaginaries did not appear until 1615. Had this result been published in the sixteenth century, it might have delayed any interest and investigation into the nature of complex numbers.

Bombelli's notation, *p. di m.* and *m. di m.*, was not subsequently adopted. The notation that would survive, $\sqrt{-a}$, was introduced by Albert Girard (1595–1632), while giving the solutions of the quartic $x^4 = 4x - 3$ as *les quatre solutions seront* $1, 1, -1 + \sqrt{-2}, -1 - \sqrt{-2}$.[17] This notation was adopted by many mathematicians, including Newton, Wallis, and Euler. As

---

[17] *Invention nouvelle en l'algèbre* (The new invention in algebra), 1629, ff. FA and FB. See Cajori, *A history of mathematical notations*, II, 1929, p. 127.

for the word, "imaginary" that we use to this day, it was originally introduced by René Descartes (1596–1650) in *La Géométrie* of 1637 as follows [p. 380]:

> Neither the true nor the false [negative] roots are always real; but sometimes only imaginary.[18]

For a long time this would be the word to describe what we now call complex numbers. Acceptance was one thing but understanding was slower in coming, and mathematicians were puzzled by the meaning of imaginaries. Newton referred to imaginary roots as *radices impossibiles*,[19] and Leibniz, in a more literary mood, expressed himself as follows:

> Therefore, an elegant and admirable way out is found in that miracle of Analysis, that portent of the ideal world, almost Amphibian between Being and non-Being, that we call an imaginary root.[20]

3.4 HIGHER-ORDER ROOTS OF COMPLEX NUMBERS

More than a century and a half passed since the publication of Bombelli's *l'Algebra* until a method to compute the nth roots of complex numbers was made available. It was developed by Abraham de Moivre (1667–1754), a French mathematician who resided in England for religious reasons since the revocation of the Edict of Nantes protecting Huguenots.

It appeared in a paper of 1739,[21] in which he considered four problems. The first does not refer to complex numbers, the second deals with the extraction of cube roots, and the third was "To extract the Root, whose Index is n, of the impossible Binomial $a + \sqrt{-b}$." Note that n must be a positive integer and that $b > 0$ if the "impossible binomial" is what we now call a complex

---

[18] "Au reste tant les vrayes racines que les fausses ne sont pas tousiours reelles; mais quelquefois seulement imaginaires." In *The geometry of René Descartes*, translated by Smith and Latham, 1954, pp. 174–175. Also in Hawking, ed., *God created the integers. The mathematical breakthroughs that changed history*, 2005, p. 345.

[19] *Arithmetica universalis*, 1707, p. 242.

[20] "Itaque elegans et mirabile effugium reperit in illo Analyseos miraculo, idealis mundi monstro, pene inter Ens et non-Ens Amphibio, quod radicem imaginariam apellamus." In "Specimen novum analyseos pro scientia infiniti, circa summas & quadraturas," 1702 = *Opera mathematica*, p. 378 = Gerhardt, *Leibnizens Mathematische Schriften*, Sec. 2, I, No. XXIV, p. 357.

[21] "De reductione radicalium ad simpliciores terminos, seu de extrahenda radice quaqunque data ex binomio $a + \sqrt{+b}$, vel $a + \sqrt{-b}$, Epistola," 1739.

ABRAHAM DE MOIVRE IN 1736

Portrait by Joseph Highmore, engraving by John Faber.
From Florence Nightingale David, *Games, Gods and Gambling.*
Charles Griffin, London and Hafner Publishing Co., New York, 1962.

number. Then he stated his solution, which needs more than interpretation, as
follows [p. 475]:

Let that Root [the nth root of $a+\sqrt{-b}$] be $x+\sqrt{-y}$; then making $\sqrt[n]{aa + b} =$
m; and also making $\dfrac{n-1}{n} = p$, describe, or imagine describing, a circle,
whose Radius is $\sqrt{m}$, in which take some arc A, whose Cosine is $\dfrac{a}{m^p}$; let
C be the whole circumference. Take to the same Radius, the Cosines of the
Arcs

$$\frac{A}{n}, \frac{C-A}{n}, \frac{C+A}{n}, \frac{2C-A}{n}, \frac{2C+A}{n}, \frac{3C-A}{n}, \frac{3C+A}{n}, \&c.$$

Until the number of them [the Cosines] equals the number n; this done,
thereupon; these Cosines will be so many values of the quantity x; as for the
quantity y, it will always be $m - xx$.

Let us admit at the outset that this is just a statement without proof, possibly because de Moivre had just presented a fully detailed discussion of the particular case $n = 3$ as his second problem, and felt entitled to omit the details in the general case.

In any event, the equation defining p seems to contain a typo. If, instead, we define $p = (n - 1)/2$, we can interpret the result in present-day notation. If we write $z = a + i\sqrt{b}$ and recall the definition of m, then

$$|z| = \sqrt{a^2 + b} = \sqrt{m^n}$$

and

$$\arg z = \arccos \frac{a}{|z|} = \arccos \frac{a}{\sqrt{m^n}}.$$

For de Moivre, as for many of his predecessors and contemporaries, the Cosine of an arc A was what we now denote by $R \cos A$, where R is the radius of the circle, in this case $\sqrt{m}$. Then, his choice of A is such that

$$\sqrt{m} \, \cos A = \frac{a}{m^p},$$

and it follows that

$$A = \arccos \frac{a}{m^p \sqrt{m}} = \arccos \frac{a}{m^{n/2}} = \arg z.$$

This clears up the nature of the arc A, and by C de Moivre must have meant 2π, not the whole circumference of the circle with radius $\sqrt{m}$.

There is a shortcoming and a flaw in de Moivre's solution. Only complex numbers in the upper half-plane are of the form $a + \sqrt{-b}$ if $b > 0$, so that the roots of $a - \sqrt{-b}$ must be found by a similar method. This is the shortcoming, and the flaw is to assume that all the roots of a number in the upper half-plane are of the form $x + \sqrt{-y}$ with $y > 0$. Actually, some of the roots are of the form $x - \sqrt{-y}$, as shown by the simple case of $a + \sqrt{-b} = 0 + \sqrt{-1}$. In this case, $A = \frac{1}{2}\pi$, and $n = 3$, and the three arcs in de Moivre's method are

$$\frac{\frac{1}{2}\pi}{3} = \frac{\pi}{6}, \qquad \frac{2\pi - \frac{1}{2}\pi}{3} = \frac{3\pi}{6}, \qquad \text{and} \qquad \frac{2\pi + \frac{1}{2}\pi}{3} = \frac{5\pi}{6},$$

The cosine of the second arc is zero, which yields $x = 0$ and $y = 1 - x^2 = 1$. This gives the complex number $0 + \sqrt{-1}$, which is not a cube root of $0 + \sqrt{-1}$. Instead, the desired cube root is $0 - \sqrt{-1}$ (in our terms, $-i$ is a cube root of i). De Moivre could have easily avoided these problems with a change of notation; for instance, using $a + b\sqrt{-1}$ instead of $a + \sqrt{-b}$ and allowing b to be negative.

The preceding explanation notwithstanding, it is a fact that de Moivre never stated the nth roots of a complex number in a fully explicit form. Euler got closer when ten years later, in 1749, he submitted a paper in which he posed a more ambitious problem than that of de Moivre:[22]

PROBLEM 1

79. *An imaginary quantity* [meaning complex] *being raised to a power whose exponent is any real quantity, find the imaginary value that results.*

His method of solution is based on a formula that he had proved in his *Introductio*,[23] so we present this first. Once he was done with series expansions for exponentials and logarithms in Chapter VII, he turned his attention to "transcendental quantities derived from the circle" in the next chapter. In Articles 127 to 131, Euler introduced the notation that he would use—essentially the modern notation—and then gave a number of trigonometric identities. Next he evaluated the product [Art. 132]

$$(\cos.y + \sqrt{-1}.\sin.y)(\cos.z + \sqrt{-1}.\sin.z)$$
$$= \cos.y\cos.z - \sin.y\sin.z + \sqrt{-1}.(\sin.y\cos.z + \cos.y\sin.z)$$
$$= \cos.(y+z) + \sqrt{-1}.\sin.(y+z),$$

which we have reproduced keeping his notation but abbreviating his writing, and similarly,

$$(\cos.y - \sqrt{-1}.\sin.y)(\cos.z - \sqrt{-1}.\sin.z) = \cos.(y+z) - \sqrt{-1}.\sin.(y+z).$$

These give [Art. 133]

$$(\cos.z \pm \sqrt{-1}.\sin.z)^2 = \cos.2z \pm \sqrt{-1}.\sin.2z,$$

and, in general, the formula

$$(\cos.z \pm \sqrt{-1}.\sin.z)^n = \cos.nz \pm \sqrt{-1}.\sin.nz$$

---

[22] "Recherches sur les racines imaginaires des équations," 1749, pub. 1751. References will be by article number so that any of the references listed in the bibliography can be used. The stated problem and its solution are in Art. 79. In Art. 93 he considered the general problem of raising a complex number to a complex power.

[23] *Introductio in analysin infinitorum*, 1748. We shall refer to the *Introductio* by article number rather than by page, so that the reader can refer to any available edition.

for any positive integer n.

If we are permitted a digression that will tie some loose ends in our earlier discussion of trigonometry, Euler had stated the following in the second paragraph of Chapter VIII [Art. 126]:

> Let us therefore stipulate that the Radius of the Circle or total Sine be $= 1$, ... let π = the Semicircumference of the Circle whose Radius $= 1$, or [equivalently] π will be the length of the Arc of 180 degrees.[24]

It was with these words that Euler's imposing mathematical authority sealed the value of the trigonometric *sinus totus* to be 1 from that point on.[25] However, he was not being original in this selection, for it had already been made by Abu'l Wafa and Joost Bürgi, but it is almost certain that Euler was not aware of this. With this choice of the radius equal to 1, the trigonometric lengths effectively became trigonometric ratios, and soon, in the rest of Euler's work, trigonometric functions. In this manner, trigonometry disappeared as an independent branch of mathematics to become, as in the rest of this section, a part of mathematical analysis.

Before moving ahead, several comments are in order about the last equation. First, it should be pointed out that Euler was the first to consistently write $\sqrt{-1}$ for the imaginary unit, and that later he introduced the current symbol i for it.[26] It will be convenient to use this new symbol from now on even if Euler himself did not until much later, and, since the periods in the abbreviations for sine and cosine are no longer in use, we shall omit them from now on and rewrite the previous equation as

$$(\cos z \pm i \sin z)^n = \cos nz \pm i \sin nz.$$

Finally, it turns out that this formula was known to de Moivre, although he never wrote it explicitly like this. The closest he got is as follows (to be explained below):[27]

---

[24] "Ponamus ergo Radium Circuli seu Sinum totum esse $= 1$, ... ita ut sit π = Semicircumferentiæ Circuli, cujus Radius $= 1$, seu π erit longitudo Arcus 180 graduum."

[25] A complete translation of Articles 126, 127, 132 and 133 is included in Calinger, *Classics of Mathematics*, pp. 494–495.

[26] With the words "formulam $\sqrt{-1}$ littera i in posterum designabo" (in what follows I denote the formula $\sqrt{-1}$ by the letter i) in "De formulis differentialibus angularibus maxime irrationalibus, quas tamen per logarithmos et arcus circulares integrare licet," presented to the Academy of Saint Petersburg on May 5, 1777, but first published in his book *Institvtionvm calcvli integralis*, 2nd. ed., IV, 1794, p. 184.

[27] "De sectione anguli," 1722, p. 229.

Let x be the Versed Sine of any Arc whatever.

t the Versed Sine of another Arc.

1 the Radius of the Circle.

And let the former Arc be to the latter as 1 to n, Then, assume two Equations which may be regarded as known,

$$1 - 2z^n + z^{2n} = -2z^n t$$
$$1 - 2z + zz = -2zx.$$

Eliminate z, there will arise an Equation by which the Relation between x & t will be determined.

Regardless of the possible meaning of z (it is not essential for our purposes), recall that the versed sine of φ is $1 - \cos\varphi$. Then let $x = 1 - \cos\varphi$ and $t = 1 - \cos n\varphi$. Solving the quadratic equations

$$1 - 2z^n + z^{2n} = -2z^n(1 - \cos n\varphi) \quad \text{and} \quad 1 - 2z + z^2 = -2z(1 - \cos\varphi)$$

for z^n and z gives

$$z^n = \cos n\varphi \pm \sqrt{\cos^2 n\varphi - 1} \quad \text{and} \quad z = \cos\varphi \pm \sqrt{\cos^2\varphi - 1}.$$

It follows that

$$(\cos\varphi \pm i\sin\varphi)^n = \cos n\varphi \pm i\sin n\varphi.$$

Because it was implicitly contained in de Moivre's work, this equation has been known as *de Moivre's formula* since the nineteenth century.

But of course, Euler went further to generalize the validity of this equation. After restating the fundamental equation as

$$(\cos\varphi + i\sin\varphi)^m = \cos m\varphi + i\sin m\varphi$$

in his 1749 paper [Art. 85], he added:

> But that the same formula is valid also, when m is any [real] number, differentiation after having taken logarithms will show it beyond a doubt.

What this shows is that both sides of de Moivre's formula have the same derivative with respect to φ, and since they are equal for $\varphi = 0$ they must be equal for all φ. Thus, de Moivre's formula is valid for any real exponent, and, what is more, Euler used it.

He had begun the solution of Problem 1 in Art. 79 by writing $(a + bi)^m = M + Ni$, and his goal was to find M and N. First he put

$$\sqrt{(aa + bb)} = c,$$

and then he stated: "let us look for an angle φ such that its sine is $= b/c$ and the cosine $= a/c$," which if found would allow us to write

$$a + bi = c(\cos \varphi + i \sin \varphi).$$

Here φ can be replaced by $\varphi + 2k\pi$, k any integer (Euler did not write the k, he just spelled out a few of these angles and then wrote "etc."), because their sines and cosines "are the same" as for φ. Then, using de Moivre's formula (whose general proof he postponed until Art. 85), the proposed power becomes

$$(a + bi)^m = c^m (\cos \varphi + i \sin \varphi)^m = c^m (\cos m\varphi + i \sin m\varphi).$$

From this he concluded that

$$M = c^m \cos m\varphi \qquad \text{and} \qquad N = c^m \sin m\varphi.$$

After finding this solution, Euler stated five corollaries. The first two are useless today, and the third stated that if m is an integer then $(a + bi)^m$ has only one value because all the angles involved (namely, $\varphi + 2k\pi$) have the same sine and cosine [Art. 82]. The fourth [Art. 83] is of interest to us, asserting that if m is of the form μ/ν, where μ and ν are positive integers, then $(a + bi)^m$ will have exactly ν values, which essentially solves the problem of finding the roots of complex numbers. If $m = 1/n$, where n is a positive integer, and if we combine all the possible values of M and N in a single formula (which Euler did not do), we have

$$\sqrt[n]{a + bi} = \sqrt[n]{c} \left(\cos \frac{\varphi + 2k\pi}{n} + i \sin \frac{\varphi + 2k\pi}{n} \right).$$

Equivalently, this formula can be written using $\varphi - 2k\pi$ instead of $\varphi + 2k\pi$. In either case, the only distinct values are for $k = 0, 1, \ldots, n - 1$. Finally, the fifth corollary [Art. 84] states that if m is irrational, $(a + bi)^m$ "will have an infinite number of different values."

Now we can return to Cardano's formula, but choosing the cubic from Bombelli's proof for the *Cubo equale a Tanti e numero*, namely $x^3 = 6x + 4$.

This is easier for hand calculation than $x^3 = 15x + 4$, and when solved by Euler via Cardano's rule yielded the solution [28]

$$x = \sqrt[3]{2 + 2i} + \sqrt[3]{2 - 2i}.$$

At this point, Euler added: "which cannot be expressed otherwise." This is surprising because it is easy to express it "otherwise" using the formula just derived to find complex roots. If $a + bi = 2 + 2i$, it is clear that $c = \sqrt{4 + 4} = \sqrt{8}$ and $\cos\varphi = \sin\varphi = 2/\sqrt{8} = 1/\sqrt{2}$, so that $\varphi = \pi/4$. Then

$$\sqrt[3]{c} = \sqrt[3]{\sqrt{8}} = \sqrt{\sqrt[3]{8}} = \sqrt{2},$$

and the three cube roots of $a + bi$ are

$$\sqrt{2}\left(\cos\frac{\pi}{12} + i\sin\frac{\pi}{12}\right),$$

$$\sqrt{2}\left(\cos\frac{\pi/4 + 2\pi}{3} + i\sin\frac{\pi/4 + 2\pi}{3}\right) = \sqrt{2}\left(\cos\frac{3\pi}{4} + i\sin\frac{3\pi}{4}\right),$$

and

$$\sqrt{2}\left(\cos\frac{\pi/4 + 4\pi}{3} + i\sin\frac{\pi/4 + 4\pi}{3}\right) = \sqrt{2}\left(\cos\frac{17\pi}{12} + i\sin\frac{17\pi}{12}\right).$$

Similarly, but using $\varphi = -\pi/4$, $\varphi - 2\pi$, and $\varphi - 4\pi$, the three cube roots of $2 - 2i$ are

$$\sqrt{2}\left[\cos\left(-\frac{\pi}{12}\right) + i\sin\left(-\frac{\pi}{12}\right)\right] = \sqrt{2}\left(\cos\frac{\pi}{12} - i\sin\frac{\pi}{12}\right),$$

$$\sqrt{2}\left[\cos\left(-\frac{3\pi}{4}\right) + i\sin\left(-\frac{3\pi}{4}\right)\right] = \sqrt{2}\left(\cos\frac{3\pi}{4} - i\sin\frac{3\pi}{4}\right),$$

and

$$\sqrt{2}\left[\cos\left(-\frac{17\pi}{12}\right) + i\sin\left(-\frac{17\pi}{12}\right)\right] = \sqrt{2}\left(\cos\frac{17\pi}{12} - i\sin\frac{17\pi}{12}\right).$$

Hence, the three solutions provided by Cardano's formula are

$$\sqrt{2}\left(\cos\frac{\pi}{12} + i\sin\frac{\pi}{12}\right) + \sqrt{2}\left(\cos\frac{\pi}{12} - i\sin\frac{\pi}{12}\right) = 2\sqrt{2}\cos\frac{\pi}{12},$$

---

[28] *Vollständige Anleitung zur Algebra*, 1770, Second Part, First Section, Chapter 12, Article 188.

$$\sqrt{2}\left(\cos\frac{3\pi}{4}+i\sin\frac{3\pi}{4}\right)+\sqrt{2}\left(\cos\frac{3\pi}{4}-i\sin\frac{3\pi}{4}\right)=2\sqrt{2}\cos\frac{3\pi}{4}=-2,$$

and

$$\sqrt{2}\left(\cos\frac{17\pi}{12}+i\sin\frac{17\pi}{12}\right)+\sqrt{2}\left(\cos\frac{17\pi}{12}-i\sin\frac{17\pi}{12}\right)=2\sqrt{2}\cos\frac{17\pi}{12}.$$

If we now use the formula $\cos(\alpha\pm\beta)=\cos\alpha\cos\beta\mp\sin\alpha\sin\beta$, we obtain

$$\cos\frac{\pi}{12}=\cos\left(\frac{\pi}{3}-\frac{\pi}{4}\right)=\frac{1}{2}\frac{\sqrt{2}}{2}+\frac{\sqrt{3}}{2}\frac{\sqrt{2}}{2}=\frac{1+\sqrt{3}}{4}\sqrt{2}$$

and

$$\cos\frac{17\pi}{12}=\cos\left(\frac{\pi}{4}+\frac{7\pi}{6}\right)=\frac{\sqrt{2}}{2}\left(-\frac{\sqrt{3}}{2}\right)-\frac{\sqrt{2}}{2}\left(-\frac{1}{2}\right)=\frac{1-\sqrt{3}}{4}\sqrt{2},$$

and then the three solutions of $x^3=6x+4$ become $1+\sqrt{3}$, -2, and $1-\sqrt{3}$. All three are real, and this represents a triumph for imaginary numbers. From this moment on they could not be denied.

3.5 THE LOGARITHMS OF COMPLEX NUMBERS

To lay the ground for obtaining the logarithms of complex numbers, we return to the *Introductio*. In Article 134 Euler deduced the following series expansions for the sine and the cosine, using the letter v for the variable:

$$\cos v=1-\frac{v^2}{1\cdot 2}+\frac{v^4}{1\cdot 2\cdot 3\cdot 4}-\frac{v^6}{1\cdot 2\cdot 3\cdot 4\cdot 5\cdot 6}+\&\text{c.},\quad\&$$

$$\sin v=v-\frac{v^3}{1\cdot 2\cdot 3}+\frac{v^5}{1\cdot 2\cdot 3\cdot 4\cdot 5}-\frac{v^7}{1\cdot 2\cdot 3\cdot 4\cdot 5\cdot 6\cdot 7}+\&\text{c.}[29]$$

A few paragraphs later, Euler gave formulas for $\cos v$ and $\sin v$ in terms of exponentials, and from these he deduced his now famous formula [Art. 138]

$$e^{iv}=\cos v+i\sin v,[30]$$

---

[29] This was unnecessary, since these series were obtained by Newton in 1665 and first made public in 1669 in *De analysi per æquationes numero terminorum infinitas*. The sine and cosine series are on page 17 of Jones' edition and on pages 236 and 237 of Whiteside's 1968 edition. They appeared in *De analysi* for the first time in Europe, although they were known in India since the fourteenth century (see Section 4.3).

[30] Since it is famous, here it is in Euler's original notation: $e^{+v\sqrt{-1}}=cos.v+\sqrt{-1}.sin.v$.

which is extremely useful in many practical applications. It is universally known today as *Euler's formula* or *Euler's identity*. In proving this result Euler was original because he was not aware of an equivalent result that was obscurely embedded in the one and only paper published by the English mathematician Roger Cotes (1682–1716),[31] the man of whom Newton said: "If He [Cotes] had lived we might have known Something."[32]

The argument that Euler used to deduce this formula in the *Introductio* is rather dirty. Perhaps no less dirty but definitely briefer is a third proof (his second requires knowledge of integral calculus) that he gave in 1749.[33] Here he put an imaginary number $i\varphi$ in place of the real variable x in the series previously obtained for the exponential function, taking it as an act of faith that [p. 166; 219]

$$e^{i\varphi} = 1 + i\varphi + \frac{(i\varphi)^2}{1 \cdot 2} + \frac{(i\varphi)^3}{1 \cdot 2 \cdot 3} + \frac{(i\varphi)^4}{1 \cdot 2 \cdot 3 \cdot 4} + \frac{(i\varphi)^5}{1 \cdot 2 \cdot 3 \cdot 4 \cdot 5} + \text{etc.}$$

$$= 1 + i\varphi - \frac{\varphi^2}{1 \cdot 2} - i\frac{\varphi^3}{1 \cdot 2 \cdot 3} + \frac{\varphi^4}{1 \cdot 2 \cdot 3 \cdot 4} + i\frac{\varphi^5}{1 \cdot 2 \cdot 3 \cdot 4 \cdot 5} - \text{etc.}$$

Actually, Euler did not write $e^{i\varphi}$ as we have done. Rather, in the vein of what

---

[31] "Logometria," 1714. The words "obscurely embedded" are explained by the form in which the formula appears on page 32, as a small statement in the midst of a number of geometric constructions: "Nam si quadrantis circuli quilibet arcus, radio CE descriptus, sinun habeat CX sinumque complementi ad quadrantem XE; sumendo radium CE pro Modulo, arcus erit rationis inter $EX + XC\sqrt{-1}$ & CE mensura ducta in $\sqrt{-1}$." This can be translated as follows: "For instance if any arc of a quadrant of a circle, described with radius CE [with the center at E], has sinus CX and sinus of the complement to the quadrant [cosinus] XE; taking the radius CE as modulus, the arc will be the measure of the ratio between $EX + XC\sqrt{-1}$ & CE multiplied by $\sqrt{-1}$." For Cotes, if A, B, and M are positive numbers, "the measure of the ratio A over B modulo M" meant $Ml(A/B)$, where l denotes the natural logarithm. Thus, if $R = CE$ and if the arc is denoted by $R\theta$, the ratio $(EX + XC\sqrt{-1})/CE$ equals $\cos\theta + i\sin\theta$, and Cotes' statement becomes $iRl(\cos\theta + i\sin\theta) = R\theta$. Except for the obvious sign error, this is Euler's formula. This paper is combined with other previously unpublished works in Cotes' posthumous publication *Harmonia mensurarum*, 1722, pp. 4–41. The stated quotation is on page 28 of *Harmonia mensurarum* and page 170 of Gowing's translation.

[32] According to a handwritten note by Robert Smith, editor of *Harmonia mensurarum*, on his own copy of this book.

[33] "De la controverse entre Mrs. Leibnitz et Bernoulli sur les logarithmes négatifs et imaginaires," 1749, pub. 1751. Page references below are to the original paper first and, after a semicolon, to the *Opera omnia*.

he had written in Art. 115 of the *Introductio* (see page 139), he wrote

$$\left(1 + \frac{i\varphi}{n}\right)^n,$$

where n is infinitely large. Putting this little detail aside, not only did he dare to replace the variable in the series for the exponential with an imaginary number, but he rearranged the infinitely many resulting terms at will. Implicitly gathering all the real terms together and then all the imaginary ones, and in view of the infinite series previously obtained for the sine and the cosine, he obtained

$$e^{i\varphi} = \cos\varphi + i\sin\varphi.$$

We should say in passing that, besides its usefulness, Euler's formula has an interesting consequence that has always fascinated mathematicians. Putting $\varphi = \pi$, we obtain $e^{i\pi} = -1$, which is usually rewritten as

$$e^{i\pi} + 1 = 0.$$

This is said to be the most beautiful formula of mathematics, relating the five most important numbers, 0 and 1 from arithmetic, π from geometry, i from algebra, and e from analysis, by three of the most important operations: addition, multiplication, and exponentiation

As for the logarithms of complex numbers, Euler already knew that there was an infinitude of them for any number, as mentioned at the end of Chapter 2, and now posed the problem as follows [p. 165; 218]:

PROBLEM III

Determine all the logarithms of any imaginary [complex] *quantity.*

His solution consisted in taking a nonzero complex number $a + bi$, letting $\sqrt{(aa + bb)} = c$, and choosing an angle φ whose sine is b/c and cosine a/c. Then,

$$a + bi = c(\cos\varphi + i\sin\varphi),$$

"or, since c is a positive number, let C be its real logarithm, and we will have"

$$l(a + bi) = C + l(\cos\varphi + i\sin\varphi).$$

At this point we must depart from Euler's own presentation, which has become archaic in that it does not use the exponential function or his famous formula stated above, lacks rigor by using an "infinite number," and is way too long. If we use Euler's formula instead, as written in the *Introductio* and as is written today, and if p is any even integer, then

$$l(a + bi) = C + l[\cos(\varphi + p\pi) + i\sin(\varphi + p\pi)]$$
$$= C + l\left[e^{(\varphi + p\pi)i}\right]$$
$$= C + (\varphi + p\pi)i.$$

This last line is the result as stated by Euler [p. 167; 220]. To rewrite it in current terminology, put $z = a + bi$, in which case $c = |z|$, $C = l(c)$, and $\varphi = \arg z$. Then we have

$$l(z) = l(|z|) + (\varphi + p\pi)i,$$

where p is any even integer, giving infinitely many values of the logarithm of every complex number z.

If z is real and positive, only one value of its logarithm is real (the one for $p = 0$, since $\varphi = 0$). But Euler easily concluded that "all the logarithms of an imaginary [complex] quantity are also imaginary" [p. 167; 220]. In particular,

$$l(-1) = l(1) + (\pi + p\pi)i = (1 + p)\pi i,$$

or, as Euler wrote it (but without using the parentheses) [pp. 168; 221–222],

$$l(-1) = q\pi i,$$

where q is "any odd number."

In June 1746 Euler wrote a letter to his friend Christian Goldbach, at the very end of which he revealed an even more surprising find than the logarithms of imaginaries:[34]

that the expression $(\sqrt{-1})^{\sqrt{-1}}$ has a real value, which in decimal fraction = 0,2078795763, which to me appears to be remarkable.

The problem of evaluating i^i had a long history. The Italian Giulio Carlo de'Toschi di Fagnano (1682–1766), Johann Bernoulli, and Euler himself before this time had made some preliminary progress. In 1746 he limited himself

---

[34] In Fuss, *Correspondence mathématique et physique de quelques célèbres géomètres du XVIIIème siècle*, tome 1, 1843, p. 383; and on p. 182 of the reprint by Johnson Reprint Corporation.

to state the one value given above, but in 1749 he gave the complete solution. It is very simple to obtain it if we put $a = 0$ and $b = 1$ in Euler's formula for the logarithm, for this gives $C = l(1) = 0$ and $\varphi = \pi/2$, and then

$$l(i) = \left(\tfrac{1}{2}\pi + p\pi\right)i.$$

Therefore,

$$l(i^i) = il(i) = -\left(\tfrac{1}{2}\pi + p\pi\right),$$

so that

$$i^i = e^{-\frac{1}{2}\pi - p\pi},$$

"what is all the more remarkable because it is real and includes an infinitude of different real values."[35]

But in spite of Euler's enormous success with complex numbers, their acceptance was slow, very slow. Solving the cubic was no longer fashionable, several other trigonometric series had been obtained without the help of complex numbers and—more importantly—what were these numbers and what could their infinitely many logarithms mean?

3.6 CASPAR WESSEL'S BREAKTHROUGH

It may be said that complex numbers made their most important inroads into mathematical thought through the back door; not through their algebraic prowess but through their geometric representation. After all, if they can be represented geometrically they must exist. The first to try his hand at such a task was John Wallis (1616–1703), in Chapters LXVI through LXVIII of his *Algebra* of 1685, but his efforts did not amount to much of any use.

---

[35] The last formula and the quoted statement are from Art. 97 of "Recherches sur les racines imaginaires des équations" (except that Euler wrote $\sqrt{-1}$ instead of i and 2λ in place of p), but we have not presented his own proof because the method of using logarithms directly is much briefer and more relevant at this moment. His own proof begins by setting [Art. 93]

$$(a + bi)^{m+ni} = x + yi,$$

where a, b, m, and n are assumed to be given. Then taking logarithms, finding the total differentials of both sides, and separating real and imaginary parts leads to two equations in the unknowns x and y. Solving them gives

$$x = c^m e^{-n\varphi} \cos(m\varphi + nl(c)) \qquad \text{and} \qquad y = c^m e^{-n\varphi} \sin(m\varphi + nl(c)),$$

where c and φ are above. The result follows by putting $a = m = 0, b = n = c = 1$, and $\varphi = \tfrac{1}{2}\pi + p\pi$.

On March 10, 1797, Caspar Wessel (1745–1818), a land surveyor (born in Vestby, in what today is Norway but was then part of Denmark), presented a paper to the Royal Danish Academy of Sciences, *On the analytic representation of direction*, which was published two years later, essentially containing the idea of representing each complex number by a vector in the plane anchored at the origin. On this basis, he then gave laws of addition and multiplication that were geometric in nature. The first is just the familiar rule for vector addition in the plane [§ 1]:[36]

> Two right lines [vectors] are added if we write them in such a way that the second line begins where the first one ends, and then pass a right line from the first to the last point of the united lines; this line is the sum of the united lines.

He thereby initiated the vector calculus. But the multiplication law, which we shall explore next, was Wessel's main contribution [§ 4]:

> Firstly, the factors shall have such a direction that they both can be placed in the same plane with the positive unit.
>
> Secondly, as regards length, the product shall be to one factor as the other factor is to the unit. And,
>
> Finally, if we give the positive unit, the factors, and the product a common origin, the product shall, as regards its direction, lie in the plane of the unit and the factors and diverge from the one factor as many degrees, and on the same side, as the other factor diverges from the unit, so that the direction angle of the product, or its divergence from the positive unit, becomes equal to the sum of the direction angles of the factors.

Since this write-up is beyond our modern endurance for verbosity (or perhaps it is a shortage of modern patience with detailed description), we can rephrase it as follows:

- We consider two vectors from the origin coplanar with the one from $(0, 0)$ to $(0, 1)$.

- The length of their product is defined to be the product of their lengths, and

---

[36] The quotations from Wessel's paper included here are from the translation by Martin A. Nordgaard in Smith, *A source book in mathematics*, pp. 55–66. They are by section number rather than by page. For the original paper and other translations, see the bibliography.

- The angle of their product (counterclockwise from the positive x-axis) is the sum of the angles of the factors.

So far this looks like the beginning of vector calculus—which it is—but then Wessel stated the following [§5]:

> Let $+1$ designate the positive rectilinear unit and $+\epsilon$ a certain other unit perpendicular to the positive unit and having the same origin; then the direction angle of $+1$ will be equal to $0°$, that of -1 to $180°$, that of $+\epsilon$ to $90°$, and that of $-\epsilon$ to $-90°$ or $270°$. By the rule that the direction angle of the product shall equal the sum of the angles of the factors, we have: $(+1)(+1) = +1$; $(+1)(-1) = -1$; $(-1)(-1) = +1$; $(+1)(+\epsilon) = +\epsilon$; $(+1)(-\epsilon) = -\epsilon$; $(-1)(+\epsilon) = -\epsilon$; $(-1)(-\epsilon) = +\epsilon$; $(+\epsilon)(+\epsilon) = -1$; $(+\epsilon)(-\epsilon) = +1$; $(-\epsilon)(-\epsilon) = -1$.
>
> From this it is seen that $\epsilon = \sqrt{-1}$; and the divergence of the product is determined such that not any of the common rules of operation are contravened.

In this fashion, Wessel introduced what would eventually become known as the complex plane, while using the letter ϵ instead of i, and he also proved the formula

$$(a + \epsilon b)(c + \epsilon d) = ac - bd + \epsilon(ad + bc)$$

from his multiplication rule [§10] and the usual rule for division [§12]:

$$\frac{a + \epsilon b}{c + \epsilon d} = \frac{ac + bd + \epsilon(bc - ad)}{c^2 + d^2}.$$

We should observe that Wessel's geometric multiplication rule has the advantage over the usual formula $(a + bi)(c + di) = ac - bd + (ad + bc)i$, reverting now to i instead of ϵ, from a computational point of view. For instance, which rule would you choose to evaluate $(0.6 + 0.5i)^{20}$, even if a hand-held calculator were allowed? Using the latter formula nineteen times seems rather forbidding, but noting that the length of $0.6 + 0.5i$ is

$$\sqrt{0.6^2 + 0.5^2} = 0.781024968$$

and that its direction angle is

$$\text{arccos}\, \frac{0.6}{0.781024968} = \text{arccos}\, 0.76822128 = 39.8055711°,$$

we have

$$0.781024968^{20} = 0.007133429$$

and

$$20 \times 39.8055711° = 796.111422°.$$

Then

$$(0.6 + 0.5i)^{20} = 0.007133429 \,(\cos 796.111422° + i \sin 796.111422°)$$
$$= 0.001712269 + i0.006924879.$$

This illustrates dramatically the power of Wessel's multiplication law.

In the same manner, Wessel discovered that the direction angle corresponding to the mth root of a complex number is the original direction angle divided by m, and gave the already known formula for the mth roots [§15]. In the process, he rediscovered de Moivre's formula once more, even for fractional exponents [§13]. In his isolation, as a nonmathematician, he must have been unaware of these known facts. The reader is invited to use this method to evaluate the cube roots contained in the solution of Bombelli's cubic, $x^3 = 15x + 4$.

Although his paper was published, in Danish, in the Academy's memoirs two years after Wessel's oral presentation, it remained largely ignored until it was rediscovered and its value recognized in 1895.

Meanwhile, a similar geometric representation of complex numbers was privately printed, but not published, by an amateur mathematician residing in Paris, Robert Argand (1768–1822), in a booklet that did not even include the author's name.[37] It also remained ignored until, by a happy series of circumstances, Argand's work was rediscovered by Jacques Français (1775–1883), a professor at Metz, who published a summary in 1813 together with a plea for the unknown author to reveal his name.[38] Argand answered the plea, and his name was printed in the next issue.[39]

Argand posed the problem and the difficulty in solving it as follows [Art. 3]: to find [40]

---

[37] *Essai sur une manière de représenter les quantités imaginaires, dans les constructions géométriques*, privately printed in 1806 (according to Argand). The often repeated particulars about Argand's name, biographical details, and occupation are now questionable. In this regard see Schubring, "Argand and the early work on graphical representation: New sources and interpretations," 2001, pp. 130–134.

[38] "Philosophie mathématique. Nouveaux principes de géométrie de position, et interprétation géométrique des symboles imaginaires," 1813, p. 71.

[39] "Philosophie mathématique. Essay sur une manière de représenter les quantités imaginaires dans les constructions géométriques," 1813, p. 133.

[40] This translation is from Hoüel's second edition, 1874, p. 6. There is another translation in Hardy, *Imaginary quantities: Their geometrical interpretation*, 1881, p. 23.

the quantity x that satisfies the proportion

$$+1 : +x :: +x : -1$$

[that is, $+1/x = x/(-1)$]. We stop here ... because x cannot be made equal to any number, either positive or negative ...

He provided the solution [Art. 4] with the aid of the following figure (from which we have eliminated twelve additional radii that are not necessary for our

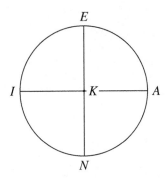

purposes), in which the directed distance $\overline{KA}$ represents $+1$ and $\overline{KI}$ represents -1. Then he stated that

> the condition [the proportion mentioned above] to be satisfied will be met by KE, perpendicular to the above and with the direction from K to E, expressed in like manner by $\overline{KE}$. For the direction of $\overline{KA}$ is to that of $\overline{KE}$ as is the latter to that of $\overline{KI}$. Moreover, we see that this same condition is equally met by $\overline{KN}$, as well as by $\overline{KE}$, these two last quantities being related to each other as $+1$ and -1. They are, therefore, what is ordinarily expressed by $+\sqrt{-1}$, and $-\sqrt{-1}$.

In other words, it is possible to turn 1 into -1 if we rotate 1 counterclockwise in the plane by $90°$ and then by another $90°$. The same is accomplished if we multiply 1 by $\sqrt{-1}$ and then multiply the result again by $\sqrt{-1}$. In this way we can think of $\sqrt{-1}$ as a counterclockwise rotation of 1 by $90°$.

Argand's ideas were ignored as much as Wessel's by the mathematical community of the times, but once complex numbers were accepted there was frequent mention of *le diagramme d'Argand* in France.

Two additional contributions of 1828, one French[41] and one English,[42] have remained largely ignored since their publication even to this day.

3.7 GAUSS AND HAMILTON HAVE THE FINAL WORD

Carl Wilhelm Friedrich Gauss (1777–1855) was a more influential exponent of the geometric representation of complex numbers. In a letter of December 18,

CARL FRIEDRICH GAUSS IN 1828
Portrait by Siegfried Detlev Bendixen.

---

[41] Mourey, *La vrai Théorie des quantités négatives et des quantités prétendues imaginaires*, 1828, 1861.

[42] Warren, *A treatise on the geometrical representation of the square roots of negative numbers*, 1828.

1811, to the German astronomer Friedrich Wilhelm Bessel (1784–1846), of the Königsberg observatory, he stated that [43]

> we can get a feeling for the total Realm of both Quantities, real and imaginary Quantities by means of an infinite Plane, wherein each Point, determined by an Abscissa $= a$ and an Ordinate $= b$, represents as it were the Quantity $a + bi$.

But he published this view for the first time in a paper of 1832, stating that[44]

> any complex quantity can be represented by some point in an infinite plane, by which it is readily referred to real quantities, one may know the complex quantity $x + iy$ by the point, whose abscissa $= x$, ordinate $\ldots = y$.

On both occasions he described a complex number as a point, not as a vector as Wessel and Argand had done, and on the same page of the 1832 paper he defined subtraction and multiplication in the expected way. He had already defined division algebraically.[45]

It was at this time that Gauss started "by habit denoting by i the imaginary quantity $\sqrt{-1}$," [46] and that, referring to numbers of the form $a+bi$, stated "such numbers we call complex numbers," [47] as opposed to imaginary numbers. The name of Gauss carried sufficient authority to establish the geometric ideas on complex numbers in mathematics, and, since then and with justifiable pride, many German mathematicians have made reference to *die Gaussische Ebene*. In the words of Einar Hille, the "Norwegians with becoming modesty avoid claiming *det Wesselske planet*." [48]

The geometric approach to complex numbers was rejected by Sir William Rowan Hamilton (1805–1865), of Dublin, who felt that algebra must be sepa-

---

[43] *Briefwechsel zwischen Gauss und Bessel*, 1880, pp. 156–157 $=$ *Werke*, **8**, 1900, pp. 90–91 $=$ *Werke*, **101**, 1917, p. 367.

[44] "Theoria residuorum biquadraticorum. Commentatio secunda," 1832, Art. 38 $=$ *Werke*, **2**, 1863, p. 109.

[45] "Theoria residuorum biquadraticorum. Commentatio secunda," Art. 32 $=$ *Werke*, **2**, p. 104.

[46] ... *denotantibus i, pro more quantitatem imaginariam* $\sqrt{-1}$, ... "Theoria residuorum biquadraticorum. Commentatio secunda," Art. 30 $=$ *Werke*, **2**, p. 102. Possibly, the first appearance of i in print since Euler introduced this symbol was in Gauss' *Disquisitiones arithmeticæ*, 1801, Art. 337 $=$ Werke, **1**, 1863, p. 414, where he stated *scribendo brevitatis caussa i pro quantitate imaginaria* $\sqrt{-1}$ (for the sake of brevity writing i for the imaginary quantity $\sqrt{-1}$).

[47] *Tales numeros vocabimus numeros integros complexos* [*integros* meaning complete], "Theoria residuorum biquadraticorum. Commentatio secunda," Art. 30 $=$ *Werke*, **2**, p. 102.

[48] Hille, *Analytic function theory*, **1**, 1959, p. 18.

SIR WILLIAM ROWAN HAMILTON
From Smith, *Portraits of Eminent Mathematicians*, II.

rate from geometry. In a paper written and read at the Irish Academy in 1833 and 1835, but published two years later,[49] he differed from other authors

> in not introducing the sign $\sqrt{-1}$ until he has provided for it, by his Theory of Couples, a possible and real meaning, as a symbol of the couple (0, 1).[50]

Hamilton preferred to think of a complex number as a *couple* of real numbers (a, b).[51] In the second part of his paper, "Theory of conjugate functions, or algebraic couples" (the part written in 1833, which starts on page 393; 76; 87),

---

[49] "Theory of conjugate functions or algebraic couples; with a preliminary and elementary essay on algebra as a science of pure time," 1837. Page references are to the three references given in the bibliography in that order, separated by semicolons.

[50] End of the last footnote in the "General introductory remarks."

[51] Actually, Hamilton worked with several kinds of couples: *moment-couples*, *step-couples*, *number couples*, *limit-couples*, *logarithmic function-couples*, etc. We shall concern ourselves with number couples only.

he set down the following basic definitions [Art. 6, p. 403; 83; 95]:

$$(b_1, b_2) + (a_1, a_2) = (b_1 + a_1, b_2 + a_2);$$
$$(b_1, b_2) - (a_1, a_2) = (b_1 - a_1, b_2 - a_2);$$
$$(b_1, b_2)(a_1, a_2) = (b_1, b_2) \times (a_1, a_2) = (b_1 a_1 - b_2 a_2, b_2 a_1 + b_1 a_2);$$
$$\frac{(b_1, b_2)}{(a_1, a_2)} = \left(\frac{b_1 a_1 + b_2 a_2}{a_1^2 + a_2^2}, \frac{b_2 a_1 - b_1 a_2}{a_1^2 + a_2^2} \right).$$

He added that these definitions "are really *not arbitrarily chosen*, and that though others might have been assumed, no others would be equally proper." Of course, the definition of multiplication, for instance, must have been inspired by what we want to obtain through the traditional manipulations:

$$(b_1 + b_2 i)(a_1 + a_2 i) = b_1 a_1 + b_2 a_1 i + b_1 a_2 i + b_2 a_2 i^2$$
$$= b_1 a_1 + (b_2 a_1 + b_1 a_2)i - b_2 a_2$$
$$= b_1 a_1 - b_2 a_2 + (b_2 a_1 + b_1 a_2)i,$$

where we have used the fact that $i^2 = -1$.

Then Hamilton showed that addition and multiplication are commutative and that multiplication is distributive over addition. With these definitions and laws, he may have started what would later be called modern algebra.

Clearly, the couples of the form $(a, 0)$ can be put in a one-to-one correspondence with the real numbers, and we may write, as Hamilton did, $a = (a, 0)$ [p. 404; 83; 95]. Then, with his definition of multiplication, $(0, 1)^2 = (-1, 0) = -1$, so that one can think of $(0, 1)$ as representing the dreaded $\sqrt{-1}$. Then we have

$$(a_1, a_2) = (a_1, 0) + (0, a_2) = a_1 + a_2(0, 1) = a_1 + a_2\sqrt{-1}.$$

Or, to explain it in his own emphatic words [Art. 13, pp. 417–418; 93; 107],

... and then we shall have the particular equation

$$\sqrt{(-1, 0)} = (0, 1);$$

which may ... be concisely denoted as follows,

$$\sqrt{-1} = (0, 1).$$

In the THEORY OF SINGLE NUMBERS, the symbol $\sqrt{-1}$ is *absurd*, and denotes an IMPOSSIBLE EXTRACTION, or a merely IMAGINARY NUMBER; but in the THEORY OF COUPLES, the same symbol $\sqrt{-1}$ is *significant*, and denotes a POSSIBLE

EXTRACTION, or a REAL COUPLE, namely (as we have just now seen) the *principal square-root of the couple* $(-1, 0)$. In the latter theory, therefore, though not in the former, this sign $\sqrt{-1}$ may be properly employed; and we may write, if we choose, for any couple (a_1, a_2) whatever,

$$(a_1, a_2) = a_1 + a_2\sqrt{-1},$$

interpreting the symbols a_1 and a_2, in the expression $a_1 + a_2\sqrt{-1}$, as denoting the pure primary couples $(a_1, 0)$ $(a_2, 0), \ldots$

This formulation as ordered couples really clears any metaphysical worries posed by complex numbers. It places these numbers on a solid foundation that the geometric interpretation could not provide.

In 1837, Wolfgang Bolyai, and old university friend of Gauss, wrote him a letter chastising him for having propagated the geometric theory of complex numbers, while their foundation should be the real numbers, whose arithmetic was known. In his reply, Gauss agreed with Bolyai, and claimed that he had had the same idea and had regarded complex numbers as ordered couples since 1831. While the world at large was readier to accept the geometric approach than Hamilton's algebraic one, Bolyai and Gauss knew something that few others did at the time: that there are geometries other than Euclid's.

Ten years after the publication of Hamilton's work on algebraic couples, Augustin-Louis Cauchy—the creator of complex function theory—stamped the seal of approval on complex numbers with his adoption of the letter i for $\sqrt{-1}$, approximately 300 years after these numbers originated. On the second page of a paper on "a new theory of imaginaries"[52] he stated:

> There was no need to torture the spirit to seek to discover what the symbol $\sqrt{-1}$, for which the German geometers substitute the letter i, could represent. This symbol or this letter was, if I can so express myself, a tool, an instrument of calculation whose introduction into formulas permitted us to arrive more quickly at the very real solution of questions that had been posed.

---

[52] "Mémoire sur une nouvelle théorie des imaginaires, et sur les racines symboliques des équations et des équivalences," 1847 p. 1121 = *Œuvres*, Ser. 1, X, p. 313.

4

INFINITE SERIES

4.1 THE ORIGINS

In the preceding chapters we have seen Newton, Gregory, Euler, and others freely using infinite series. This terminology, as applied to the object of our study, was introduced in the last third of the seventeenth century. It probably appeared in print for the first time in the second of Gregory's *Exercitationes geometricæ*, published in London in 1668. Here, on pages 10 to 12, we read "infinitæ seriei" twice and "series infinita" four times, but Gregory had already used "series in infinitum" the year before.[1] Newton already used *Infinitam terminorum Seriem* (infinite series of terms) in 1669,[2] and Jakob Bernoulli later wrote a *Tractatus de seriebus infinitis* that became a classic on the subject.

But medieval mathematicians, writing in Latin, used the word *progressio*. Among the progressions studied at that time, but whose origins go back to antiquity, there are three that have come to us with names that are still in use. In current notation, if a and r are constants, these are the *arithmetic progression*

$$a, \; a + r, \; a + 2r, \; a + 3r, \; \dots,$$

the *geometric progression*

$$a, \; ar, \; ar^2, \; ar^3, \; \dots,$$

---

[1] In *Vera circvli et hyperbolæ qvadratvra, in propria sua proportionis specìe, inuenta & demonstrata*, 1667, pp. 34–39.

[2] On the first page of *De analysi per æquationes numero terminorum infinitas*, 1669.

and the *harmonic progression*

$$\frac{1}{1}, \frac{1}{2}, \frac{1}{3}, \frac{1}{4}, \cdots .$$

The name of the last one is due to the interest that the Pythagoreans showed in music and to their realization that the pitch of a vibrating string doubles, triples, etc., as the length of the string is shortened to one half, one third, etc., of its original length.

We have written these progressions as sequences (using the modern word), but of course there was an early interest in adding their terms and finding their sum. Already in an Egyptian papyrus, transcribed from an older work by the scribe Ahmoses c. 1650 BCE, there is a collection of 84 problems, of which the 79th simply states

From Robins and Shute, *The Rhind mathematical papyrus,
an ancient Egyptian text*, 1987, 1998.

which can be translated as [3]

The one scale		Houses	7
1	2801	Cats	49
2	5602	Mice	343
4	11204	Spelt	2301
Together	19607	*Hekat*	16807
		Together	19607

On the right (left in the papyrus) we have a finite geometric progression, with $a = r = 7$ (Ahmoses wrote 2301 (ⲓⲩⲗ) erroneously since $7^4 = 2401$), and its

---

[3] The first line ends with a 7 (ⲍ), both in the translation and in the original, because the Egyptians wrote from right to left. Spelt is a hard-grain variety of wheat, and a *hekat* is a measure of grain just short of 5 liters.

sum:

$$7 + 7^2 + 7^3 + 7^4 + 7^5 = 19607.$$

The words written to the left of the terms of the progression have no interest for us and can be disregarded (but originally they may have been intended as part of a cute riddle, such as "each of seven houses has seven cats, each cat caught seven mice ... ," and so on). The most interesting part of this problem is the sum of multiples of 2801 in the left part of this problem. Where did the number 2801 come from, so conveniently chosen that seven $(1 + 2 + 4)$ times this number is the sum of the progression? If we look at the previous sum,

$$7 + 7^2 + 7^3 + 7^4 = 2800,$$

it shows that the original author of this problem knew that $19607 = 7(1 + 2800)$, and suggests that he obtained the terms of the progression in this manner:

$$7 = 7$$
$$7 + 7^2 = 7(1 + 7) = 56$$
$$7 + 7^2 + 7^3 = 7(1 + 56) = 399$$
$$7 + 7^2 + 7^3 + 7^4 = 7(1 + 399) = 2800$$
$$7 + 7^2 + 7^3 + 7^4 + 7^5 = 7(1 + 2800) = 19607.$$

To us, who can use notation developed over millennia, this shows that if s_n denotes the sum of the first n terms of the progression, then $s_{n+1} = 7(1 + s_n)$. That is, $s_n + 7^{n+1} = 7 + 7s_n$, from which

$$s_n = 7\frac{7^n - 1}{7 - 1}.$$

A general formula, containing this as a particular case, was actually demonstrated in Egypt, and has come to us from the pen of Eucleides of Alexandria (c. 325 BCE–c. 265 BCE), usually known as Euclid in an English-language context. Euclid, the great compiler of all Hellenic mathematical knowledge up to his time, flourished in the pampering conditions at the *Museum* of Alexandria and wrote at least ten scientific works, most of which have been lost. His most famous work was *The Elements*, a work in thirteen books, of which the first six served as the standard text in geometry for the next two millennia. Books VII to IX are devoted to arithmetic, and it is in this last book that we find the following statement:

PROPOSITION 35.

If as many numbers as we please be in continued proportion, and there be subtracted from the second and the last numbers equal to the first, then, as the excess of the second is to the first, so will the excess of the last be to all those before it.

It is not difficult to translate Euclid's verbal description into a form that is familiar to us and would have been so foreign to him. What Euclid says is that, if $a, ar, ar^2, ar^3, \ldots, ar^{n+1}$ is a finite geometric progression, then

$$\frac{ar - a}{a} = \frac{ar^{n+1} - a}{a + ar + ar^2 + ar^3 + \cdots + ar^n},$$

which, if we denote the denominator on the right by s_{n+1}, is equivalent to

$$s_{n+1} = a\frac{r^{n+1} - 1}{r - 1}.$$

But we are more interested in infinite sums. The first infinite series ever summed was

$$1 + \frac{1}{4} + \left(\frac{1}{4}\right)^2 + \cdots + \left(\frac{1}{4}\right)^n + \cdots,$$

a feat accomplished by Archimedes of Syracuse (c. 287 BCE–212 BCE) in the process of performing the quadrature of the parabola, that is, of finding the area of a parabolic segment such as QPq in the next figure. Archimedes, who

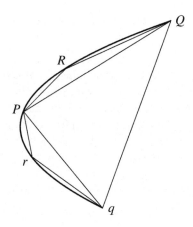

There is no existing portrait that beyond a doubt represents Archimedes.

had spent quite some time at Alexandria and can be considered a member of this school, was one of the greatest mathematicians of all times if not the greatest. His life, on return to Syracuse, was entirely devoted to mathematical research and to the development of some mechanical inventions. He summed the series stated above in a series of propositions that he included in a letter to Dositheos of Pelusium, probably at Alexandria. We shall follow the translation by Heath.[4]

By the segment QPq in the previous figure we mean the area between the arc of parabola QPq and the chord Qq. Archimedes chose a point P so that the tangent to the parabola at P is parallel to Qq and then considered two points, R and r in Heath's notation, located on the parabola at the points where its tangents (not drawn) are parallel to the chords PQ and Pq, respectively. Then he proved, in Proposition 21 [p. 248], that the sum of the areas of the smaller triangles PRQ and Prq is one quarter of the area of the triangle QPq.

---

[4] In *The works of Archimedes with The method of Archimedes*; reprinted by Dover, 1953, pp. 233–252. Page references are to this reprint.

Next he drew, in each smaller parabolic segment left over (such as the one subtended by the arc from P to R) another triangle whose base is the chord joining the endpoints of the arc (for instance, the segment PR) and with a vertex located on the subtending arc at the point where the tangent to the parabola is parallel to the chosen base (for instance, at the point where the tangent to the parabola is parallel to PR). As before, the sum of the areas of these smaller triangles with bases PR and RQ is one quarter of the area of the triangle PRQ. Continuing in this way, he "exhausted" the original parabolic segment.[5] In our terms, if A denotes the area of the triangle QPq, then the area of the parabolic segment QPq is the sum

$$A + \frac{1}{4}A + \left(\frac{1}{4}\right)^2 A + \cdots + \left(\frac{1}{4}\right)^n A + \cdots .$$

But of course, Archimedes did not express himself in this way. Instead, he evaluated first a partial sum of this series as follows [p. 249]:

PROPOSITION 23.

Given a series of areas A, B, C, D, ..., Z, of which A is the greatest, and each is equal to four times the next in order, then

$$A + B + C + D + \cdots + Z + \tfrac{1}{3}Z = \tfrac{4}{3}A.$$

Translated into our notation, this means that for any n,

$$A + \frac{1}{4}A + \left(\frac{1}{4}\right)^2 A + \cdots + \left(\frac{1}{4}\right)^n A + \frac{1}{3}\left(\frac{1}{4}\right)^n A = \frac{4}{3}A,$$

so that

$$1 + \frac{1}{4} + \left(\frac{1}{4}\right)^2 + \cdots + \left(\frac{1}{4}\right)^n = \frac{4 - \left(\frac{1}{4}\right)^n}{3}.$$

It is clear, from our present point of view, that as n grows indefinitely,

$$1 + \frac{1}{4} + \left(\frac{1}{4}\right)^2 + \cdots + \left(\frac{1}{4}\right)^n + \cdots = \frac{4}{3},$$

---

[5] This *method of exhaustion* to find areas, which Archimedes perfected and used countless times, was introduced by Eudoxos of Cnidos (408–355 BCE).

which yields the sum of the geometric series for $r = \frac{1}{4}$. Without disregarding the first area A of the triangle QPq, Archimedes phrased his conclusion as follows [p. 251]:

PROPOSITION 24.

Every segment bounded by a parabola and a chord Qq is equal to four-thirds of the triangle which has the same base as the segment and equal height.

But a long time would pass before the general formula for the sum of the geometric series was obtained, and this was accomplished first in India by Nilakantha Somayaji (1444–1545). A native of Sri-Kundapura (Trkkantiyur in the local Malayalam language) in Kerala, on the southern coast of India, he is chiefly known for his astronomical work *Tantrasangraha* (A digest of scientific knowledge), c. 1500, but his masterpiece is the *Aryabhatiyabhasya*, a commentary on the *Aryabhatiya* of Aryabhata.[6] Here [1930, p. 106] he stated and proved the sum of an infinite geometric progression:

> Thus the sum of an infinite series, whose later terms (after the first) are got by diminishing the preceding one by the same divisor, is always equal to the first term divided by one less than the common mutual divisor.[7]

In other words,

$$a + \frac{a}{r} + \frac{a}{r^2} + \cdots + \frac{a}{r^n} + \cdots = \frac{a}{r-1}.$$

European scholars were unaware of these developments in India, and François Viète rediscovered this formula in 1593.[8] First he stated a theorem about the sum of a finite number of terms, but without any notation. Next he gave an explanation using notation, as follows:

> Let there be magnitudes [numbers] in continued proportion, of which D is the largest, X the smallest, & the composition [sum] of all of them is F, and let the ratio of a larger to the next smaller term be as D is to B. I say that D is to B, as F minus X is to F minus D.

---

[6] Sastri, *Aryabhatiya with the bhashya of Nilakantha*, 1930, 1931.

[7] Quoted from Sharma, *Hindu Astronomy*, 2004, p. 222.

[8] In Chapter XVII, entitled *Progressio Geometrica*, of *Variorvm de rebvs mathematicis responsorvm, liber VIII*, 1593 = Schooten, *Francisci Vietæ opera mathematica*, 1646, pp. 397–398. The theorem and equations discussed here are from page 397.

With these words Viète referred to the progression

$$D + B + \cdots + X = F$$

(notice that this is the proper order of its terms, since they must decrease from left to right for it to have a finite sum), and then concluded that

$$\frac{D}{B} = \frac{F - X}{F - D}. \text{ [9]}$$

Then Viète drew four additional conclusions, of which the first was

Itaque datis D, B, X, *dabitur* F. *Enimvero* $\frac{B\text{quad.}-B\text{ in }X}{D-B}$ *aquabitur* F.

This equation has been reproduced photographically to be true to the original, but it contains a typo. The fraction's numerator should start with D^2, not B^2, and then the equation on the right should be

$$\frac{D^2 - BX}{D - B} = F,$$

as is easy to deduce from the previous equation.

Next, Viète considered the case in which the number of terms goes to infinity:

> In truth when the magnitudes are in continued proportion to infinity, X goes away to nothing.

This means that the term BX can be considered zero when the number of terms is infinite, and we obtain

$$\frac{D^2}{D - B} = F, \text{ [10]}$$

where F is now the sum of the series. If we write $D = a$ and $B = ar$, in the notation established above, we obtain the sum of the geometric series as

$$\frac{a}{1 - r}.$$

---

[9] He gave no proof of this, and none is necessary because it follows immediately from Euclid's formula (in its present-day equivalent, just replace a with D, r with B/D, ar^n with X, and the sum in the second denominator with F).

[10] Viète stated this equation in the form $(D-B)/D = D/F$ with the following words: "Vt differentia terminorum rationis at terminum rationis majorem, ita maxima ad compositam ex omnibus."

The harmonic series did not fare as well. Using the modern word, it is divergent. The first to give a written proof of this fact was Nicole Oresme (c. 1323–1382), a professor at the *Collège de Navarre* in Paris who would later become dean of the Rouen Cathedral and Bishop of Lisieux in Normandy. About 1360, he gave the following argument, which we write first in modern symbols, in *Quæstiones super geometriam Euclidis*:

$$1 + \frac{1}{2} + \frac{1}{3} + \frac{1}{4} + \frac{1}{5} + \frac{1}{6} + \frac{1}{7} + \frac{1}{8}$$

$$+ \frac{1}{9} + \frac{1}{10} + \frac{1}{11} + \frac{1}{12} + \frac{1}{13} + \frac{1}{14} + \frac{1}{15} + \frac{1}{16} + \cdots$$

$$> 1 + \frac{1}{2} + \frac{1}{4} + \frac{1}{4} + \frac{1}{8} + \frac{1}{8} + \frac{1}{8} + \frac{1}{8}$$

$$+ \frac{1}{16} + \frac{1}{16} + \frac{1}{16} + \frac{1}{16} + \frac{1}{16} + \frac{1}{16} + \frac{1}{16} + \frac{1}{16} + \cdots$$

$$= 1 + \frac{1}{2} + \frac{1}{2} + \frac{1}{2} + \frac{1}{2} + \cdots .$$

Clearly, the sum on the right is not finite. Oresme's original argument was:[11]

> Add to a magnitude of 1 foot: $\frac{1}{2}$, $\frac{1}{3}$, $\frac{1}{4}$ foot, etc.; the sum of which is infinite. In fact, it is possible to form an infinite number of groups of terms with sum greater than $\frac{1}{2}$. Thus, $\frac{1}{3} + \frac{1}{4}$ is greater than $\frac{1}{2}$ [*quia 4ᵃ et 3ᵃ sunt plus quam una medietas*], similarly $\frac{1}{5} + \frac{1}{6} + \frac{1}{7} + \frac{1}{8}$ is greater than $\frac{1}{2}$, $\frac{1}{9} + \frac{1}{10} + \frac{1}{11} + \cdots + \frac{1}{16}$ is greater than $\frac{1}{2}$, and so *in infinitum*.

4.2 THE SUMMATION OF SERIES

Gottfried Wilhelm Leibniz (1646–1716), a lawyer turned diplomat, arrived in Paris in 1672 on a (failed) diplomatic mission on behalf of the elector of Mainz. When in Paris, where he stayed until 1676, he met the Dutch scientist Christiaan Huygens (1629–1695) and developed an interest in mathematics. Huygens was Leibniz's mentor in Paris and, possibly wanting to test

---

[11] Quoted from Struik, *A source book in mathematics, 1200–1800*, 1969, p. 320. Note also that Oresme did not write fractions as we do. The inset in Latin illustrates his original writing.

the ability of the young man, proposed to him the following problem in 1672 [pp. 14; 404; 49; 16]:[12]

> To find the sum of a decreasing series of fractions, of which the numerators are all unity and the denominators are the triangular numbers; of which he [Huygens] said that he had found the sum among the contributions of Hudde on the estimation of probability.

Triangular numbers are so called because you can form a triangle with as many dots as each number, as shown in the next figure. Thus, the triangular

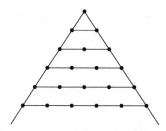

numbers are 1, 3, 6, 10, 15, 21, ... or, as we would express this today, those of the form $n(n + 1)/2$ for each positive integer n.

Leibniz found the answer to this problem very quickly, and it was in agreement with the one that Huygens already had. He wrote the result in a 1674 manuscript never published during his life, entitled *Theorema arithmeticæ infinitorum*,[13] first in narrative form and then as follows:

Series $\odot$ $\frac{1}{1} + \frac{1}{3} + \frac{1}{6} + \frac{1}{10} + \frac{1}{15} + \frac{1}{21}$ etc. in infinitum $\sqcap$ 2.

Two of these symbols need an explanation. Leibniz used $\sqcap$ as an equal sign, but the symbol $\odot$ is just a marker that he used instead of numbering the equation, and has no mathematical meaning. Thus, Leibniz' statement is that the sum of the series $\odot$ is 2. Of this fact he included a demonstration, which begins by considering two more series. First the one that we have called harmonic:

---

[12] This is as Leibniz himself recalled the problem in his manuscript "Historia et origo calculi differentialis." This quotation is from the printed references given in the bibliography, in this order.

[13] This extract is from f. 1^r of the manuscript.

Series ☽ $\frac{1}{1} + \frac{1}{2} + \frac{1}{3} + \frac{1}{4} + \frac{1}{5} + \frac{1}{6}$ etc. in infinitum.

Then another whose terms are half the terms of ☉:

Series ☿ $\frac{1}{2} + \frac{1}{6} + \frac{1}{12} + \frac{1}{20} + \frac{1}{30} + \frac{1}{42}$ etc. in infinitum.

This series, he asserted, has sum 1. He began his proof by subtracting ☿ from ☽ to obtain $\frac{1}{2} + \frac{4}{12} + \frac{9}{36} + \frac{16}{80} + \frac{25}{150} + \frac{36}{252}$ etc., which, after simplifying, becomes

series ♀ $\frac{1}{2} + \frac{1}{3} + \frac{1}{4} + \frac{1}{5} + \frac{1}{6} + \frac{1}{7}$ etc. in infinitum.

That is, Leibniz had shown that ☽ − ☿ = ♀, and then he pointed out that ♀ = ☽ − 1. It follows that ☿ = 1 and therefore twice the series ☿, or the series ☉, equals two. In his own words:

Ergo dupla series ☿ sive series ☉ erit æqualis binario.

Today's readers may be shocked by Leibniz' use of divergent series with such flagrant disregard for the appropriate mathematical etiquette. He thought nothing of it, and neither did most of his contemporaries. They were used to rigor in geometry, where it belonged, but what many other mathematical fields wanted was exploration. Obtaining results was what counted, and the niceties of mathematical rigor would have to wait for a better time. In this particular case, what counted is that the sum of the series in question is actually 2. However, modern presentations of Leibniz' result are usually "sanitized" in the following manner:

$$1 + \frac{1}{3} + \frac{1}{6} + \frac{1}{10} + \frac{1}{15} \cdots = 2\left[\frac{1}{2} + \frac{1}{6} + \frac{1}{12} + \frac{1}{20} + \frac{1}{30} + \cdots\right]$$

$$= 2\left[\left(1 - \frac{1}{2}\right) + \left(\frac{1}{2} - \frac{1}{3}\right) + \left(\frac{1}{3} - \frac{1}{4}\right) + \left(\frac{1}{4} - \frac{1}{5}\right) + \left(\frac{1}{5} - \frac{1}{6}\right) \cdots\right],$$

which, after all the obvious cancellations, equals 2.

After this result and the invention of a calculating machine that could perform addition, subtraction, multiplication, division, and the extraction of roots, Leibniz went on to achieve immortal fame in mathematics with his development of the calculus.

However, he failed to solve a problem in infinite series that had originally been posed to him by Henry Oldenburg, in a letter of April 1673:[14] to sum the reciprocals of the squares of all the natural numbers, a task at which Pietro Mengoli had already failed. A little over one month later, Leibniz confessed that he had not yet found the sum.[15] He never found it.

Next to succeed at summing infinite series was Jakob Bernoulli (1655–1705), of Basel, who included some known results—such as the sum of the geometric series, the divergence of the harmonic series, and the preceding series summed by Leibniz—in his *Tractatus de seriebus infinitis*, which was a collection of five papers, written between 1689 and 1704, published as an appendix to his *Ars conjectandi*,[16] all put together by his nephew Niklaus. In this appendix, Bernoulli also set out to sum some new series with a considerable degree of success, but the procedure was essentially the one already worked out by Leibniz and illustrated in the preceding equation: the decomposition of each term into a difference and the cancellation of most of the terms. Then he attempted to sum the reciprocals of the squares of the natural numbers. Bernoulli made some progress, for he realized that each term of this series is smaller than the corresponding term of the series that Leibniz had summed, and therefore this series must also have a sum. But as for finding it, Bernoulli had to admit that [p. 254]

> when [the denominators] are purely Quadratic, as in the series
>
> $$1 + \frac{1}{4} + \frac{1}{9} + \frac{1}{16} + \frac{1}{25} \quad \&c.$$
>
> it is difficult, more than one would have expected, to search for its sum, ...

So he could not find the sum, concluding the paragraph with the following plea [p. 254]:

> If someone discovers and communicates to us what has eluded our efforts thus far, he will earn our deep gratitude.

Much later, Euler started with some numerical approximations and found the approximate value 1.644934 for the sum of the series [Art. 22],[17] but this

---

[14] Gerhardt, *Der Briefwechsel von Gottfried Wilhelm Leibniz mit Mathematikern*, 1899, p. 86.

[15] "summan nondum fateor reperi," Gerhardt, *Der Briefwechsel von Gottfried Wilhelm Leibniz mit Mathematikern*, p. 95.

[16] Page references are to the 1713 edition.

[17] "De summatione innumerabilium progressionum," 1730/31, pub. 1738. References to this and the next paper by Euler are by article number.

JAKOB BERNOULLI IN 1687
Portrait by his brother Nikolaus.

figure was not sufficient to make him guess the exact answer. However, in 1735, he was able to report his success in obtaining the exact value as follows [Art. 2]:[18]

> Recently, however, and altogether unexpectedly I have found an elegant expression for the sum of this series
>
> $$1 + \frac{1}{4} + \frac{1}{9} + \frac{1}{16} + \text{etc.,}$$
>
> which depends on the quadrature of the circle ... In fact, I have found that six times the sum of this series is equal to the square of the circumference of a circle whose diameter is 1, ...

In current terminology, this means that the sum of the series is $\pi^2/6$. If we wonder about the strange description of Euler's result in words, note that at

[18] "De svmmis seriervm reciprocarvm," 1734/35, pub. 1740.

that time there was no universally accepted symbol for π.[19] In 1735, Euler used p instead of π, and, because of the lack of a standard symbol, he must have thought it appropriate to state the result in unequivocal words. However, it was Euler's use of the letter π, starting in 1736,[20] that made this symbol popular.

Euler began with the series for $y = \sin s$ (see Section 4.3),

$$y = s - \frac{s^3}{1 \cdot 2 \cdot 3} + \frac{s^5}{1 \cdot 2 \cdot 3 \cdot 4 \cdot 5} - \frac{s^7}{1 \cdot 2 \cdot 3 \cdot 4 \cdot 5 \cdot 6 \cdot 7} + \text{etc.}$$

[Art. 3], and then found the sum of the stated series [Art. 11] by a long method that we shall not present here. Fortunately, he then did it again by a shorter method that starts by putting $y = 0$ and dividing by s if $s \neq 0$ [Art. 16]:

$$0 = 1 - \frac{s^2}{1 \cdot 2 \cdot 3} + \frac{s^4}{1 \cdot 2 \cdot 3 \cdot 4 \cdot 5} - \frac{s^6}{1 \cdot 2 \cdot 3 \cdot 4 \cdot 5 \cdot 6 \cdot 7} + \text{etc.}$$

The sine of s, and therefore this series, vanishes for $s = \pm k\pi$ (we shall use π while Euler still used p). Therefore, the right-hand side can be written as the product of the factors

$$1 - \frac{s}{\pi}, \quad 1 + \frac{s}{\pi}, \quad 1 - \frac{s}{2\pi}, \quad 1 + \frac{s}{2\pi}, \quad \text{etc.,}$$

and combining them two by two we have

$$1 - \frac{s^2}{1 \cdot 2 \cdot 3} + \frac{s^4}{1 \cdot 2 \cdot 3 \cdot 4 \cdot 5} - \frac{s^6}{1 \cdot 2 \cdot 3 \cdot 4 \cdot 5 \cdot 6 \cdot 7} + \text{etc.}$$

$$= \left(1 - \frac{s^2}{\pi^2}\right)\left(1 - \frac{s^2}{4\pi^2}\right)\left(1 - \frac{s^2}{9\pi^2}\right)\left(1 - \frac{s^2}{16\pi^2}\right) \text{etc.}$$

From this Euler concluded [Art. 17]:

> It is now clear from the nature of this equation, that the coefficient of ss or $\frac{1}{1 \cdot 2 \cdot 3}$ is equal to
>
> $$\frac{1}{\pi^2} + \frac{1}{4\pi^2} + \frac{1}{9\pi^2} + \frac{1}{16\pi^2} + \text{etc.}$$

---

[19] The symbol π for 3.14159 ... was introduced by Sir William Jones in his *Synopsis palmariorum matheseos*, 1706, p. 236, but had not yet become popular.

[20] Starting in *Mechanica sive motvs scientia analytice exposita*, 1736, vol. 1, p. 119, with the statement "denotante 1:π rationem diametri ad peripheriam" = *Opera omnia*, Ser. 2, **1**, 1912, p. 97.

In short,

$$1 + \frac{1}{4} + \frac{1}{9} + \frac{1}{16} + \text{etc.} = \frac{\pi^2}{6}.$$

Upon learning of this result, Johann Bernoulli (1667–1748) wrote: "if only my brother were alive."[21] In the same paper [p. 133] Euler found the sum of the p-series for $p = 4, 6, 8, 10,$ and 12 [Art. 18], and later, in 1740, he found many more sums by similar methods.[22]

Euler was 23–24 years of age when he obtained his first results on the summation of series, but he would be from this moment forever to be known in the field of mathematics. However, in a letter to Euler of April 2, 1737, Johann Bernoulli pointed out some defects in the preceding method, which Euler admitted in his reply of August 27, 1737. Also, Niklaus Bernoulli (1687–1759), a nephew of both Jakob and Johann, wrote him a letter in 1743 to point out that in expanding an infinite series into a product using its roots, he was treating the series as if it were a polynomial, a criticism that was well founded. But Euler was not to be distracted from his path, for he believed in the validity of his result even in the absence of proof. He devoted Chapter X of the *Introductio* to the sum of infinite series, including once more the p-series previously obtained, on the basis of the following statement [Art. 165]:

If we have $1 + Az + Bz^2 + Cz^3 + Dz^4 + \&\text{c.} = (1 + \alpha z)(1 + \beta z)(1 + \gamma z)(1 + \delta z) \&\text{c.}$, these Factors, whether they are finite or infinite in number, if they are actually multiplied by each other, must produce that expression $1 + Az + Bz^2 + Cz^3 + Dz^4 + \&\text{c.}$. Therefore, the coefficient A will be equal to the sum of all the quantities $\alpha + \beta + \gamma + \delta + \epsilon + \&\text{c.}$. The coefficient B will be equal to the sum of their products two at a time, and it will be $B = \alpha\beta + \alpha\gamma + \alpha\delta + \beta\gamma + \beta\delta + \gamma\delta + \&\text{c.}$. Then the coefficient C will be equal to the sum of the products three at a time, namely it will be $C = \alpha\beta\gamma + \alpha\beta\delta + \beta\gamma\delta + \alpha\gamma\delta + \&\text{c.}$ And so on ....[23]

---

[21] The original statement, "utinam frater superstes esset", is in *Johannis Bernoulli opera omnia*, 1742, vol. 4, p. 22.

[22] "De seriebus quibusdam considerationes," 1740, pub. 1750.

[23] A similar result for the roots of a polynomial had already been given by Newton in his *Arithmetica universalis, sive de compositione et resolutione arithmetica liber*, 1707. This was written as a text for Newton's lectures from 1673 to 1683. The statement in question is on page 20 of the English translation by Raphson as *Universal arithmetick*, London, 1720, which can also be seen in Struik, *A source book in mathematics, 1200–1800*, p. 94. Also in Whiteside, *The mathematical papers of Isaac Newton*, V, p. 359. Newton did not provide a proof, and, in fact, no proof was yet available at the time that Euler's *Introductio* was published.

Then he defined [Art. 166]

$$P = \alpha + \beta + \gamma + \delta + \epsilon + \&c.$$
$$Q = \alpha^2 + \beta^2 + \gamma^2 + \delta^2 + \epsilon^2 + \&c.$$
$$R = \alpha^3 + \beta^3 + \gamma^3 + \delta^3 + \epsilon^3 + \&c.$$
$$S = \alpha^4 + \beta^4 + \gamma^4 + \delta^4 + \epsilon^4 + \&c.$$
$$\&c.$$

and asserted that

$$P = A$$
$$Q = AP - 2B$$
$$R = AQ - BP + 3C$$
$$S = AR - BQ + CP - 4D$$
$$\&c.$$

"the truth of which formula is easy to know by simple inspection: until it is proved with maximum rigor in the differential calculus." Dispensing then with a proof, as Euler did, we shall be content with a simple inspection of the second equation in the last quartet. If we change Euler's notation $\alpha, \beta, \gamma, \ldots$ to $\alpha_1, \alpha_2, \alpha_3, \ldots$, so that we can indulge in the use of friendly subscripts, and if we write

$$A = P = \sum \alpha_i, \qquad B = \sum_{i<j} \alpha_i\alpha_j \qquad \text{and} \qquad Q = \sum \alpha_i^2,$$

it is possible to believe that

$$AP = \left(\sum \alpha_i\right)\left(\sum \alpha_i\right) = \sum \alpha_i^2 + 2\sum_{i<j} \alpha_i\alpha_j = Q + 2B,$$

so that $Q = AP - 2B$.

Accepting Euler's equations on this narrow basis, we now return to the basic identity [24]

$$1 - \frac{s^2}{1\cdot2\cdot3} + \frac{s^4}{1\cdot2\cdot3\cdot5} - \frac{s^6}{1\cdot2\cdot3\cdot5\cdot7} + \&c.$$
$$= \left(1 - \frac{s^2}{\pi^2}\right)\left(1 - \frac{s^2}{4\pi^2}\right)\left(1 - \frac{s^2}{9\pi^2}\right)\left(1 - \frac{s^2}{16\pi^2}\right) \&c.$$

---

[24] Euler did not do this, but began Art. 167 with another identity that he had proved in Art. 156. We shall stay with his older expansion to avoid proving the new one, but otherwise will proceed exactly as he did.

and make the substitution $s^2 = \pi^2 z$ to obtain

$$1 - \frac{\pi^2}{1 \cdot 2 \cdot 3} z^2 + \frac{\pi^4}{1 \cdot 2 \cdot 3 \cdot 5} z^4 - \frac{\pi^6}{1 \cdot 2 \cdot 3 \cdot 5 \cdot 7} z^6 + \&c.$$

$$= (1 - z)\left(1 - \frac{1}{4}z\right)\left(1 - \frac{1}{9}z\right)\left(1 - \frac{1}{16}z\right) \&c.$$

"Therefore, applying the rules found above to this case, it will be"

$$A = -\frac{\pi^2}{6}, \quad B = \frac{\pi^4}{120}, \quad C = -\frac{\pi^6}{5040}, \quad D = \frac{\pi^8}{362880} \quad \&c..$$

Therefore if we put

$$P = -1 - \frac{1}{4} - \frac{1}{9} - \frac{1}{16} - \&c.$$

$$Q = 1 + \frac{1}{4^2} + \frac{1}{9^2} + \frac{1}{16^2} + \&c.$$

$$R = -1 - \frac{1}{4^3} - \frac{1}{9^3} - \frac{1}{16^3} - \&c.$$

$$S = 1 + \frac{1}{4^4} + \frac{1}{9^4} + \frac{1}{16^4} + \&c.$$

we obtain

$$1 + \frac{1}{2^4} + \frac{1}{3^4} + \&c. = Q = AP - 2B = A^2 - 2B = \frac{\pi^4}{36} - \frac{\pi^4}{60} = \frac{\pi^4}{90},$$

$$1 + \frac{1}{2^6} + \frac{1}{3^6} + \&c. = -R = -AQ + BP - 3C = \frac{\pi^6}{540} - \frac{\pi^6}{720} + \frac{3\pi^6}{5040} = \frac{\pi^6}{945},$$

and

$$1 + \frac{1}{2^8} + \frac{1}{3^8} + \&c. = S = AR - BQ + CP - 4D$$

$$= \frac{\pi^8}{5670} - \frac{\pi^8}{10800} + \frac{\pi^8}{30240} - \frac{\pi^8}{90720} = \frac{\pi^8}{9450}.$$

Euler concluded [Art. 168]:

It is thus clear that all the infinite Series contained in this general form

$$1 + \frac{1}{2^n} + \frac{1}{3^n} + \frac{1}{4^n} + \&c.,$$

every time that n is an even number, can be produced by means of the Periphery of the Circle π; and in fact the sum of the Series will always have a rational ratio to π^n.

After this he gave the sum of the p-series for all even values of p up to $p = 26$.

However, while he was able to give approximations to the sum for some odd values of p in his 1740 paper [Art. 29], Euler was unable to sum a p-series for p odd. Nobody has ever been able to sum such a series.

4.3   THE EXPANSION OF FUNCTIONS

Obtaining the sums of particular series of numbers is of relatively little interest in comparison with expanding a function in a series whose terms depend on the variable x. Such series can be used to compute values of the expanded functions, such as in the case of the binomial theorem and the expansion of the logarithm, both already covered in Chapter 2, and obtained by various ingenious methods in the 1660s.

As for the series for the sine function, which we saw Euler using in the preceding section, it was first obtained by Madhava of Sangamagrama—the present-day Irinjalakuda, in Kerala—who lived from about 1340 to 1425. His writings on this and other infinite series have not survived, but they are quoted in two works. One is the *Yuktibhasa* (An exposition of the rationale), c. 1550, of Jyesthadeva (c. 1500–1575) and the other is the anonymous *Tantrasangrahavyakhya*, a commentary on Nilakantha's *Tantrasangraha* written by a student of Jyesthadeva, probably while his teacher was still alive. Both works contain statements for the series for the sine, cosine, and arctangent, but only the *Yuktibhasa* shows the demonstrations.

However, the statements in the *Tantrasangrahavyakhya* are important because they contain a second version of the sine series, which was quoted in Nilakantha's *Aryabhatiyabhasya* and attributed to Madhava.[25] It can be assumed, then, that the versions in the *Yuktibhasa* and the *Tantrasangrahavyakhya* are also by Madhava. The statement of the sine series in this last work, both in Sanskrit and in English translation, is as follows.[26]

---

[25] See Sastri, *Aryabhatiya with the bhashya of Nilakantha*, No. 101, 1930, p. 113.

[26] These are taken from Rajagopal & Rangachari, "On an untapped source of medieval Keralese mathematics," 1978, 95–96. The authors based their translation on an original manuscript in the Government Oriental Manuscript Library at Madras. This article also contains Nilakantha's version in the *Aryabhatiyabhasya*, p. 97.

निहृत्य चापवर्गेण -चापं तत्तन्फलानि -च
इरेत्समूल युग्वर्गे स्त्रिज्यावर्गहतैः क्रमात्
-चापं फलानि चाधोऽधो व्यस्योपर्युपरि त्यजेत्
जीवाप्यै संग्रहोऽस्यैवं विद्वानित्यादिना कृतः ।

The arc is to be repeatedly multiplied by the square of itself and is to be divided [in order] by the square of each even number increased by itself and multiplied by the square of the radius. The arc and the terms obtained by these repeated operations are to be placed in sequence in a column, and any last term is to be subtracted from the next above, the remainder from the term then next above, and so on, to obtain the *jyva* of the arc. It was this procedure which was briefly mentioned in the verse starting with *vidvan*.

Since we are temporarily back to sixteenth-century mathematics, we must explain a statement made in narrative form. If s denotes the arc corresponding to a central angle θ and R is the radius, the first sentence refers to the terms

$$s, \quad s \cdot \frac{s^2}{(2^2+2)R^2}, \quad s \cdot \frac{s^2}{(2^2+2)R^2} \cdot \frac{s^2}{(4^2+4)R^2},$$

and so on. The rest of the statement means that we have to alternately add and subtract these terms to obtain the *jyva* of θ. That is,

$$\mathrm{jya}\,\theta = s - \frac{s^3}{6R^2} + \frac{s^5}{6R^2 \cdot 20R^2} - \frac{s^7}{6R^2 \cdot 20R^2 \cdot 42R^2} + \cdots,$$

and replacing $\mathrm{jya}\,\theta$ with $R \sin \theta$, dividing by R, and writing $s = R\theta$, this equation becomes

$$\sin \theta = \theta - \frac{\theta^3}{3!} + \frac{\theta^5}{5!} - \frac{\theta^7}{7!} + \cdots,$$

which is the familiar infinite series for $\sin \theta$.

These discoveries could have reached Europe through many Portuguese explorers and mathematically savvy Jesuit missionaries that were in Kerala in the sixteenth century. But if Indian mathematics was imported into Europe at that time it made no significant impact. European mathematicians carried on as if they knew nothing about it, which is probably the case. They rediscovered

Madhava's series more than 250 years after they were originally proved, and then they found new series expansions.

Newton found the following expansion for arcsin, among others, "in winter between the years 1664 & 1665," which he did via integration and the binomial theorem as in the case of the logarithm.[27] If z denotes what we call arcsin x, then

$$z = x + \frac{x^3}{6} + \frac{3x^5}{40} + \frac{5x^7}{112} \ \&c.$$

Newton restated this result, exactly in this way, in *De analysi*, along with the series for the sine and cosine, which were new to him [pp. 17; 236–237].[28] John Collins, librarian of the Royal Society, received a copy of this work through Barrow, Newton's colleague at Cambridge, and communicated these results to James Gregory (1638–1675), of the universities of Padua (1664–1668), Saint Andrews (1668–1674), and Edinburgh (1674–1675), on 24 December 1670 [p. 155; 54].[29]

In his response of 15 February 1671,[30] Gregory said: "I thank you for the serieses you sent me, & send ye these following in requital" [p. 170; 62]. Two of these are, in current notation (they will later be seen in Gregory's),

$$\tan \theta = \theta + \frac{\theta^3}{3} + \frac{2\theta^5}{15} + \frac{17\theta^7}{315} + \cdots$$

and

$$\sec \theta = 1 + \frac{\theta^2}{2} + \frac{5\theta^4}{24} + \frac{61\theta^6}{720} + \cdots.$$

---

[27] Reproduced in Whiteside, *The mathematical papers of Isaac Newton*, I, 1967, pp. 108–109. The result is not explicitly written in the original as it is here. Columns 1 and 3 of the table on page 109 must be multiplied to obtain it. Note Newton's trivial error on page 108, as explained in Whiteside's footnote.

[28] The first page refers to Jones' edition, and those after the semicolon to Whiteside's translation.

[29] Any double reference to Gregory's work or correspondence in this chapter is to the works by Turnbull, *James Gregory tercentenary memorial volume*, 1939 and *The Correspondence of Isaac Newton*, I, 1959, in this order. If only one page reference is given, it is to the *memorial volume*. Most of Gregory's papers were lost or put away and ignored soon after his death, and thus the man who was second only to Newton in the early 1670s was almost forgotten, and his very important contributions to mathematics were not rediscovered until the publication of Turnbull's memorial volume. It contains reproductions and translations of Gregory's manuscripts, a summary of the contents of his books, and Turnbull's extensive researches and interpretation. In this and the next section, I rely heavily on Turnbull's interpretations.

[30] For the sake of authenticity, dates of letters and manuscripts mentioned in the rest of this book are given in the style in which they were originally written by their authors: small unit (day), middle unit (month), and large unit (year).

JAMES GREGORY
His only known portrait, at Marischal College.
University of Aberdeen.

Gregory did not explain how he obtained the expansions. But finding out is important because it will turn out that while previous series expansions by others had been found by disparate methods, these were obtained by a general procedure that has been in use ever since. This method can be obtained on the basis of a very valuable interpolation formula first found by Gregory and communicated to Collins in a letter of 23 November 1670. The same kind of work was carried out by Newton, independently, in his 1676 manuscript *Regula differentiarum*. It was published in 1711 with the title *Methodus differentialis* (Method of differences), but a modified version had already appeared in Lemma V of Book III in the *Principia*.[31]

The function concept and the now usual notation for functions did not exist at the time of this discovery, but it will add clarity if we present it in modern

---

[31] *Philosophiæ naturalis principia mathematica*, 1686, pp. 481–483 = Motte's translation *The mathematical principles of natural philosophy*, 1729, pp. 333–336

notation and terminology in addition to the original notation by Gregory. This will allow an easy comparison of the main result of the next section with its usual statement at present. Assume then that c is a positive constant and that $y = f(x)$ is a function whose values are known at the points of an evenly spaced collection of points $x_i = ci, i = 0, \ldots, n$. Our problem is to determine the values of f at every other point x or, as Newton put it, *Given any number of points, to describe a curve which shall pass through all of them.*[32] To present the solution, we define the first- and higher-order increments

$$\Delta f(x_i) = f(x_i + c) - f(x_i)$$
$$\Delta^2 f(x_i) = \Delta f(x_i + c) - \Delta f(x_i)$$
$$\Delta^3 f(x_i) = \Delta^2 f(x_i + c) - \Delta^2 f(x_i)$$
$$\Delta^4 f(x_i) = \Delta^3 f(x_i + c) - \Delta^3 f(x_i)$$

$$\cdots$$

But instead of using these, Gregory, in his letter to Collins, and in reference to a figure like the next one,[33] said [p. 119; 46]:

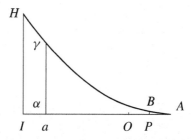

In the end of my Geometrical exercitations I fail exceedinglie ... and hence in place of any thing I have described there; putting $AP = PO = c$, $PB = d$, ...

Thus, with the origin at A and the direction of the positive x-axis to the left, $d = \Delta f(0) = f(c) - f(0)$. Then he defined the higher-order increments in

---

[32] "Datis quotcunque punctis Curvan describere quæ per omnia transibit," in Whiteside, *The mathematical papers of Isaac Newton*, IV, pp. 60–61.

[33] Fig. 8.va, *Exercitationes geometricæ*, 1668. I have omitted some inscribed rectangles that are irrelevant in this context.

verbal form, denoting them by f ("primam ex differentiis primis"), h, i, and so on, which in our notation can be expressed as

$$f = \Delta^2 f(0) = \Delta f(c) - \Delta f(0)$$
$$h = \Delta^3 f(0) = \Delta^2 f(c) - \Delta^2 f(0)$$
$$i = \Delta^4 f(0) = \Delta^3 f(c) - \Delta^3 f(0)$$
$$\cdots$$

With all this, he was able to state his result as follows [pp. 119–120; p. 46].

> In the eighth figure of my exercitationes imagine any straight line [segment] $A\alpha$ in the straight line AI to which $\alpha\gamma$ is perpendicular, with γ on the curve ABH [the segment $\alpha\gamma$ has been added above to the original figure], which remains as before, consider the infinite series
>
> $$\frac{a}{c}, \quad \frac{a-c}{2c}, \quad \frac{a-2c}{3c}, \quad \frac{a-3c}{4c}, \quad \text{etc.,}$$
>
> let the product of the first two terms of the series be b/c, that of the first three k/c, that of the first four l/c, that of the first five m/c, etc., to infinity; then the straight line [segment]
>
> $$\alpha\gamma = \frac{ad}{c} + \frac{bf}{c} + \frac{kh}{c} + \frac{li}{c} + \text{etc.}$$
>
> to infinity.

To interpret the last equation replace the quotients b/c, k/c, l/c, $\ldots$ with the products that they stand for and then replace d, f, h, i, $\ldots$ with the expressions given above. Then, since $\alpha\gamma = f(a)$ in current notation, where a is the variable, we have

$$f(a) = \frac{a}{c}\Delta f(0) + \frac{a(a-c)}{c \cdot 2c}\Delta^2 f(0) + \frac{a(a-c)(a-2c)}{c \cdot 2c \cdot 3c}\Delta^3 f(0)$$
$$+ \frac{a(a-c)(a-2c)(a-3c)}{c \cdot 2c \cdot 3c \cdot 4c}\Delta^4 f(0) + \text{etc.}$$

This is Gregory's formula,[34] and it can be considered only an approximation if a is arbitrary. Of course, if $f(0) \neq 0$ then we can add $f(0)$ to the beginning of the right-hand side.

---

[34] Newton did not originally state a formula. In 1676—in what may be considered a memo to himself rather than a paper—he just gave a table of divided differences (our increments divided by Gregory's c), hoping perhaps that the rest would be obvious. But his notation

We can further generalize the formula by replacing 0 with a; a with $a + h$ in the left-hand side; and any a in the numerators of the right-hand side, where it has the meaning $a - 0$, with h. If we also replace c with Δx, it becomes

$$f(a + h) = f(a) + h\frac{\Delta f(a)}{\Delta x} + \frac{h(h - \Delta x)}{1 \cdot 2}\frac{\Delta^2 f(a)}{\Delta x^2} + \cdots .$$

It must be pointed out that this formula can be used to obtain the binomial theorem by considering the special case

$$f(x) = \left(1 + \frac{d}{b}\right)^x,$$

a function whose values are known at $x = 0, 1, 2, \ldots$ In this case, $f(0) = 1$,

$$\Delta f(0) = f(1) - f(0) = 1 + \frac{d}{b} - 1 = \frac{d}{b},$$

$$\Delta^2 f(0) = \Delta f(0 + 1) - \Delta f(0) = f(0 + 2) - f(0 + 1) - \Delta f(0)$$

$$= \left(1 + \frac{d}{b}\right)^2 - \left(1 + \frac{d}{b}\right) - \frac{d}{b} = \frac{d^2}{b^2},$$

and so on. In this manner, putting $a = 0$, $h = a/c$, and $\Delta x = 1$,

$$\left(1 + \frac{d}{b}\right)^{a/c} = 1 + \frac{a}{c}\frac{d}{b} + \frac{\frac{a}{c}\left(\frac{a}{c} - 1\right)}{1 \cdot 2}\frac{d^2}{b^2} + \frac{\frac{a}{c}\left(\frac{a}{c} - 1\right)\left(\frac{a}{c} - 2\right)}{1 \cdot 2 \cdot 3}\frac{d^3}{b^3} + \cdots .$$

This is the binomial theorem, but we have worked it out in current notation.

In an enclosure sent with his letter to Collins of 23 November 1670, Gregory posed a problem and then gave its solution by what is essentially recognized as the last equation, but did not explain how he arrived at it. It is a

---

is very foreign to modern tastes. Suffice it to say that he denoted his x-values by $A + p$, $A + q$, $A + r \ldots$, the function values at these points by $\alpha, \beta, \gamma \ldots$, the first-order divided differences by $\zeta, \eta, \theta \ldots$, the second-order divided differences by $\lambda, \mu \ldots$, the third-order divided differences by $\xi \ldots$, and so on. Pity he had not discovered the use of subscripts. In 1711 Newton included some explanations from which the formula can be reconstructed. This was done—in Newton's own notation—by Whiteside in *The mathematical papers of Isaac Newton*, IV, pp. 63–64 and VIII, p. 248. However, the curious reader is warned that a certain amount of work would still be required to transform Whiteside's formula into the one stated here. There is a formula in the *Principia*'s version: "RS will be $= a + bp + cq + dr + es + ft, + \&c,$" but it also requires careful interpretation.

permissible conjecture that he may have obtained this result from his interpolation formula as shown above, but in his own notation. Here is a translation of Gregory's Latin passage [pp. 131–133]:

> To find the number of a logarithm.
>
> Given two numbers, the first b, the second $b + d$, let the logarithm of b be e and the logarithm of $b + d$ be $e + c$, it is desired to find the number whose logarithm is $e + a$.
>
> Consider a series of continued proportions,
>
> $$b, \quad d, \quad \frac{d^2}{b}, \quad \frac{d^3}{b^2}, \quad \&c.;$$
>
> and another series
>
> $$\frac{a}{c}, \quad \frac{a - c}{2c}, \quad \frac{a - 2c}{3c}, \quad \frac{a - 3c}{4c}, \quad \&c.;$$
>
> let the product of the first two terms of this series be f/c, that of the first three g/c, that of the first four h/c, that of the [first] five i/c, & c.; the number whose logarithm is $e + a$ will be
>
> $$= b + \frac{ad}{c} + \frac{fd^2}{cb} + \frac{gd^3}{cb^2} + \frac{hd^4}{cb^3} + \frac{id^5}{cb^4} + \frac{kd^6}{cb^5} + \&c.^{35}$$
>
> Hence with some work but no trouble any pure equation whatever may be solved.

If the last equation does not look like the binomial formula stated above, it is a matter of moments to recognize it as such. Just replace f/c to k/c with their definitions, divide throughout by b, and the last series becomes

$$1 + \frac{a}{c}\frac{d}{b} + \frac{a(a - c)}{c \cdot 2c}\frac{d^2}{b^2} + \frac{a(a - c)(a - 2c)}{c \cdot 2c \cdot 3c}\frac{d^3}{b^3} + \&c,$$

which, as we have seen above, is the binomial series of

$$\left(1 + \frac{d}{b}\right)^{a/c}.$$

---

[35] Actually, Gregory omitted the factor d/b in several of the terms, but corrected the mistake in a new letter of 19 December [p. 148; 50].

Thus, the last series in the quotation from Gregory corresponds to the product of this expression by b. The logarithm of this product is, using Euler's notation,

$$l(b) + \frac{a}{c}[l(b+d) - l(b)] = e + \frac{a}{c}(e + c - e) = e + a,$$

as stated by Gregory. Thus, we see that Gregory essentially discovered the binomial series in 1670 independently of Newton. He knew what he had in his hands, for he gave an example to find the daily compound interest rate equivalent to 6% per annum [pp. 131–133], that is, what we now express as

$$100\left(1 + \frac{6}{100}\right)^{\frac{1}{365}}.$$

Then in a letter to Collins of 19 December he used the binomial theorem to perform several term-by-term integrations, giving the area under a circle, the expansion of the arcsine, and the length of the curve $y = r \log x$ [pp. 148–149; 50–51]. Of course, Newton put the binomial theorem to greater use, not only in the case of the quadrature of the hyperbola, but also making it an essential tool in his development of the calculus, as we shall see in the next chapter.

4.4 THE TAYLOR AND MACLAURIN SERIES

Returning now to Gregory's interpolation formula, essentially the same result was obtained by Brook Taylor (1685–1731), who had been secretary of the Royal Society from 1714 to 1718, and is contained in a letter dated 26 July 1712, to John Machin (1680–1751),[36] professor of astronomy at Gresham College. Then he published it in 1715,[37] giving proper credit to Newton's work in the *Principia*, and including a "demonstration" and two corollaries. It was in the second corollary, after the proposal: "If we substitute for evanescent increments the fluxions proportional to them," that he obtained the famous *Taylor series*, as it was called by Colin Maclaurin (1698–1746) and as it has been known ever since. What Taylor meant by substituting "evanescent increments" is better expressed now as "letting $\Delta x \to 0$" in the modernized Gregory formula, which gives, in today's notation,

$$f(a + h) = f(a) + f'(a)h + \frac{f''(a)}{2!}h^2 + \frac{f'''(a)}{3!}h^3 + \cdots.$$

---

[36] Reproduced by Bateman in "The correspondence of Brook Taylor," 1906–1907.

[37] *Methodus incrementorum directa & inversa* (Direct and reverse methods of incrementation), 1715, p. 21.

BROOK TAYLOR IN 1714 AND HIS SERIES
Portrait by an unknown artist.
Taylor's series appears in Corollary II.

If we put $a = 0$ in the Taylor series and replace h with x, we obtain

$$f(x) = f(0) + f'(0)x + \frac{f''(0)}{2!}x^2 + \frac{f'''(0)}{3!}x^3 + \cdots,$$

which is called the *Maclaurin series* because it was frequently used by Maclaurin in his *Treatise on fluxions*, published in Edinburgh in 1742. Maclaurin, of course, made it clear that the result was "given by Dr. Taylor."

Taylor, who credited Newton, did not mention Johann Bernoulli, who claimed to have published an equivalent result eleven years before. This prompted one of the many disputes in which Johann Bernoulli used to find himself embroiled.[38]

However, the whole story must be told, and to do that we must return to Gregory's series expansions for the tangent and the secant. Gregory did not

---

[38] In "Additamentum effectionis omnium quadraturarum & rectificationum curvarum per seriem quandam generalissimam," 1694, p. 438 = *Opera omnia*, I, p. 126, Bernoulli had

give them in our notation, but, instead, wrote the following (translated from the Latin) in his letter to Collins of 15 February 1671 [p. 170; 62]:

Let radius $= r$, arc $= a$, tangent $= t$, secant $= s$, it will be

$$a = t - \frac{t^3}{3r^2} + \frac{t^5}{5r^4} - \frac{t^7}{7r^6} + \frac{t^9}{9r^8}$$

and it will be

$$t = a + \frac{a^3}{3r^2} + \frac{2a^5}{15r^4} + \frac{17a^7}{315r^6} + \frac{3233a^9}{181440r^8},$$

and

$$s = r + \frac{a^2}{2r} + \frac{5a^4}{24r^3} + \frac{61a^6}{720r^5} + \frac{277a^8}{8064r^7} :$$

The last two series are the same as the ones previously quoted at the beginning of Section 4.3. To see this in the case of the tangent, note that $a = r\theta$ and that Gregory still used trigonometric lengths rather than trigonometric functions. For example, his t means $r \tan \theta$, and then, replacing t and a with the values given here, the second series in the quoted paragraph becomes

$$\tan \theta = \theta + \frac{\theta^3}{3} + \frac{2\theta^5}{15} + \frac{17\theta^7}{315} + \frac{3233\theta^9}{181440} + \cdots,$$

as given in Section 4.3.

The concept of factorial did not exist in the seventeenth century, but to write this series as we might expect it today, note that

$$3 = \frac{3!}{2}, \quad 15 = \frac{5!}{8}, \quad 315 = \frac{7!}{16}, \quad \text{and} \quad 181440 = \frac{9!}{2},$$

---

used successive integrations by parts to show that if n is a function of the variable z, then

$$Integr. \, n \, dz \left[\int_0^z n \, dz \right] = +nz - \frac{zz \, dn}{1 \cdot 2 \cdot dz} + \frac{z^3 \, ddn}{1 \cdot 2 \cdot 3 \cdot dz^2} - \frac{z^4 \, dddn}{1 \cdot 2 \cdot 3 \cdot 4 \cdot dz^3} \quad \&c.$$

It is easy to obtain the Maclaurin series from Bernoulli's formula, but not immediately. Indeed, if f is an infinitely differentiable function, apply Bernoulli's formula with $n = f'$, $n = f''$, $n = f'''$, $n = f^{(4)}$, and so on to infinity, performing the integrations on the left in each case. Then multiply the resulting equations by 1, z, $z^2/2!$, $z^3/3!$, and so on, and add the results. Canceling all terms containing $f'(z)$, $f''(z)$, $f'''(z)$, ... will yield the Maclaurin series for f. The details are left to the reader, as well as passing judgment on Bernoulli's claim.

and then Gregory's series for the tangent becomes

$$\tan\theta = \theta + \frac{2}{3!}\theta^3 + \frac{16}{5!}\theta^5 + \frac{272}{7!}\theta^7 + \frac{6466}{9!}\theta^9 + \cdots.$$

Those who own the appropriate electronic means of computation or have some free time on their hands will have no trouble verifying that this is the Maclaurin expansion of $\tan\theta$ (except for a numerical error in the last coefficient, which we shall discuss later).

But in 1670 neither Taylor nor Maclaurin had been born, and it is clear that Gregory could not have used their work. So, how did he find his expansions? The answer lies (and lay dormant until 1938) on some manuscript notes that he wrote about a fortnight before his letter to Collins giving these series. They fill some of the blank space of a letter that he had received from his bookseller, Gideon Shaw, dated 29 January 1671 (paper was expensive in those times and Gregory was in the habit of writing on the blank spaces of received correspondence). Gregory's notes start as follows [p. 350]:

(i)   arc $= a$ 1st 2nd 3rd

sine $= s$

radius $= r$ $m = q$ $m = \dfrac{r^3}{c^2}$ $m = \dfrac{2r^4 q}{vc^3}$

secant $= v$

sine com. $= c$ $t = \dfrac{r^2 q}{v^2}$ $t = \dfrac{vc}{2q}$ $t =$

tang. $= q$

They are not particularly intelligible from a current point of view, but they can be interpreted, and what follows is an elaboration of Turnbull's interpretation. It is clear that the left column just establishes the notation for several trigonometric items, "sine com." being the sine complement ,or cosine, and since we are dealing with lengths, $c = r\cos\theta$. Note that the tangent (meaning $r\tan\theta$) is now q, and not t as before, and we are interested in the values of m given on the third line. These are the values of the ordinate of the curve $q = r\tan\theta$ in the 1st column, what we now call its first derivative with respect to θ in the 2nd column, and its second derivative in the 3rd column. We are not concerned with the values of t, which now stands for the subtangent.

To verify the statements about the derivatives of q, note that (in current notation)

$$\frac{dq}{d\theta} = \frac{r}{\cos^2\theta} = \frac{r^3}{(r\cos\theta)^2} = \frac{r^3}{c^2},$$

and then, from the first of these equations and using the quotient rule,

$$\frac{d^2q}{d\theta^2} = -\frac{r[2\cos\theta(-\sin\theta)]}{\cos^4\theta} = \frac{2r\sin\theta}{\cos^3\theta} = \frac{2r^4(r\tan\theta)}{(r\sec\theta)(r^3\cos^3\theta)} = \frac{2r^4q}{vc^3}.$$

Gregory may have thought that computing "higher derivatives" in this way was time-consuming. In any event, he modified his approach and then wrote

$$m = q \qquad m = r + \frac{q^2}{r} \qquad m = 2q + \frac{2q^3}{r^2},$$

which is another way to write q and its first two derivatives. Indeed, this form of the first derivative is obtained as follows:

$$\frac{dq}{d\theta} = \frac{r}{\cos^2\theta} = r(1 + \tan^2\theta) = r + \frac{r^2\tan^2\theta}{r} = r + \frac{q^2}{r},$$

and the second can be found using the chain rule

$$\frac{d^2q}{d\theta^2} = \frac{d}{dq}\left(r + \frac{q^2}{r}\right)\frac{dq}{d\theta} = \frac{2q}{r}\left(r + \frac{q^2}{r}\right) = 2q + \frac{2q^3}{r^2}.$$

But did Gregory know the chain rule? Before we venture an opinion, consider the first seven derivatives of $q = r\tan\theta$, which he gave later in the manuscript under discussion in the following form [p. 352]:

(xiii)   1st   2nd   3rd   4th

$$m = q \quad m = r + \frac{q^2}{r} \quad m = 2q + \frac{2q^3}{r^2} \quad m = 2r + \frac{8q^2}{r} + \frac{6q^4}{r^2}$$

5th   6th

$$m = 16q + \frac{40q^3}{r^2} + \frac{24q^5}{r^4} \qquad m = 16r + \frac{136q^2}{r} + \frac{240q^4}{r^3} + \frac{120q^6}{r^5}$$

7th

$$m = 272q + 987\frac{q^3}{r^2} + 1680\frac{q^5}{r^4} + 720\frac{q^7}{r^6}$$

8th

$$m = 272r + 3233\frac{q^2}{r} + 11361\frac{q^4}{r^3} + 13440\frac{q^6}{r^5} + 5040\frac{q^8}{r^7}$$

Now, if Gregory did not know the chain rule he must have needed a lot of additional scrap paper before he wrote these derivatives down on the scrap paper that contains them. Be that as it may, the derivatives—for that is what they are called now—are correct, except for the coefficient ~~978~~ in the sixth, which should be 1232. The error carries on to the last line, in which the second and third coefficients should be 3968 and 12096 instead of 3233 and 11361. It seems to be the consequence of a trivial mistake, which is, for our purposes, very significant. Computing the sixth derivative from the fifth using the chain rule requires the addition of the terms

$$\frac{2(136)q}{r}\frac{q^2}{r} \quad \text{and} \quad \frac{4(240)q^3}{r^3}r$$

among others. Thus, the coefficient of q^3/r^2 in the seventh derivative should be $272 + 960 = 1232$, but, in an obvious slip, Gregory added $27 + 960 = 987$.

Putting this mistake aside, but only for a moment, the rest of Gregory's manuscript contains the same type of successive differentiations for six other functions, one of them being $q = r \sec \theta$. And six of these seven are functions whose power series expansions he gave in the subsequent letter to Collins. Thus, we have the following partial answer to the question that we posed before: yes, Gregory knew how to compute, and in fact did compute, the derivatives necessary for a Maclaurin expansion of those six functions. Still, he could have obtained the expansions in some other manner. But besides the obvious fact that the repeated differentiations evaluated on Shaw's letter would then have been pointless, there is a clincher. The error that Gregory made in computing the last two derivatives of $r \tan \theta$ appears again, two weeks later, in the second series that he sent to Collins: the numerator 3233 on the fifth term of the series for t (page 222). The evidence is pretty conclusive, and then the conclusion inescapable. Gregory used the successive differentiations on Shaw's letter to obtain six of the series expansions that he sent to Collins. In other words, Gregory discovered the Maclaurin series expansion before Taylor and Maclaurin were born, 41 years before Taylor discovered it.

Why did Gregory not publish his results? His intention to publish was clearly expressed in his letter to Collins of 23 November 1670 [p. 118; 45]:

> I have almost readie for the presse another edition of my quadratura circuli
> & hyperbolæ, wherein (if I be not much mistaken) I demonstrat my intent
> many & several ways; I purpose also to add to it several universal methods,
> as I imagine, as yet unheard of in Mathematicks, both in geometrie and
> analyticks: I am afrayed I can have it but naughtilie don here; and therfor I
> humblie desire your concurrence to try how I can have it don in London, and
> advertise me with the first occasion.

Those "unheard of universal methods" must have been the interpolation theorem, the binomial theorem, and the Taylor–Maclaurin series, already in his
mind. It may be doubted that this quotation refers to the Maclaurin series,
which was first given later in the Shaw letter the following January, but Gregory had actually announced it in November 1670 [p. 120; p. 47]:

> I have a methode also whereby I turn ageometrick problems (at least al I have
> yet considered) into an infinit series; and amonge others Keplers probleme
> ... I resolve with great ease.

So he was enthusiastic about his discoveries and ready to publish. What happened that deprived him of so much deserved credit? In November, Gregory
had received only one of Newton's series from Collins, but then Collins wrote
on 24 December 1670, expressing himself as follows [p. 154; 54]:

> I have since had some few Series more out of Newtons generall method ...
> the method is universall and performes all Quadratures ... in this method,
> the curved lines of all figures that have a common property, are streightened,
> their tangents and Centers of Gravitie discovered ... the length of the Curved
> line being given an Ordinate is found and the Converse.

In short, Newton was ready to unveil the universal panacea. As a sample,
Collins included four of those series that he got out of Newton, including
those for the sine, cosine, and arcsine [p. 155; 54]. To this, Gregory replied
on 15 February 1671 [p. 170; 62]:

> As for Mr Newton his universal method, I imagine I have some knowledge
> of it ...[39]

And it was at this point that he included the seven series already discussed.
But it is possible to read an undercurrent of disappointment in this statement.
Gregory might have assumed that Newton already knew everything that he
had recently discovered himself and thought it was pointless to publish. The
fact is that he changed his mind about publishing, expressing his decision as
follows in the same letter of 15 February 1671 [p 171; 63]:

---

[39] As we shall see in Chapter 5, Gregory had already published how to find "the length of
the Curved line being given an Ordinate" in his *Geometriæ pars vniversalis* of 1668.

I thank you werie hertilie for your good advice, as to the publication of my notions, & for your civil profer; I would be werie sorrie to put you to so much trouble. I have no inclination to publish any thing, safe only to reprint my quadrature of the circle, and to add some little trifles to it.

Was this loss of interest in publishing compounded by Collins' warnings about the dire publishing situation in London? In the same letter of 24 December, Collins said [pp. 156; 55–56]:

There is not any Printer now in London accustomed to Mathematicall worke, or indeed fitted with all convenient Characters, and those handsome fractions but Mr Godbid where your Exercitationes were printed, and at present he is full of this kind of worke to wit ...

And here he proceeded to give a long, exhaustive list of all those works in the queue to be printed. Was this the trouble that Gregory hesitated to put Collins through? Whatever the reasons, the three most important pieces of Gregory's lifetime work remained unpublished.

Newton was not in possession of such methods. Of the new results that Gregory had obtained and could have published, Newton had discovered only, at that time, the binomial series, as shown in Section 2.5. But Newton was a giant, and five years later he carried out, and eventually even published, his work on interpolation as mentioned in Section 4.3. The next step took him a little longer, but he discovered the Taylor series twenty years after Gregory but still twenty years before Taylor. This, however, he never published. The Maclaurin and Taylor expansions are the most important achievement in Newton's manuscript *De quadratura curvarum*.[40] At some point Newton stated the following result [p. 93].

PROPOSITION XII.

Out of an equation involving two fluent quantities, either alone or together with their fluxions,[41] to extract one or other quantity in an unterminated converging series.

---

[40] This is the original manuscript of November–December 1691. Page references are to its printing in Whiteside, *The mathematical papers of Isaac Newton*, VII, 1976, pp. 48–129. For the many versions of this work, see the bibliography.

[41] We are getting ahead of the story here. Fluent quantities (that is, quantities that vary with time) and their fluxions (or the rates at which they vary) will take center stage in the next chapter.

Then, after considering three cases, he gave four corollaries, of which we reproduce here the third and fourth, and then a translation of the third, admittedly out of context. Let us just say, for the benefit of today's reader, that a dot on top of a variable denotes its derivative with respect to time, two dots the second derivative, and so on. But it is neither possible nor necessary to fully understand Newton's corollaries at this point [pp. 97–99].

Newton's discovery of the Taylor series.
From Whiteside, *The mathematical papers of Isaac Newton*, VII, Facing page 98.

Corol. 3. Hence, indeed, if the series proves to be of this form

$$y = az + bz^2 + cz^3 + dz^4 + ez^5 + \&c$$

(where any of the terms $a[z]$, $b[z^2]$, $c[z^3]$, $d[z^4]$, ... can either be lacking or be negative), the fluxions of y, when z vanishes, are had by setting $\dfrac{\dot{y}}{\dot{z}} = a$,

$$\frac{\ddot{y}}{\dot{z}^2} = 2b, \; \frac{\dddot{y}}{\dot{z}^3} = 6c, \; \frac{\ddddot{y}}{\dot{z}^4} = 24d, \; \frac{\dddddot{y}}{\dot{z}^5} = 120e.$$

That is,

$$y = \frac{\dot{y}}{\dot{z}}z + \frac{\ddot{y}}{2\dot{z}^2}z^2 + \frac{\dddot{y}}{6\dot{z}^3}z^3 + \frac{\ddddot{y}}{24\dot{z}^4}z^4 + \frac{\dddddot{y}}{120\dot{z}^5}z^5 + \&c,$$

which, if we set $z = t$, so that $\dot{z} = 1$, and insert the omitted term $y_0 = y(0)$, gives the Maclaurin series

$$y = y_0 + \dot{y}z + \frac{\ddot{y}}{2!}z^2 + \frac{\dddot{y}}{3!}z^3 + \frac{\ddddot{y}}{4!}z^4 + \frac{\dddddot{y}}{5!}z^5 + \cdots.$$

Corollary 4 is similar but gives the full Taylor series, as shown in the photographic reproduction of Newton's manuscript.

There has been too much talk about derivatives and integrals in this section. It is clear that the calculus had already been discovered when these general expansions were obtained, but we have not talked about it. We must, therefore, interrupt the story on infinite series to present the development of the calculus before we embark on the thorny question of convergence.

5

THE CALCULUS

5.1 THE ORIGINS

We have seen some of the roots of the integral calculus in the method of exhaustion to perform quadratures, at least as exemplified by Archimedes' quadrature of the parabola. These ideas were to be resurrected and perfected in the seventeenth century.

An ancestor of future ideas about what we now call the differential calculus can be found in the fourteenth-century study of the *latitude of forms*. To put it in very simple terms, a form—a concept of Aristotle—was what we now call a function, for it was thought of as a quality subject to variability, such as velocity, temperature, or density. John Duns Scotus (1266–1308) was one of the first to consider the increase and decrease (*intensio* and *remissio*) of the intensity of forms. For the intensity of a form must be distinguished from its extension: velocity, temperature, and density are intensities, while distance traveled, heat, and mass are extensions. But fourteenth-century Scholastic scholars, especially at Merton College, Oxford—in particular, William Heytesbury and Richard Suiseth (or Swineshead)—also studied such fine points as the rate of change and the rate of change of the rate of change of a form, of course in a verbal manner and not always separate from philosophy or mysticism.

In a manuscript entitled *De motu* (On motion), attributed to Suiseth, he classified all motions into two kinds: uniform [pp. 243; 245][1]

---

[1] The quotations in this section are from Clagett, *The science of mechanics in the Middle Ages*, 1959. This work contains the Latin texts of the manuscripts by the Merton College

in which in every equal part of the time an equal distance is described.

and difform:

> Difform motion [*difformis motus*] is that in which more space is acquired in
> one part of the time and less in another equal part of the time.

In turn, *difformis* motion was classified into *uniformiter difformis* and *difformiter difformis*, according to whether the instantaneous rate of change of
the rate of change was constant or not, and further subdivisions were considered. For instance, in 1335 Heytesbury gave the following definition in
Regule solvendi sophismata (Rules for solving sophisms), Part VI: Local Motion [pp. 237; 241]:

> For any motion whatever is uniformly accelerated [*uniformiter intenditur*]
> if, in each of any equal parts of the time whatsoever, it acquires an equal
> increment [*latitudinem*] of velocity.

The scholars at Merton College arrived at a theorem for the mean intensity
of a *uniformiter difformis* form. It was stated first by Heytesbury [pp. 270;
277], and then by Suiseth in *De motu* as follows [pp. 244; 245–246]:

> Wherever there is uniform increase [*intensio*] of local motion, the local motion is uniformly difform motion. Since local motion uniformly difform corresponds to its mean degree [of velocity] in regard to effect, so it is evident
> that in the same time so much is traversed by means of [a uniform movement]
> at the mean degree as by means of the uniformly difform movement.

More precisely, if an object travels a distance d in time t, starting with initial
velocity v_i, moving with constant acceleration, and reaches a final velocity v_f,
then "in the same time [t] so much [d] is traversed" as if the object moved "at
the mean" constant velocity

$$\frac{v_i + v_f}{2}.$$

In short, the average velocity of a body moving with constant acceleration is
the mean of its initial and final velocities.

---

scholars and translations of large passages from them. Page references given here are to
this edition; the page on the left of the semicolon refers to the English translation, and the
one on the right to the original in Latin.

Several proofs were provided, some based on manipulations with infinite series. The leading fourteenth-century treatise on the latitude of forms was the *Liber calculationum* (probably composed in the second quarter of the century) of Richard Suiseth, who became known as *Calculator*. It contains the *Regule de motu locali* (Rules of local motion), in which Suiseth gave four proofs, of which the third is the most interesting of all the Merton College proofs.[2]

When this subject reached the Continent, Nicole Oresme clarified the picture by introducing graphs to distinguish the different types of variation in the latitude of a form.[3] He made this approach an integral part of his work *De configurationibus qualitatum*, which was probably written before 1361. His geometric representation is proposed in the following way in Chapter 1 [pp. 347; 368].

> Hence every intension which can be acquired successively is to be imagined by means of a straight line erected perpendicularly on some point or points of the [extensible] space or subject of the intensible thing.[4]

In short, we have a system of coordinates. On the horizontal axis (the *longitudo* axis in geographic terms) we represent the extension of the form and on the vertical axis the *latitudo* or intensity of the form.

Later, opening Chapter 11, Oresme added [pp. 352; 372]:[5]

> Thus every uniform quality is imagined by a [rectangular] quadrangle [that is, a horizontal line at a constant height], and every quality uniformly difform terminated at no degree [ending at zero] is imaginable by a right triangle. Every quality uniformly difform terminated at both ends at some degree is to be imagined as a quadrangle having right angles on the base and the other angles unequal. Moreover, every other linear quality is called "difformly difform" and is to be imagined by figures disposed in other and considerable varying ways.

---

[2] It is reconstructed in modern notation by Clagett, *The science of mechanics in the Middle Ages*, pp. 295–297.

[3] If it comes to questions of priority, Giovanni di Casali had used graphs for the same purpose before Oresme, but not to the same extent or with the same influence. In this connection see Clagett, *The science of mechanics in the Middle Ages*, pp. 332–333.

[4] "Omnis igitur intensio successive acquisibilis ymaginanda est per lineam rectam perpendiculariter erectam super aliquod punctum aut aliquot puncta spacii vel subiecti illius rei intensibilis."

[5] A slightly different translation can be found in Clagett, *Nicole Oresme and the medieval geometry of qualities and motions*, 1968; reprinted in Calinger, *Classics of Mathematics*, 1995, p. 255.

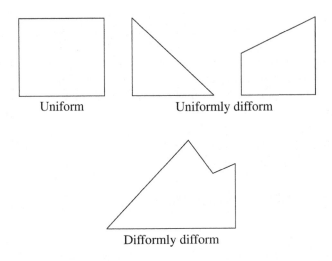

Uniform   Uniformly difform

Difformly difform

The preceding discussion notwithstanding, the first work that we can recognize as properly belonging to the subject now called the calculus is Kepler's
book on the solid geometry of wine casks.[6] In spite of its title, this is a book on
mathematics, largely devoted to evaluating the volumes of solids of revolution.
For example, a chord in a circle that is not a diameter divides the circle into
two parts of unequal areas. If the largest is rotated about the chord, it produces
a solid "which is of the form of the fruit of the apple tree"; if the smallest is
rotated it produces "what can be called the figure of a lemon" [Supplement
to Part I, p. 576]. All in all, Kepler considered the volumes of 96 solids of
revolution.[7] Later he turned to the problem of finding the best dimensions for
wine casks and was able to determine that the largest right parallelepiped with
a square base that can be inscribed in a sphere is the cube [Part II, Theorem IV,
pp. 607–609], and that of all right circular cylinders with equal diagonals the
largest is the one that has a ratio of diameter to altitude of $\sqrt{2}$ to 1 [Part II,
Theorem V and Corollary II, pp. 610–612].

After Kepler, many seventeenth-century mathematicians worked on problems of finding tangents and quadratures. The list includes, but is not limited

---

[6] *Nova stereometria doliorum vinariorum*, 1615 = *Opera omnia*, IV, 1863. Page references are to the Latin text in the *Opera*.

[7] The volume of an apple is obtained in the Corollary to Theorem XIX of Part II, which
can be seen in English translation in Struik, *A source book in mathematics, 1200–1800*,
1969, pp. 192–197; reprinted in Calinger, *Classics of Mathematics*, pp. 356–360.

to, Galileo Galilei, Bonaventura Cavalieri, and Evangelista Torricelli in Italy; Blaise Pascal, René Descartes, and Gilles Personne de Roberval in France; Frans van Schooten, René François de Sluse, Johann Hudde, and Christiaan Huygens in the Low Countries; and John Wallis in England. No history of the calculus would be complete without an account of their contributions, but it would be impossible to do that here without turning this chapter into a book,[8] so we shall restrict our attention to the work of five men in the 1600s, each of which has been called the true inventor of the calculus by someone at some time: Pierre de Fermat, James Gregory, Isaac Barrow, Isaac Newton, and Gottfried Wilhelm Leibniz.[9]

5.2 FERMAT'S METHOD OF MAXIMA AND MINIMA

Pierre de Fermat (1601–1665), a lawyer working in the *parlement* of the city of Toulouse, found some spare time to work on mathematics. The fact that he was an amateur mathematician notwithstanding, he is considered to be the best French mathematician of the seventeenth century.

About 1629 Fermat produced an algorithm to find maxima and minima, which is contained in a copy of his original manuscript sent in 1637 by Roberval to Marin Mersenne under the double title "A method to investigate maxima and minima" and "About the tangents to curved lines." [10] It was widely circulated in mathematical circles in Paris—Mersenne himself sent a copy to René Descartes—but it was not well received. It simply gives an algorithm to find maxima and minima but includes no supporting explanations whatsoever.

If Fermat's method was found unclear by his contemporaries, used as they

---

[8] There are, in fact, some comprehensive works on the history of the calculus. For instance, Baron, *The origins of the infinitesimal calculus*, 1969 (Dover Publications, 1987); Boyer, *The concepts of the calculus. A critical and historical discussion of the derivative and the integral*, 2nd. ed., 1949 (reprinted by Dover Publications as *The history of the calculus and its conceptual development*, 1959); and Edwards, *The historical development of the calculus*, 1979.

[9] I'll exhibit my own prejudice on this matter at the outset. The calculus was not invented—not in the sense that many gadgets have been invented—but whatever word one wants to apply to its development, it came about as the cumulative result of work by many people—much as in the case of geometry and algebra—some more insightful, some more extensive. Some represented a breakthrough but some did not. Moreover, the calculus as taught today is a product of the nineteenth century, owing much to the rigorization efforts of da Cunha, Cauchy, Riemann, Weierstrass, and others.

[10] "Methodus ad disquirendam maximam et minimam" and "De tangentibus linearum curvarum."

PIERRE DE FERMAT
From D. E. Smith, *Portraits of Eminent Mathematicians*,
Portfolio II, Scripta Mathematica, New York, 1938.

were to the terminology of the times, it is to be expected that today's readers would fare no better. Thus, we shall adopt the following course of action: first we include a translation of the first part of this manuscript—a printed copy of which can be seen in the next photograph—up to and including the example; then we attempt to clarify Fermat's steps and interpret them in today's terminology; and finally, we shall include a summary of his own explanations as to how he got the idea for the method.

The translation (from the printed copy reproduced on the next page) is as follows, but we have inserted roman numerals, which are not in the original, to number the steps in the algorithm so that we can refer to these steps later.

The whole science of the discovery of maxima & minima, rests on the position of two unknown [expressions], & this unique rule; (*i*) let *A* be any unknown of the problem, whether plane, or a solid, or a length, according to the proposition to be satisfied, & (*ii*) express the maximum or minimum

METHODUS

Ad diſquirendam maximam & minimam.

 MNIS de inventione maximæ & minimæ doctrina, duabus poſitioni
bus ignotis innititur, & hac unica præceptione ſtatuatur quilibet quæſtio
nis terminus eſſe A, ſive planum, ſive ſolidum, aut longitudo, prout pro
poſito ſatisfieri par eſt, & inventa maxima aut minima in terminis ſub A,
gradu ut libet inuolutis; Ponatûr rurſus idem qui prius eſſe terminus A,
+ E, iterumque inveniatur maxima aut minima in terminis ſub A & E, gradibus ut
libet coefficientibus. Adæquentur, ut loquitur Diophantus. duo homogenea maximæ
aut minimæ æqualia & demptis communibus (quo peracto homogenea omnia ex parte
alterutra (ab E, vel ipſius gradibus afficiuntur) applicentur omnia ad E, vel ad elatio
rem ipſius gradum, donec aliquod ex homogeneis, ex parte utravis affectione ſub E,
omnino liberetur.

Elidantur deinde utrimque homogenea ſub E, aut ipſius gradibus quomodolibet in-
voluta & reliqua æquentur. Aut ſi ex unâ parte nihil ſupereſt æquentur ſine, quod eo-
dem recidit, negata ad firmatis. Reſolutio ultimæ iſtius æqualitatis dabit valorem A,
quâ cognita, maxima aut minima ex repetitis prioris reſolutionis veſtigiis innoteſcet.
       Exemplum ſubijcimus

Sit recta A C, ita dividenda in E, ut rectang. A E C, ſit maximum ; Recta A C, di-
catur B.

A   E   C

ponatur par altera B, eſſe A, ergo reliqua erit B, — A, & rectang. ſub ſegmentis erit
B, in A, — A² quod debet inueniri maximum. Ponatur rurſus pars altera ipſius B,
eſſe A, + B, ergò reliqua erit B, —, A — E, & rectang. Sub. ſegmentis erit B, in A
, A² + B, in E, 'E in A, — E, quod debet adæquati ſuperiori rectang. B, in A,
     A², demptis communibus B, in E, adæquabitur A, in E' + E', & omnibus per
E, diviſis B, adæquabitur 'A + E, elidatur E, B, æquabitur 'A, igitur B, bitariam
eſt dividenda, ad ſolutionem propoſiti, nec poteſt generalior dari methodus.

❧❧❧❧❧❧❧❧❧❧❧❧❧❧❧

De Tangentibus linearum curvarum.

A D ſuperiorem methodum inventionem Tangentium ad data puncta in lineis
quibuſcumque curvis reducimus

H 4

First page of the printed copy of Fermat's *Methodus*
in his *Varia opera mathematica*, 1679.

in terms of *A*, using terms that can be of any degree whatever; (*iii*) Posit in
turn that the same term as before is $A + E$, and express again the maximum
or minimum in terms of *A* & *E*, using terms of any degree. (*iv*) Adequate,
as Diophantos would say.[11] the two homogeneous [expressions] giving the
maximum or minimum & (*v*) remove the common terms (when done all terms
in the homogeneous [expressions] on either side will contain *E*, or powers
of it) (*vi*) divide all terms by *E*, or by a higher power of it, until any of the
homogeneous [expressions], on one side or the other, is completely free of *E*.

---

[11] Fermat was borrowing here a word and a meaning from a translation of Diophantos'
Arithmetica by Holtzman, 1575. He owned a copy of its republication with added commentary by Claude-Gaspard Bachet de Méziriac, 1621.

(*vii*) Remove then from both sides all the terms involving E or a power of it in any manner, and (*viii*) adequate the rest. Or, if nothing remains on one of the sides reasonably adequate, which amounts to the same thing, the negative terms to the positive. The solution of this last equation will give the value of A, which once known, the maximum or minimum will become known by repeating the steps of the preceding solution.[12]

Now, with the aid of the inserted roman numerals, we can simplify and rewrite Fermat's algorithm in our own terms as follows:

(*i*) Let A be the variable in the problem, for which we shall write x.

(*ii*) Express the quantity to be maximized or minimized as a function of the variable. We shall denote it by $f(x)$.

(*iii*) Increment the variable as $A + E$, for which we shall write $x + h$, and evaluate $f(x + h)$.

(*iv*) "Adequate" (in the sense of setting approximately equal) the two preceding expressions:

$$f(x + h) \approx f(x).$$

(*v*) Simplify by removing terms that appear in both $f(x + h)$ and $f(x)$.

(*vi*) Cancel any power of h higher than the first that is a common factor of both $f(x + h)$ and $f(x)$ and then divide by h, which is equivalent to setting

$$\frac{f(x + h) - f(x)}{h} \approx 0.$$

(*vii*) Disregard or strike out any terms that still contain h or a power of h.

(*viii*) "Adequate" the rest (to 0 since we had moved $f(x)$ to the left-hand side).

With this interpretation, part of Fermat's algorithm is what we do today, at least to the evaluation of what we call the difference quotient, but then we take the limit as $h \to 0$ and actually equate it to 0. However, the concept of limit was not available to Fermat and he did not create it. Thus, he had to make do with disregarding higher powers of h and working with an approximation. Of

---

[12] The Latin version of this quotation, as well as Fermat's example given below, can also be seen in *Œuvres de Fermat*, I, 1891, pp. 133–136. There are two somewhat different English translations of the preceding paragraphs that the reader can consult, and the detailed references are given in the bibliography. The one by Struik includes the translation of Fermat's complete manuscript.

course this is valid only if h is very small, a claim that he did not make at any stage of this procedure, but seems to have had in mind.

An example would be welcome at this point, so we consider the problem of splitting a straight line segment AC of length B into two portions by a point E,

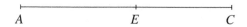

$$A \qquad\qquad E \qquad\qquad C$$

and the problem is to find E such that the product $AE \times EC$ is a maximum. Here are the problem and its solution as stated by Fermat.

> We exhibit an example.
>
> Divide the straight line [segment] AC at E, so that the rectang. AEC [meaning the area $AE \times EC$] is a maximum; Denote the segment AC by B. Let A be one part [segment] of B, so that the rest will be $B - A$,[13] & the rectang. [product] of the segments will be $BA - A^2$ which must become a maximum. Let in turn $A+E$ be one part of B, so that the rest will be $B-A-E$, & the rectang. [product] of the segments will be $BA - A^2 + BE - 2EA - E^2$, which must be adequated to the rectang. given above $BA - A^2$, remove the common terms [to obtain] BE, adequal to $2AE + E^2$, & divide all by E [to obtain] B, adequal to $2A + E$, remove E, therefore adequate $2A$ to B, [and] divide by 2, to [obtain] the proposed solution, it is not possible to give a more general method.

And here it is restated in our own terms.

(i) Let $x = AE$.

(ii) $f(x) = x(B - x) = Bx - x^2$.

(iii) $f(x + h) = (x + h)(B - x - h) = Bx - x^2 + Bh - 2hx - h^2$.

(iv) "Adequate": $Bx - x^2 + Bh - 2hx - h^2 \approx Bx - x^2$.

(v) Simplify: $Bh \approx 2xh + h^2$.

(vi) No power of h other than the first is a common factor of all the terms, so just divide by h: $B \approx 2x + h$.

(vii) Disregard the term that still contains h: $B \approx 2x$.

($viii$) The rest is already "adequated," and the solution is $x = B/2$.

---

[13] Here Fermat is being naughty in denoting by A one of the segments in which AC is divided by E. Also, notice from the photograph that we have changed the notation $B, -A$ to $B - A$ and we shall also write BA for the product of B and A rather than B, in A.

Today we would obtain the same answer by setting $f(x) = Bx - x^2$ and then

$$\lim_{h \to 0} \frac{f(x+h) - f(x)}{h} = \lim_{h \to 0} \frac{Bh - 2hx - h^2}{h} = B - 2x = 0,$$

and it is clear to us that Fermat had the seed of the present method.

To his contemporaries, in particular to Descartes, he was at best obscure and at worst accused of having found the algorithm by trial and error (*à tâtons* and *par hasard*). Fermat was stung, and reacted by producing several explanations of his algorithm. The most important for our purposes is in a manuscript that begins with the words "Dum *syncriseos* et *anastrophes* Vietæ methodum expenderem,"[14] probably written about 1639 or 1640 [15] to directly refute Descartes' allegations. This paper, which we shall simply call *Syncriseos*, begins with an acknowledgment of Fermat's source of inspiration:

> When considering Viète's method of *syncrisis* and *anastrophe*, ... a new method derived from it came to my mind for finding maxima and minima
> . . .

Viète was fond of creating Greek neologisms, and "syncrisis" was one of them. The key element in the method of syncrisis is the assumption that a polynomial of order n has up to n roots and the relation between those roots and the order of the polynomial.[16]

In the second paragraph of this paper Fermat revealed his second source of inspiration to be a result of Pappos of Alexandria, which he later identified more precisely [p. 151] as Proposition 61 from Book VII of the *Synagoge* or *Mathematical collection*.[17] The opening statement of this lemma is this:

> Given three straight lines [segments] AB, BC, CD if one makes the rectangle ABD to the rectangle ACD, as the square on BE to the square on EC, the

---

[14] It appears in Fermat, *Œuvres*, I, p. 147, with the title "Methodus de maxima et minima," chosen by the editors. Page references are to this work.

[15] For the reasons for this dating see Mahoney, *The mathematical career of Pierre de Fermat 1601–1665*, 1994, p. 145.

[16] Syncrisis is explained in Chapter XVI of *De recognitione æquationum* = *Francisci Vietæ opera mathematica, in unum volumen congesta, ac recognita*, 1646, pp. 104ff; English translation in Witmer, *The analytic art by François Viète*, 1983, pp. 207ff.

[17] This is a work on classic Alexandrian geometry, with some new proofs and additions, written in the late 200s. This proposition can be found in Commandino's translation: *Mathematicæ collectiones*, 1588, f. 196^r.

ratio is unique, & smallest of rectangle AED to rectangle BEC.[18]

This statement is to be interpreted with reference to the following figure, which

is a small part of Pappos' figure. Although Commandino, the translator, confessed to not understanding the end of this statement—which he had just rendered as "singularis proportio, & minima est rectanguli AED ad rectangulum BEC"—Fermat did understand. Unfortunately, he changed Commandino's notation in his explanation. Using the original notation to avoid unnecessary confusion, the problem that the proposition poses is [p. 151]:

> Let $ABCD$ be a straight line, in which the points A, B, C, D are given. Choose a point E between the points B and C, such that the rectangle AED to the rectangle BEC has minimum ratio.

The solution given by Pappos in Proposition 61 is to choose E such that

$$\frac{AB \times BD}{AC \times CD} = \frac{BE^2}{EC^2}.$$

Fermat had solved this problem in a manuscript written before *Syncriseos*, and this led him to a quadratic equation.[19] Such an equation normally has two

---

[18] Recall that "the rectangle ABD" means the area $AB \times BD$, and similarly for the other rectangles.

[19] In publication, the title of the manuscript is "Ad eamdem methodum." It is possible to obtain both the quadratic equation and Pappos' solution rather quickly using today's differentiation. Write (as Fermat wrote in *Syncriseos*, except that we must use lowercase letters, while he used uppercase ones) $BC = b$, $BD = z$, $AB = d$, and $BE = a$. As is frequently the case in Fermat's writing, a is the variable. Then, according to his interpretation, we have to minimize

$$\frac{AE \times ED}{BE \times EC} = \frac{(d+a)(z-a)}{a(b-a)} = \frac{dz - da + za - a^2}{ba - a^2}.$$

Then equating the derivative of the right-hand side with respect to a to zero and simplifying gives

$$(b + d - z)a^2 - 2dza + bdz = 0$$

["Ad eamdem methodum," in *Varia opera*, p. 67 = *Œuvres*, I, p. 144; *Syncriseos*, p. 152]. This is the quadratic equation, which Fermat obtained by syncrisis (to be explained shortly) rather than by differentiation. Multiplying it by b, and then adding and subtracting dza^2 to

solutions. But then he focused attention on the words that Commandino did not understand and realized that at the minimum there is only one solution. What happened to the other solution?

To find out, Fermat started by considering a simpler problem, and found it in Euclid's Proposition 27 from Book VI of *The Elements*. It is the one he presented in the *Methodus*: split a segment of length B into two parts A and E so that the product of A and E is largest. He knew the solution because it was given by Euclid as "that parallelogram is greatest which is applied to half of the straight line" [20] and its area is $B^2/4$.

But in *Syncriseos* Fermat was in search of the two solutions of a quadratic, and modified the problem by asking how the original segment should be split so that the product of the resulting segments is a constant $Z < B^2/4$. In his own words [p. 148]:

> But, if it is proposed that the *same straight line B is to be cut under the condition that the rectangle* [product] *of the segments is equal to an area Z* (which is supposed to be smaller than one fourth of B square), then two points satisfy the proposition, and the point of maximum rectangle is between them.

He meant that there are two solutions of the quadratic that we would write now as $A(B - A) = Z$ (but he wrote in Viète's style as "B in $A - A$ quad. æquale Z plano"), and denoted these solutions by A and E. Here is Fermat's view of what happens to these solutions as Z approaches its maximum value $B^2/4$ [pp. 148–149]:

> If, in place of the area Z, we take another that is larger than the area Z, but smaller than one-fourth of B square, then the straight lines [segments] A and E will differ from each other by less than the ones above, as the points of division more nearly approach the point constitutive of the maximum rectangle,[21] always, as the rectangles of the divisions increase, the difference

---

the left-hand side, we obtain

$$(b^2 + db - bz - dz)a^2 - 2dza + b^2dz + dza^2 = 0,$$

which, after factoring and rearranging, can be written as

$$\frac{dz}{(d + b)(z - b)} = \frac{a^2}{(b - a)^2}.$$

This is Pappos' solution if the lowercase letters are replaced by the segments they represent.

[20] Quoted from the Dover Publications edition of *The Elements*, vol. 2, 1956, p. 257.

[21] This is Fermat's intuition, for he offered no proof of this or of the fact the "the point of maximum rectangle lies between" A and B.

of these A and E will decrease, until with the last division of the maximum rectangle it vanishes, in which case a $\mu o \nu \alpha \chi \grave{o} \varsigma$ [Pappos' word] or unique solution will be produced, since the two quantities [become] equal, that is, A will equal E.

Now, if

$$A(B - A) = Z \qquad \text{and} \qquad E(B - E) = Z,$$

then, by the method of Viète ("syncrisis"),

$$A(B - A) = E(B - E)$$

or, rearranging,

$$B(A - E) = A^2 - E^2.$$

Dividing by $A - E$, we obtain $B = A + E$ (which in this very simple example is geometrically obvious). Fermat drew the following conclusion [p. 149]:

> Since, therefore, in the two correlate equations above, by the method of Viète, B will equal $A + E$, if E is equal to this A (which appears to hold always at the point constitutive of a maximum or a minimum), then, in the proposed case, B will equal $2A$: that is, if the straight line B is cut in half, the rectangle [product] of its segments will be a maximum.

This is the essence of Fermat's method to find maxima and minima, which he then applied to another example [p. 149]:

> To cut the straight line B, in such a manner that the square of one of its segments [call it A again] times the other is a maximum.

In this case we have to maximize $A^2(B - A)$. Then consider the equation $BA^2 - A^3 = Z$, where Z is a constant smaller than the maximum value. If E is another solution of this equation, then, by syncrisis,

$$BA^2 - A^3 = BE^2 - E^3,$$

or

$$B(A^2 - E^2) = A^3 - E^3,$$

and dividing by $A - E$,

$$B(A + E) = A^2 + AE + E^2.$$

At the maximum, $A = E$, and then

$$2BA = 3A^2,$$

so that $A = 2B/3$.

But Fermat noted at this point that division by binomials was exceedingly laborious and generally troublesome ("operosa nimis et plerumque intricata") [p. 149], and put forth the following idea [p. 150]:

> However, since E (as well as A) is an uncertain quantity, nothing prevents us from calling it $A + E$.

Then the difference of the two solutions is just E, a much easier quantity to divide by. While Fermat showed his new approach by a new example—and went on to crown the *Syncriseos* with the solution of Pappos VII, 61—we shall not do that. Instead, to make a long story a little shorter, we shall apply it to his first example. Then, by syncrisis,

$$A(B - A) = (A + E)[B - (A + E)],$$

or, simplifying and rearranging,

$$BE = 2AE + E^2,$$

just as in the example included in the *Methodus*. Dividing by E and recalling that at the maximum "E will give nothing" [*Syncriseos*, p. 150], we obtain $B = 2A$. The solution is, as before, $A = B/2$. However, since he was hiding any explanation of his methods in the *Methodus*, he needed an explanatory gimmick, and he found it in Diophantos' adequality.

The *Methodus* also included a procedure to find tangents, and posing the problem is quite simple:[22]

> Let there be given, for example, a Parabola BDN, whose vertex is D, the diameter DC, & a point B given on it, to extend a straight line BE, tangent to the parabola, & meeting the diameter at a point E.

To solve the problem, Fermat took a point O on BE located over the segment CD and drew OI parallel to the ordinate BC. In Fermat's time there was no equation of a parabola as we know it today, but if we think of a rectangular

---

[22] *Varia opera*, p. 64; *Œuvres*, I, p. 135. English translation in Stedall, *Mathematics emerging*, 2008, pp. 73–74.

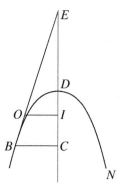

coordinate system with the origin at D, the positive x-axis in the direction DC, and the positive y-axis horizontally to the left, it is clear from what he wrote next that he was dealing with the parabola $x = y^2$. Then, if the ordinate at $x = DI$ is denoted by y_I, we see that $y_I < OI$, and then

$$\frac{CD}{DI} = \frac{BC^2}{y_I^2} > \frac{BC^2}{OI^2}.$$

But from similar triangles,

$$\frac{BC^2}{OI^2} = \frac{CE^2}{IE^2},$$

and therefore

$$\frac{CD}{DI} > \frac{CE^2}{IE^2}.$$

At this point Fermat managed to be naughty again by using some of the old letters with new meanings as follows:

> But since the point B is given, the applied line BC is given, hence the point C; is given as well as CD; therefore let CD be equal to a given D. Let CE be A; let CI be E.

Why not? After all, the alphabet is limited. In the new notation the last inequality becomes (translating once more from Fermat's narrative form to present-day usage)

$$\frac{D}{D - E} > \frac{A^2}{A^2 + E^2 - 2AE},$$

or

$$DA^2 + DE^2 - 2DAE > DA^2 - A^2 E.$$

And now we shall let Fermat have the last word to avoid the blame for any unclarity in his statement.

> Adequate therefore according to the method above: with common terms removed, $DE^2 - 2DAE \approx -A^2 E$ [$\approx$ meaning "will be adequal to"], or what is the same, $DE^2 + A^2 E \approx 2DAE$. Let everything be divided by E, hence $DE + A^2 \approx 2DA$, remove DE, hence $A^2 = 2DA$, and thus, $A = 2D$, therefore we have proved that CE is twice CD, which indeed holds true.

In this manner, almost in one breath, an inequality turned first into an adequality and then into an equality. Fermat never gave any subsequent explanations.

But today's readers deserve some interpretation in current terminology, and to provide it we turn to the equation that Fermat obtained from similarity of triangles, which can be rewritten as follows:

$$\frac{OI}{BC} = \frac{IE}{CE}.$$

Referring now to the rectangular coordinate system with origin at D that was mentioned above, define $x = DC$, $h = x - DI$ and write the arc $\overset{\frown}{BD}$ of the parabola as $y = \sqrt{x}$. Then $BC = \sqrt{x}$, $IE = CE - h$, and $DI = x - h$. If we use the adequality $OI \approx \sqrt{x-h}$, the equation of similarity can be replaced with

$$\frac{\sqrt{x-h}}{\sqrt{x}} \approx \frac{CE - h}{CE}.$$

Subtracting 1 from each side and simplifying,

$$\frac{\sqrt{x-h} - \sqrt{x}}{\sqrt{x}} \approx -\frac{h}{CE},$$

or

$$\frac{\sqrt{x-h} - \sqrt{x}}{-h} \approx \frac{\sqrt{x}}{CE}.$$

Now multiply numerator and denominator of the fraction on the left-hand side by $\sqrt{x-h} + \sqrt{x}$ and then simplify to obtain

$$\frac{1}{\sqrt{x-h} + \sqrt{x}} \approx \frac{\sqrt{x}}{CE}.$$

At this point, we would be happy to let $h \to 0$, but the last step in Fermat's procedure was to discard h (his E) and to turn the adequality into an equality:

$$\frac{1}{2\sqrt{x}} = \frac{\sqrt{x}}{CE},$$

and it follows that the subtangent is $CE = 2x$, the same result that in his notation was $A = 2D$.

Descartes had quite a number of objections to Fermat's method, and it is quite clear that some of them are as justified as being impossible to address with the mathematical knowledge available at that time. Nevertheless, in response to these criticisms Fermat elaborated and improved on the method. But Descartes also thought that this method was applicable only to functions given in what we call explicit form and challenged Fermat to find the tangent to the curve that we now write as $x^3 + y^3 = Pxy$,[23] then known as *la galande* and now as the *folium of Descartes*. Rising to this challenge, Fermat promptly found the tangent that Descartes demanded, sending it to him (via Mersenne) in a manuscript entitled "Méthode de maximis et minimis expliquée et envoyée par M. Fermat a M. Descartes."[24] He considered a curve through the origin

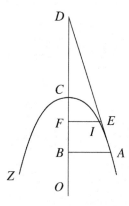

like the one shown in the figure, in which the origin is at C, DO is the x-axis, and DA is tangent to the curve at A. Fermat started [p. 154] by renaming

[23] Descartes, who gave no graph for this curve, expressed its equation as follows: "les deux cubes des deux lignes BC [our y] et CD [our x] soient ensemble égaux au parallélépipède des deux lignes BC, CD et de la ligne donnée P," Fermat, *Œuvres*, II, 1894, p. 130.

[24] Fermat, *Œuvres*, II, pp. 154–176.

some segments as follows (again reusing some letters as before): $BA = B$, $BC = D$, $BD = A$, and, for any point E on the tangent, $BF = E$. Then the coordinates of the point E on the tangent are [p. 155] $CF = D - E$, and, from the similarity of the triangles DFE and DBA,

$$FE = \frac{B(A - E)}{A} = \frac{BA - BE}{A}.$$

Next Fermat considered FI adequal to FE if E is close to A, and proceeded as if E were on the curve. Thus, replacing Descartes' P with N, the property of *la galande* is $CF^3 + FE^3 = N \cdot CF \cdot FE$. With the expressions obtained above for CF and FE in terms of single letters, the sum of the cubes is [p. 156]

$$D^3 - E^3 - 3D^2E + 3DE^2 + \frac{B^3A^3 - B^3E^3 - 3B^3A^2E + 3B^3AE^2}{A^3},$$

while $N \cdot CF \cdot FE$ is [p. 157]

$$N(D - E)\frac{BA - BE}{A} = \frac{NDBA - NDBE - NBAE + NBE^2}{A}.$$

Multiplying both expressions by A^3, we must compare by adequality

$$D^3A^3 - E^3A^3 - 3D^2EA^3 + 3DE^2A^3 + B^3A^3 - B^3E^3 - 3B^3A^2E + 3B^3AE^2$$

with

$$NDBA^3 - NDBEA^2 - NBEA^3 + NBE^2A^2.$$

Since the equation for *la galande* is $D^3 + B^3 = NDB$, the terms $D^3A^3 + B^3A^3$ and $NDBA^3$ can be dropped after adequation. Fermat completed this example as follows:

> Divide the rest by E and then drop all the terms that still contain E; it will remain [after division by $-A^2$]
>
> $$3D^2A + 3B^3 = NDB + NBA$$
>
> and we will have
> $$\frac{NDB - 3B^3}{3D^2 - NB} = A,$$
>
> which is what we had to find.

That is, Fermat had found the subtangent. If we recall that the origin is at C and if we write $D = CB = x$ and $B = BA = y$, then $A = y/y'$, and the reader should easily verify the correctness of Fermat's result using today's implicit differentiation.

5.3 FERMAT'S TREATISE ON QUADRATURES

Fermat developed an interest in the problem of the quadrature of the higher parabolas and hyperbolas, and eventually found a solution, but he did not publish anything on this subject during his lifetime, which makes it difficult to date his work. From indirect evidence, it appears that he may have obtained some results by the 1640s and even by the 1630s.[25] Much later, in 1658 or 1659, he was motivated to write a treatise on quadratures by the appearance in 1655 of the *Arithmetica infinitorum* by John Wallis (1616–1703). Fermat claimed to have obtained the same results long ago and showed some criticisms of Wallis' methods.[26]

In his treatise "On the transformation and amendment of local equations for the manifold comparisons of curvilinear figures among themselves or to rectilinear figures, to which is annexed the use of geometric progressions for the quadrature of an infinite number of parabolas and hyperbolas,"[27] Fermat started with the following assertion [p. 44; 255]:

> Archimedes used geometric progressions only for the quadrature of the parabola ... I have recognized and proved this sort of progression to be very useful for quadratures, and I willingly communicate to modern geometers my invention, which performs the quadratures of parabolas and hyperbolas by an altogether similar method.

He worked out the quadrature of the hyperbola $x^2 y = 1$ first (he chose the right-hand side to be an arbitrary constant, but there is no loss of generality in taking it to be 1), then that of the parabola $y^2 = x$, and finally generalized

---

[25] For a discussion of these dates, see Boyer, *The history of the calculus and its conceptual development*, 1959, p. 127; Bortolotti, "La scoperta e le successive generalizzazioni di un teorema fondamentale di calcolo integrale," 1924, p. 215; Walker, *A study of the Traité des Indivisibles of Gilles Persone de Roberval*, 1932, pp. 142–164; or Mahoney, *The mathematical career of Pierre de Fermat, 1601–1665*, p. 244.

[26] Wallis found the area under an arc of $y = x^{p/q}$, where p/q is any rational number other than -1, using a method that he called *investigatio per modum inductionis* [Proposition XIX], which essentially means guessing what we would call some limits. His descriptions are purely verbal, without any variable or function notation [Propositions LV to LVII and CII]. The stated propositions can be seen in his *Opera mathematica*, **1**, 1695, pp. 373, 390–391 and 408.

[27] *De æquationum localium transmutatione, & emendatione, ad multimodam curvilineorum inter se, vel cum rectilineis comparationem. Cvi annectitvr proportionis geometricæ in quadrandis infinitis parabolis & hyperbolis usus.* Page references are to *Varia opera* and, after a semicolon, to *Œuvres*, I. See the bibliography for available English translations.

his method to all curves of the form $x^m y^n = 1$, m and n nonzero integers, with the exception of $xy = 1$. To give an idea of the method it will suffice to present some simple examples. Fermat considered first the hyperbola $DSEF$

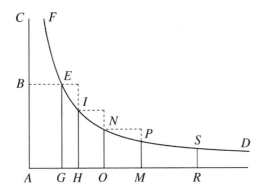

shown in the figure (the dashed lines were not printed in *Varia opera* but added in *Œuvres* to help visualize some rectangles), and since he did not have our equation for it, began the procedure as follows [pp. 44; 256–257]:

> Consider, if you please, the hyperbola defined by the property
>
> $$\frac{AH^2}{AG^2} = \frac{EG}{HI} \quad \text{and} \quad \frac{AO^2}{AH^2} = \frac{HI}{NO}, \quad \text{etc.}$$
>
> [that is, $x^2 = 1/y$]. I say that the infinite space whose base is EG, and one of its sides is the curve ES, and the other the infinite asymptote GOR, is equal to a given rectilinear area.

We shall show this and find the "given rectilinear area" in two ways. First, using present-day notation, to give a quick idea of what is going on, and then in Fermat's terminology, so that we know what he actually did. In the preceding figure, let the origin be at A and define $a = AG$. Now choose a real number $r > 1$ and then points $G, H, O, M, \ldots$ with abscissas

$$a, \quad ar, \quad ar^2, \quad ar^3, \quad \ldots .$$

Then the segments $GH, HO, OM, MR, \ldots$ have lengths

$$a(r-1), \quad ar(r-1), \quad ar^2(r-1), \quad ar^3(r-1), \quad \ldots ,$$

and the heights $EG, HI, ON, PM, \ldots$ have lengths

$$\frac{1}{a^2}, \quad \frac{1}{a^2 r^2}, \quad \frac{1}{a^2 r^4}, \quad \frac{1}{a^2 r^6}, \quad \ldots ,$$

so that the rectangles with diagonals $EH, IO, NM, PR, \ldots$ have areas

$$\frac{1}{a}(r-1), \quad \frac{1}{ar}(r-1), \quad \frac{1}{ar^2}(r-1), \quad \frac{1}{ar^3}(r-1), \quad \ldots.$$

These form a geometric progression with ratio $1/r$, whose sum, according to Viète's formula, is

$$\frac{\frac{1}{a}(r-1)}{1-\frac{1}{r}} = \frac{r}{a}.$$

It approaches $1/a$ as $r \to 1$, which is, then, the area of the region $DNEGR$.

Fermat, to whom all this notation was unavailable, chose the segments $AG, AH, AO, AM, \ldots$ so that they formed a geometric progression, which he expressed as [p. 45; 257]

$$\frac{AG}{AH} = \frac{AH}{AO} = \frac{AO}{AM} = \ldots, ^{28}$$

and this is equivalent to

$$\frac{AG}{AH} = \frac{GH}{HO} = \frac{HO}{OM} = \ldots.$$

He did not give any reasons, but those who prefer an explanation should note that

$$\frac{AG}{AH} = \frac{AH}{AO} = \frac{AG+GH}{AH+HO},$$

which, after dismissal of the middle fraction, cross multiplication of the remaining fractions and simplification, leads to

$$\frac{AG}{AH} = \frac{GH}{HO}.$$

Similarly for the other fractions.

Then Fermat used this to examine the ratio of rectangular areas. For example [p. 45; 257],

$$\frac{EG \times GH}{HI \times HO} = \frac{GE}{HI} \cdot \frac{GH}{HO} = \frac{GE}{HI} \cdot \frac{AG}{AH}.$$

---

[28] In quoting or describing Fermat's work I use modern fractions and dots as a convenience. Fermat did not use either, expressing himself in verbal form. For example, he gave this line as follows: "as AG, is to AH, so is AH, [to] AO, & so is AO to $AM \ldots$"

On the other hand, by the defining property of the hyperbola (see the preceding quotation) and the fact that the proportionality of abscissas gives $AH^2 = AG \cdot AO$, we have [p. 45; 258]

$$\frac{GE}{HI} = \frac{HA^2}{GA^2} = \frac{AO}{GA}.^{29}$$

Therefore,

$$\frac{EG \times GH}{HI \times HO} = \frac{AO}{GA} \cdot \frac{AG}{AH} = \frac{AO}{AH} = \frac{AH}{AG}.$$

Then he arrived at the following conclusion [pp. 45–46; 258]:

> Similarly we can prove that
>
> $$\frac{HI \times HO}{ON \times OM} = \frac{AO}{HA}.$$
>
> But the segments AO, HA, GA that make up the parallelogram ratios are proportional by construction [that is, $GA/HA = HA/AO$, so that $AO/HA = HA/AG$]; hence the [areas of the] parallelograms $GE \times GH$, $HI \times HO$, $ON \times OM$, etc., assumed to be infinitely many, will form a continued geometric progression, the ratio of which will be HA/AG.

Fermat had already remarked that the terms of the original progression (the abscissas) can be chosen close enough to each other that one can "adequate the rectilinear parallelogram $GE \times GH$ and the mixed [curvilinear] quadrilateral GHE" [p. 45; 257]. Thus the sum of the areas of all the rectangles can be adequated to the area under the hyperbola to the right of EG. To sum the progression of rectangular areas Fermat could have used Viète's formula, with which he must have been familiar. However, he chose to state this main tool in the following manner, which is equivalent to Viète's, at the start of his paper [pp. 44; 255–256]:

> Given any geometric progression, whose terms decrease indefinitely, the difference of two consecutive terms of this progression is to the smaller of the two as the largest of all the terms of the progression is to the sum of all the others to infinity.

---

[29] Fermat frequently changed the order of the letters in a segment as he wrote.

If S denotes the sum of the entire progression of rectangular areas, the application of this rule in this particular case gives [p. 46; 258]

$$\frac{EG{\times}GH - IH{\times}HO}{IH{\times}HO} = \frac{EG{\times}GH}{S - EG{\times}GH}.$$

If we recall that

$$IH{\times}HO = \frac{GA}{HA}\,EG{\times}GH,$$

then the left-hand side in the previous equation simplifies to and then equals

$$\frac{HA}{GA} - 1 = \frac{GH}{GA} = \frac{GE{\times}GH}{GE{\times}GA}.$$

On the other hand, if the progression ratio is close enough to 1, the denominator $S - EG{\times}GH$ is adequal to the area of the figure $DNIHR$ under the hyperbola, which is in turn adequal to the area $DNEGR$ because the segment GH is very small. Thus, Fermat obtained the adequation

$$\frac{GE{\times}GH}{GE{\times}GA} \approx \frac{EG{\times}GH}{\text{area } DNEGR}.$$

Since the numerators are equal, adequating the denominators, Fermat expressed this result as follows [p. 46; 259]:

> ... the parallelogram AE, in this kind of hyperbola, is equal to the area of the figure contained by the base EG, the asymptote GR and the curve ED infinitely extended.

Since the parallelogram AE has area

$$AG{\times}GE = a\,\frac{1}{a^2} = \frac{1}{a},$$

this is the result already obtained above in current notation.

Now we can understand Fermat's emphasis on the geometric progression in the opening statements of this treatise on quadratures, for it is precisely the fact that the areas of his rectangles form such a progression that allowed him to find the sum. He found a way to subdivide the axis so that this happens, but had he tried a subdivision by segments of equal length he would have been unable to succeed.

Then Fermat stated [p. 46; 259]: "It is not laborious to extend this discovery to all the hyperbolas of this sort, except the one that we mentioned [$xy = 1$]." He gave a brief proof that in the next case, $x^3 y = 1$ [pp. 46; 259–260],

$$AG{\times}GE = 2DNEGR,$$

or, if $a = AG$ and $GE = 1/a^3$,

$$DNEGR = \frac{1}{2a^2}.$$

After this, he claimed that the method would be the same in the remaining cases.

However, Fermat may have faced a technical difficulty while trying to use the same method to perform the quadrature of the first parabola, which we denote by $y^2 = x$, shown below,[30] over the finite interval CB. We shall

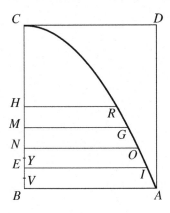

perform this quadrature in current notation first, and this will show the nature of the conjectured difficulty. The points B, E, N, M, ... were originally chosen such that

$$\frac{BC}{EC} = \frac{EC}{NC} = \frac{NC}{MC} = \cdots,$$

and then we can denote them by $a, ar, ar^2, ar^3, \ldots$, where $r < 1$ is very close to 1. Then the vertical sides of the circumscribed rectangles with diagonals $AB, IN, OM, GH, \ldots$ are

$$a(1 - r),\ \ ar(1 - r),\ \ ar^2(1 - r),\ \ ar^3(1 - r),\ \ \ldots,$$

---

[30] It was not uncommon at that time to draw the x-axis vertically with the positive direction downward. The letters Y and V were not printed in this figure in *Varia opera*.

and their horizontal sides BA, EI, NO, MG, $\ldots$ have lengths

$$a^{1/2}, \quad (ar)^{1/2}, \quad (ar^2)^{1/2}, \quad (ar^3)^{1/2}, \quad \ldots,$$

so that the rectangles have areas

$$a^{3/2}(1-r), \quad (ar)^{3/2}(1-r), \quad (ar^2)^{3/2}(1-r), \quad (ar^3)^{3/2}(1-r), \quad \ldots.$$

These form a geometric progression with ratio $r^{3/2}$, whose sum, according to Viète's formula, is

$$\frac{a^{3/2}(1-r)}{1-r^{3/2}}.$$

It may not be immediately clear what results from adequating r to 1, other than 0/0, and this may have prompted Fermat to alter his method, for he did introduce a modification, as we shall see below. However, a little algebra shows that

$$\frac{1-r}{1-r^{3/2}} = \frac{1}{r^{1/2}+\dfrac{1}{1+r^{1/2}}}.$$

Then, either letting $r \to 1$ or adequating r to 1 in the denominator yields 3/2. Therefore, the area under the parabola $y=\sqrt{x}$ from 0 to a is

$$ARCB = \frac{a^{3/2}}{3/2}.$$

Since the area of the rectangle $ABCD$ is $a\sqrt{a}=a^{3/2}$, putting $b=a^{1/2}$, we also obtain the formula

$$ARCD = a^{3/2} - \frac{a^{3/2}}{3/2} = \frac{a^{3/2}}{3} = \frac{b^3}{3}.$$

Fermat, if he had any difficulty at all, was able to deal with it by an ingenious second subdivision of the horizontal axis [p. 47; 261]:

> $\ldots$ if we take the mean proportional CV between BC and CE, and between EC and NC the mean proportional YC, $\ldots$

This means that the points V, Y, $\ldots$ are chosen so that

$$\frac{BC}{CV} = \frac{CV}{CE} = \frac{CE}{YC} = \frac{YC}{NC} = \cdots$$

and, in particular,

$$\frac{BC}{EC} = \frac{BC^2}{BC \times CE} = \frac{BC^2}{VC^2}.$$

Since the defining property of the parabola gives

$$\frac{AB^2}{IE^2} = \frac{BC}{CE},$$

it follows that

$$\frac{AB}{IE} = \frac{BC}{VC} = \frac{CE}{YC}.$$

This is one of the factors in the rectangle ratio

$$\frac{AE}{IN} = \frac{AB}{IE} \cdot \frac{BE}{EN},$$

and Fermat asserted that the remaining factor equals, "as has been shown above, BC/CE." It had not been shown above, but he may have had in mind subtracting 1 from each side of the original identity

$$\frac{BC}{EC} = \frac{EC}{NC}$$

and simplifying, which leads to

$$\frac{BE}{EC} = \frac{EN}{NC}.$$

Then

$$\frac{BE}{EN} = \frac{EC}{NC} = \frac{BC}{CE},$$

and the rectangle ratio becomes

$$\frac{AE}{IN} = \frac{CE}{YC} \cdot \frac{BC}{CE} = \frac{BC}{YC}.$$

Taking it as clear that this is also the ratio of the remaining rectangles, he proceeded to add up their areas [p. 47; 262]:

> ... and consequently from the theorem that substantiates our method, the parallelogram AE, is to the figure $IRCHE$, as the segment BY, is to the segment YC,[31] therefore the same parallelogram AE, is to the total figure $AIGRCB$, as the segment BY, is to the total diameter BC ...

---

[31] This segment appears as BC in *Varia opera* but was changed to YC in *Œuvres*

But this is going a bit too fast, so we should add some explanation. The "substantiating" theorem, stated here on page 251, gives

$$\frac{AE - IN}{IN} = \frac{AE}{\text{area } IRCHE},$$

and then using the rectangle ratio, the left-hand side simplifies to

$$\frac{BC}{YC} - 1 = \frac{BC - YC}{YC} = \frac{BY}{YC}.$$

When the rectangle ratio is very close to 1, then $YC \approx BC$ and the areas of the figures $IRCHE$ and $AIGRCB$ are approximately the same. Thus, the result of Fermat's addition can be written as

$$\frac{AE}{\text{area } AIGRCB} \approx \frac{BY}{BC} = \frac{AB \times BY}{AB \times BC} = \frac{AB \times BY}{BD},$$

and then

$$\frac{BD}{ARCB} \approx \frac{AB \times BY}{AE} = \frac{BY}{BE}.^{32}$$

But because of the fine subdivision of BC, the segments BV, VE, EY are adequal, and then the right-hand side above is approximately equal to 3/2. Fermat concluded [p. 48; 263]:

therefore the parallelogram BD, is to the figure as 3 is to 2.

That is,

$$ARCB = \frac{BD}{3/2},$$

and the quadrature has been completed.

Fermat stated that his method applies to all the other parabolas without exception, and to leave no room for doubt [p. 48; 263], he showed in full the case of the parabola $y^3 = x^2$. Now he needed two auxiliary subdivisions rather than one and was able to prove that [p. 49; 265]

in this case the parallelogram BD, is to the figure as 5 is to 3.

---

[32] In Fermat's own words [p. 47; 262]: "ut BY, ad BE, ita parallelogrammum BD, ad figuram $ARCB$;" but an error in *Varia opera* has this figure as $AROB$.

That is,

$$ARCB = \frac{BD}{5/3}.$$

This allowed him to deduce and state the following general rule:

> It is certainly clear that the parallelogram BD is always to the figure $AICB$, as the sum of the exponents of the powers of the ordinate & the abscissa is to the exponent of the power of the ordinate ...

Thus, in the general case of $y^n = x^m$,

$$AICB = \frac{BD}{\dfrac{m+n}{n}}.$$

If we denote the abscissa of B by a and define $q = m/n$, then the area of the rectangle BD is $a \cdot a^{m/n} = a^{q+1}$, and the preceding formula can be rewritten in current terminology as

$$\int_0^a x^q \, dx = \frac{a^{q+1}}{q+1}, \quad q \neq -1,$$

a general formula that Fermat found before Newton. Particular cases for a few integral values of q were known to others before Fermat.

Having stated a general rule for the quadrature of the parabolas must have inspired him to give one for hyperbolas, which he did next as follows [p. 49; 266]:

> In fact in any hyperbola it is always [the case], if you go back to the first figure [our page 249], that the parallelogram BG, is to the infinitely prolonged figure, $RGED$, as the difference of the exponents of the powers of the ordinate & the abscissa is to the exponent of the power of the ordinate.

This means that for the general hyperbola $x^m y^n = 1$ (reversing the order in Fermat's "the ordinate & the abscissa"),

$$RGED = \frac{BG}{\dfrac{m-n}{n}}.$$

Thus, if we define $a = AG$ and write $q = m/n$, then the area of BG is $a \cdot a^{-(m/n)} = a^{1-q}$. Then, the preceding formula can be rewritten in current terminology as

$$\int_a^\infty x^{-q} \, dx = \frac{a^{1-q}}{q-1}, \quad q \neq 1.$$

5.4 GREGORY'S CONTRIBUTIONS

Gregory's expertise in differentiation has already been shown in Chapter 4, where it was made clear that he was in possession of the chain rule and that he discovered Taylor's theorem. A few years before, Gregory had visited Italy, and at the end of his stay there, in 1668, he published a book entitled *Geometriæ pars vniversalis* (The universal part of geometry),[33] a work that

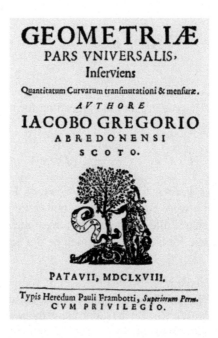

has been described as "the first attempt to write a systematic text-book on what we should call the calculus."[34] It is written in a geometrical style, as the title makes clear, and mostly in narrative form as was common at that time. It contains both original material and well-known results, and Gregory, in the

---

[33] *Geometriæ pars vniversalis, inserviens quantitatum curvarum transmutationi & mensuræ*, 1668. My discussion of this text here is based on and large portions taken from my paper "James Gregory's calculus in the *Geometriæ pars universalis*," 2007.

[34] By Adolf Prag in "On James Gregory's Geometriæ pars universalis," 1939. Prag also describes the contents of the book as follows: Propositions 1 to 11 are on arc length, area and tangent; Propositions 12 to 18 on involution and evolution; Propositions 19 to 45 on solids of revolution; and the rest applies to special figures.

second page of the introduction, left it to the reader to judge which was which (*quæ mea sunt & quæ aliena iudicet lector*).

There is very little on how to find tangents in this book, just one example given as Proposition 7 [pp. 20–22], to find the tangent to a curve that he first described in words, with reference to the next figure, as follows:

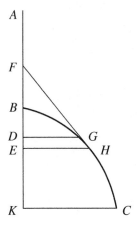

Let the curve *BHC* be a hyperbola, whose diameter is the straight line *AK* & the ordinates *EH*, *KC*, are of such nature that the solid from the square of *BE* times *AE* is to the solid from the square of *BK* times *AK* as the cube of *EH* is to the cube of *KC*.

In other words, if *A* and *B* are fixed, if *B* is the origin of coordinates, and if *BE* is an arbitrary abscissa with corresponding ordinate *EH*, then $BE^2 \times AE$ is a constant times EH^3 regardless of the position of the point *H* on the curve. Then Gregory set $AB = a$ and $BE = b$, so that $AE = a + b$, and denoted the constant that we just mentioned by a^3/c^3, where *c* is another constant. This allowed him to give the equation of the curve as

$$EH = \frac{\sqrt[3]{ab^2c^3 + b^3c^3}}{a}. \text{ 35}$$

The problem is to find the subtangent *FE*, which Gregory denoted by *z*, for the tangent *FH* at *H*. To do this, he drew a new ordinate *DG* at a vanishingly

---

35 Gregory, who omitted the equal sign, wrote $\sqrt{C}$ for cube root. But notice that he wrote b^2 instead of *bb*. In this respect, he was way ahead of his time.

small distance o from EH ("DE nihil seu serum o"), and then the product

$$DF \times EH = (z - o)\frac{\sqrt[3]{ab^2c^3 + b^3c^3}}{a}$$

is approximately equal to the area of the rectangle with sides $EF = z$ and DG. The length of DG is obtained from the equation of the curve after replacing b, which is the variable, with $b - o$, and then $DF \times EH \approx EF \times DG$, an approximation whose right-hand side is equal to

$$\frac{\sqrt[3]{c^3ab^2z^3 - 2c^3abz^3o + c^3az^3o^2 + c^3b^3z^3 - 3c^3b^2z^3o + 3c^3bz^3o^2 - c^3z^3o^3}}{a}.$$

Gregory then continued:

> & removing the denominators because they are equal, also & cubing both terms [sides] of the equation & removing from both the equal [terms] this equation results
>
> $$3b^2c^3azo^2 - 3b^2c^3az^2o - b^2c^3ao^3 - 3b^3c^3z^2o + 3b^3c^3zo^2 - b^3c^3o^3$$
> $$= c^3az^3o^2 - 2c^3abz^3o - 3c^3b^2z^3o + 3c^3bz^3o^2 - c^3z^3o^3,$$
>
> & dividing all by o
>
> $$3b^2c^3azo - 3b^2c^3az^2 - b^2c^3ao^2 - 3b^3c^3z^2 + 3b^3c^3zo - b^3c^3o^2$$
> $$= c^3az^3o - 2c^3abz^3 - 3c^3b^2z^3 + 3c^3bz^3o - c^3z^3o^2,$$
>
> & rejecting any quantities [terms] in which we find o or its powers [which is permissible as an approximation if o is very small], it remains
>
> $$-3b^2c^3az^2 - 3b^3c^3z^2 = -2c^3abz^3 - 3c^3b^2z^3,$$
>
> & whereby adding the defect [multiplying by -1] & dividing all by c^3bz^2 the equation is $3ba + 3b^2 = 2az + 3bz$, & therefore
>
> $$z = \frac{3ba + 3b^2}{a2 + 3b},$$

namely the straight line [segment] EF, which was to be found.[36]

---

[36] Since he invited the reader to judge on his originality, we must say that this is strongly reminiscent of Fermat's method. Jean de Beaugrand, a friend of Fermat and admirer of his work, made it his own business to disseminate Fermat's method of tangents, so maligned by Descartes. But he introduced some variations, one of which was the replacement of Fermat's E with o. This can be seen in his quadrature of the ellipse in the 1638 manuscript "De la manière de trouver les tangentes des lignes courbes par l'algèbre et des imperfections de celle du S. des C[artes]." In Fermat, *Œuvres*, V, *Supplément*, 1922, pp. 102–113. The letter o is introduced in page 102. Surprisingly, for an admirer, Beaugrand did not say that the method of tangents he applied in this manuscript was due to Fermat. In any event, it was widely distributed, and, since Beaugrand traveled extensively in Italy, it is quite likely that Gregory saw a copy and adopted the symbol o, along with Fermat's method.

It is easy to verify this result using current methods.[37]

There is no more on tangents per se in Gregory's *Geometriæ*, but about two years later he gave the rules for finding subtangents of composite curves. They are in manuscript form on a piece of scrap paper. He gave six rules, of which we reproduce three.[38]

> Let Ch, AK be two curves [see the next figure], and let MB be a straight line, and let GI be a curve with the property that GB is always equal to [the sum of] AB, CB; let the straight lines CD, AF touch the curves Ch, AK; put $CB = a$, $DB = b$, $AB = c$, $FB = d$; let
>
> $$BH = \frac{adb + cdb}{da + bc}$$
>
> and then join GH, which will touch the curve GI.

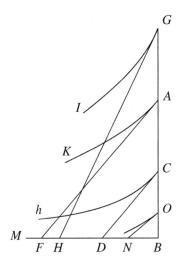

This finds the subtangent BH at an arbitrary point G of a curve GI, which is the sum of Ch and AK, in terms of the subtangents $b = DB$ and $d = FB$. In the seventeenth century mathematicians were keener on subtangents than on slopes, as we have already seen, but this result is not particularly appealing or easy to remember.

---

[37] It may be simpler to write x instead of b and y instead of EH, then find y', plug in $x = b$, and note that $z = y/y'$.

[38] The complete set in Latin, translated into English by Turnbull and with an explanatory note, can be seen in *James Gregory tercentenary memorial volume*, 1939, pp. 347–349.

Today we are interested in slopes—there is a very good reason for it to be discovered presently—and we can find that of GH from those of CD and AF using Gregory's formula. To do this denote GB, which Gregory did not name, by y, and then $y = GB = CB + AB = a + c$. The point B represents an arbitrary abscissa, and slopes and subtangents are related for our three curves by the equations (if none of the tangents are vertical)

$$y' = \frac{GB}{BH} = \frac{y}{BH}, \quad b = DB = \frac{CB}{a'} = \frac{a}{a'}, \quad \text{and} \quad d = FB = \frac{AB}{c'} = \frac{c}{c'}.$$

When the values of BH, b, and d obtained from these equations are substituted into Gregory's formula (for which he did not provide a proof in this note to himself), it becomes

$$\frac{y}{y'} = \frac{a\dfrac{c}{c'}\dfrac{a}{a'} + c\dfrac{c}{c'}\dfrac{a}{a'}}{\dfrac{c}{c'}a + \dfrac{a}{a'}c} = \frac{\dfrac{a+c}{c'a'}}{\dfrac{1}{c'} + \dfrac{1}{a'}} = \frac{a+c}{a'+c'} = \frac{y}{a'+c'},$$

and then $y' = a' + c'$. This is the derivative of a sum. It is the sum of the derivatives, while, for subtangents, $BH \neq b + d$. This is why slopes are more convenient than subtangents in calculation.

The next quotation will show that he dealt with products and quotients.

But if $GB/AB = CB/OB$,[39] and if $BH = a$, $BF = b$, $BD = c$, it will be

$$BN = \frac{acb}{ab + ac - cb}.$$

It may be annoying, but Gregory did change the notation, so that a, b, and c have new meanings now. If the four functions represented in the preceding figure are now denoted (from bottom to top) by y, u, v, and w, the relationship between slopes and subtangents is as follows:

$$y' = \frac{y}{BN}, \quad u' = \frac{u}{c}, \quad v' = \frac{v}{b}, \quad \text{and} \quad w' = \frac{w}{a}.$$

Then Gregory's formula for the subtangent becomes

$$\frac{y}{y'} = \frac{\dfrac{w}{w'}\dfrac{u}{u'}\dfrac{v}{v'}}{\dfrac{w}{w'}\dfrac{v}{v'} + \dfrac{w}{w'}\dfrac{u}{u'} - \dfrac{u}{u'}\dfrac{v}{v'}} = \frac{wuv}{u'wv + v'wu - w'uv},$$

---

[39] Actually, Gregory wrote this as a proportion, that is, as $GB : AB :: CB : OB$.

and then, since $GB \cdot OB = CB \cdot AB$ can be written as $wy = uv$,

$$y' = \frac{w(u'v + v'u) - uvw'}{w\dfrac{uv}{y}} = \frac{w(u'v + v'u) - uvw'}{w^2}.$$

This is exactly the result that we would obtain today using the quotient and product rules to differentiate $y = uv/w$. But of course, this is a modern interpretation.

After setting $FB = d$ and $HB = e$, the last of Gregory's statements, which needs interpretation, is this:

Truly generally, if the ratio GB to CB is the ratio AB to CB multiplied [by itself] in the ratio m to n, in every case it will be

$$DB = \frac{mde - nde}{me - dn}.$$

Here m and n are just positive integers, and the stated hypothesis means that

$$\frac{GB}{CB} = \left(\frac{AB}{CB}\right)^{m/n},$$

which, if we use u, v, and w as in the preceding case, becomes $w/u = (v/u)^{m/n}$. Since

$$DB = \frac{u}{u'}, \quad d = FB = \frac{v}{v'}, \quad \text{and} \quad e = HB = \frac{w}{w'},$$

Gregory's formula for the subtangent DB becomes

$$\frac{u}{u'} = \frac{m\dfrac{v}{v'}\dfrac{w}{w'} - n\dfrac{v}{v'}\dfrac{w}{w'}}{m\dfrac{w}{w'} - \dfrac{v}{v'}n} = \frac{(m - n)vw}{mwv' - vw'n}.$$

It is easy to verify that this is the result obtained using today's rules to differentiate $w/u = (v/u)^{m/n}$ (or, rather, $w^n u^{m-n} = v^m$), so that Gregory could have differentiated a power with an arbitrary rational exponent if he had not been thinking of subtangents.[40]

---

[40] At the end of the explanatory note to this manuscript, Turnbull, who had dated it to November 1670 at the earliest, states that this date "suggests that it was inspired by reading Barrows's book on *Geometry* [see Section 5.5], which reached Gregory during August

Returning to the *Geometriæ*, we find at the outset some work on arc length. The first proposition is just preparatory, but in Proposition 2 [pp. 3–8] Gregory arrived at the result that we know today. Given an arbitrary curve 79CD, which he assumed to be *simplex, seu non sinuosa* (monotonic, and in this case increasing) or else that it must be split into pieces, Gregory started by selecting an arbitrary constant X, which he viewed as the length of a segment. Now let the segment 96 in the next figure[41] be normal to the curve 79CD at an arbitrary abscissa 3, and construct another curve $PNLH$ such that its ordinate at 3 is the product of X and the quotient of the length 96 over the length 93. Then he stated a theorem, proved it, and concluded as follows [p. 8]:

> It is also clear that the [area of the] mixtilinear figure $PNLH\delta2$ is to the [area of the] rectangle X times 2δ, as the [length of the] curve 79CD is to the straight line [segment] 2δ.

Gregory's X is there for reasons of geometric homogeneity, an important concept in his time, but if we take $X = 1$, his statement reduces to

$$\text{length of } \widehat{7D} = \text{area } PH\delta2.$$

We do not present the proof, which is long, involved, and heavily multi-notational, but just interpret it in more familiar terms as follows. Let the arbitrary abscissa at 3 be denoted by x, let the arc 79CD be the graph of a differentiable function f, and let the segment 9B be tangent to the curve at 9. Then, keeping in mind that the numbers here are just labels for points and

---

1670, and which contained a systematic account of such processes of differentiation in a geometrical form." He did not give any supporting reasons. But Gregory himself shed some light on this issue when, in a letter of 5 September 1670, he wrote: "I have read over both Mr Barrows bookes of Lectures, with much pleasure and attention, wherein I find him to have infinitely transcended all that ever writt before him, I have discovered from his method of drawing Tangents, togeather with some of my owne, a generall Geometricall method without Calculation of drawing tangents to all Curves, comprehending not only Mr Barrows particular methods but also his generall Analyticall method in the end of the 10th Lecture, my method contains not above 12 Propositions." So Gregory learned from Barrow but had additional methods of his own. The differentiation of rational powers in Gregory precedes the Barrow book, a book that does not contain the differentiation of a sum or difference, in spite of some exaggerated twentieth-century claims on this issue. More on this in the next section.

[41] This is just an extract from Gregory's own figure. The same is true of all but the next of the figures from the *Geometriæ* reproduced here. Only those features are reproduced that are necessary to illustrate the statements made in this presentation. Additional lines, points, and letters used in omitted proofs have not been included.

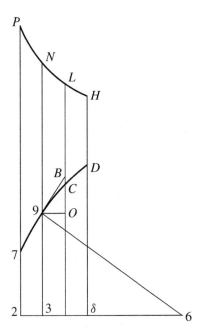

using the similarity of triangles 936 and $9OB$, we see that with $X = 1$ the ordinate of $PNLH$ at x is

$$\frac{96}{93} = \frac{\sqrt{93^2 + 36^2}}{93} = \sqrt{1 + \left(\frac{36}{93}\right)^2} = \sqrt{1 + \left(\frac{BO}{9O}\right)^2}.$$

The last quotient on the right is what we now call $f'(x)$, and then, using our definite integration for area, Gregory's discovery can be expressed in today's familiar form as

$$\text{length of } \widehat{7D} = \int_2^\delta \sqrt{1 + (f')^2}\,.$$

The most notable result involving area and arc length in the *Geometriæ* is Proposition 6 [pp. 17–19]. It is in the form of a problem to be solved, and it can be interpreted, rather than purely translated, with reference to the next figure, in which angles that appear right are indeed right angles. The problem is this: for any given curve BNS, find a curve AKQ such that, for any point I on its "axis" AO, the length of the arc AK is to that of the segment AI as the

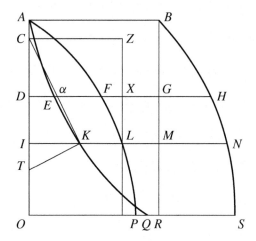

area *ABNI* bound by *BNS* and the axis is to the inscribed rectangle *ABMI*.[42]

To solve the problem, Gregory assumed that the curve *BNS* is *simplex seu non sinuosa*, and then constructed two curves in the following manner:

> Henceforth let the curve *AFLP* be of such nature, that, if any straight line [segment] *IN* is drawn perpendicular to the straight line *AO*, the curve *AFLP* cutting it at *L*, the square of the straight line [segment] *IN* is equal to those of *IL*, *IM* both; then draw the curve *AEKQ* of such nature, that, if any straight line [segment] *IM* is drawn perpendicular to the straight line *AO*, & cutting the curve *AEKQ* at *K* & *AFLP* at *L*, the rectangle *MIK* [43] [that is, the area *MI*×*IK*] is equal to that of the mixtilinear [figure] *IAFL*.

It would be convenient to interpret this in present-day notation before going ahead. Choose *A* as the origin of coordinates; let the *x*-axis contain the segment *AO* with its positive direction downward; let the *y*-axis contain the segment *AB* with the positive direction to the right; let the curves *BNS*, *AFLP*, and *AEKQ* be the graphs of functions that we denote by *f*, *g*, and *h*; and denote the length of *AB* by *a*. We assume, as in Gregory's figure, that *f* is a positive increasing function with a negative second derivative. Then Gregory's curves

---

[42] The Latin original is this: "Invenire curvam [*AKQ*], quæ ad suam axem [*AIO*] eandem habeat rationem, quam figura quælibet exhibita [*ABNSO*] ad rectangulum sibi inscriptum [*ABRO*]."

[43] The original says *MIL*, but this is just a typo that is not repeated in the rest of the proof.

are constructed so that

$$IN^2 = IL^2 + IM^2, \qquad \text{or} \qquad f^2 = g^2 + a^2,$$

and, if the abscissa of I is x,

$$MI \times IK = \text{area } IAFL, \qquad \text{or} \qquad ah(x) = \int_0^x g.$$

Once these curves were constructed, Gregory stated the solution to the problem in the following terms:

> I say that [the area of] the figure $ABSO$ is to [that of] the rectangle $ABRO$ as [the length of] the curve $AEKQ$ is to [that of] the straight line [segment] AO.

With today's terminology this can be proved rather quickly, and not just from A to O but also from A to an arbitrary point I in AO. Indeed, the area of the figure $ABNI$ is

$$\int_0^x f,$$

while the length of the arc AEK is

$$\int_0^x \sqrt{1 + (h')^2} = \int_0^x \sqrt{1 + \frac{g^2}{a^2}} = \int_0^x \sqrt{1 + \frac{f^2 - a^2}{a^2}} = \frac{1}{a} \int_0^x f.$$

Dividing both sides by AI and putting $I = O$ proves Gregory's assertion. But this proof is short and simple only because we have used the full machinery of the calculus as it is known today, including the formula for arc length and the fundamental theorem of the integral calculus to differentiate h. However, this machinery was not available to Gregory, who had to create his own results as he went. As we have seen, he had already obtained the formula for arc length in Proposition 2, and the fundamental theorem of calculus is precisely what the proof of Proposition 6 contains in disguise.

By the use of today's fractions and mathematical symbols, Gregory's proof [pp. 17–19] can be both transcribed and interpreted as follows. Assuming the curve $AEKQ$ to have been constructed, choose a point C in the segment AO, whose abscissa we denote by x_0, such that

$$\frac{IL}{IM} = \frac{IK}{IC}, \qquad \text{or} \qquad \frac{g(x)}{a} = \frac{h(x)}{x - x_0},$$

and draw the segment KC. The heart of the proof lies in showing that this segment is tangent to the curve $AEKQ$ at K, or, since this curve is the graph of the function h, that

$$h'(x) = \frac{h(x)}{x - x_0}.$$

In view of the previous equation, this is equivalent to $h'(x) = g(x)/a$. Recalling the definition of h, we observe that this is the fundamental theorem of calculus.

To prove it, Gregory showed first that KC cannot intersect DH at a point α to the right of E, as shown in the figure. To see this, we start by drawing the rectangle $ILZC$, which intersects DH at X. The equation defining C above can be rewritten as

$$MI \times IK = IL \times IC, \quad \text{or} \quad ah(x) = g(x)(x - x_0).$$

If we denote the abscissa of D by x_1 and note that the construction of the curve $AEKQ$ also gives

$$GD \times DE = \text{area } DAF, \quad \text{or} \quad ah(x_1) = \int_0^{x_1} g,$$

then we have

$$\frac{IK}{DE} = \frac{GD \times IK}{GD \times DE} = \frac{MI \times IK}{\text{area } DAF} = \frac{\text{area } IAFL}{\text{area } DAF}.$$

Since $DE < D\alpha$, using similarity of triangles and the integral representations of $ah(x)$ and $ah(x_1)$ already stated, we can write

$$\frac{\text{area } IAFL}{\text{area } DAF} > \frac{IK}{D\alpha} = \frac{IC}{DC} = \frac{\text{area } IZ}{\text{area } DZ}, \quad \text{or} \quad \frac{\int_0^x g}{\int_0^{x_1} g} > \frac{(x - x_0)g(x)}{(x_1 - x_0)g(x)},$$

which after inverting the fractions gives

$$\frac{\text{area } DAF}{\text{area } IAFL} < \frac{\text{area } DZ}{\text{area } IZ}, \quad \text{or} \quad \frac{\int_0^{x_1} g}{\int_0^x g} < \frac{(x_1 - x_0)g(x)}{(x - x_0)g(x)}.$$

Next, multiplying this inequality by -1, adding 1 to both sides, and simplifying yields

$$\frac{\text{area } IDFL}{\text{area } IAFL} > \frac{\text{area } IX}{\text{area } IZ}, \quad \text{or} \quad \frac{\int_{x_1}^{x} g}{\int_{0}^{x} g} > \frac{(x-x_1)g(x)}{(x-x_0)g(x)}.$$

The denominators of the two fractions in each of these inequalities are the same (once more, by the construction of the curve $AEKQ$ and the definition of C, or by the definition of h and the stated equation $ah(x) = g(x)(x - x_0)$), so we conclude that

$$\text{area } IDFL > \text{area } IX, \quad \text{or} \quad \int_{x_1}^{x} g > (x - x_1)g(x),$$

which is absurd.[44]

Similarly, Gregory showed that KC cannot intersect the curve $AEKQ$ on the arc KQ. Therefore, KC is tangent to $AEKQ$. As we have remarked above, this is a proof of the fundamental theorem of calculus, although it is more than likely that Gregory was not aware of this general relationship between what we now call differentiation and integration.

---

[44] To the best of my knowledge, the *Geometriæ* has not been either reprinted or translated since 1668, so here is a direct translation of the part of the proof transcribed above: "Let K be any point whatever on the curve $AFLP$, through which draw a straight line IN perpendicular to the straight line AO & cutting the curves $AFLP$, BR, $BHNS$, at points L, M, N; so that as IL is to IM so is IK to IC & draw KC: the straight line KC cuts or is tangent to the curve AQ at the point K; if it happens that it cuts in K & therefore falls inside the curve and actually inside the point E towards the vertex A: draw through the point E the straight line DH parallel to the straight line IN, cutting the curves AQ, AP, BR, BS at the points E, F, G, H, & the straight line LC [this is a typo; he means KC] in α, also complete the rectangle $ILZC$, whose side LZ cuts the straight line DH at X. Seeing that IL is to IM as IK is to IC, it will be that the rectangle MIK or the mixtilinear figure [just *mixtilineum* in Gregory, and henceforth just 'figure' here] $IAFL$ is equal to the rectangle IZ; & because the rectangle GDE is equal to the figure DAF, it will be that as IK is to DE so is the figure $IAFL$ to the figure DAF, but IK has a larger ratio to DE than to $D\alpha$; & therefore the figure $IAFL$ has a larger ratio to the figure DAF than IK has to $D\alpha$ or IC to DC; & consequently the figure $IAFL$ has a larger ratio to the figure DAF than the rectangle IZ to the rectangle DZ, & by conversion the ratio of the figure $IAFL$ to the figure $IDFL$ is smaller than the ratio of the rectangle IZ to the rectangle IX, & exchanging, the figure $IAFL$ has a smaller ratio to the rectangle IZ than the figure $IDFL$ to the rectangle IX, whereas the rectangle IZ is equal to the figure $IAFL$, it will follow that the rectangle IX is smaller than the figure $IDFL$, but it is larger, which is absurd."

Once this was established, he drew KT perpendicular to KC, and, by similarity of triangles,

$$\frac{CK}{CI} = \frac{KT}{IK}.$$

The segments IN, IM, and IL also make up a right triangle (since $IN^2 = IL^2 + IM^2$), which is similar to the triangle CIK because of the equation $IL/IM = IK/IC$ used to define C. It follows that

$$\frac{IN}{IM} = \frac{CK}{CI} = \frac{KT}{IK}. \text{[45]}$$

Without further explanations, Gregory concluded by saying that

> since this can be done in the same manner at any point of the curve AQ, it is clear from the above [Proposition] 2 that the straight line AO is to the curve AQ as the rectangle OB is to the figure $ABSO$, which was to be proved.

But a word of explanation is preferable. The conclusion of Proposition 2 on arc length, after replacing $PNLH\delta2$ with $BNIA$, X with IM, 2δ with AI, and the curve $79CD$ with the curve AEK, shows that

$$\frac{\text{area } BNIA}{IM \times AI} = \frac{\text{length of } AEK}{AI},$$

whence

$$\frac{AI}{\text{length of } AEK} = \frac{IM \times AI}{\text{area } ABNI} = \frac{\text{area } IB}{\text{area } ABNI}.$$

Gregory's conclusion follows by taking $I = O$, $K = Q$, and $N = S$.

Gregory did not have what we now call the fundamental theorem of calculus when he started his proof of Proposition 6, but he got it and he used it when he needed it, and he was the first to do so.

Proposition 11 may be another first, because, when generously interpreted in today's terms, it happens to be the first use of the method of substitution to evaluate an integral. Gregory's statement is as follows [p. 27]:

PROP. 11. THEOREM.

Consider any mixtilinear space $ABKI$ comprised between a curve BK, a straight line AI & two parallel straight lines BA, KI; and let the curve MY

---

[45] For the reason stated before, here is a direct translation of what Gregory actually wrote: "Let the perpendicular KT to the straight line CK meet the straight line AO at T; it is clear that CI is to CK as IK is to KT; but on the other hand CI is to CK as MI is to NI, because the straight lines IN, IM, IL, make a right triangle similar to the triangle CIK, whose sides IM, IN, are homologous to the sides CI, CK; & hence as IK is to KT so is IM to IN;"

be of such nature that (having taken any point C on the curve BK & from

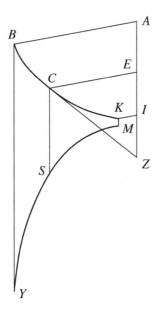

this drawn a straight line CE parallel to the straight line AB, and the straight line CZ tangent to the curve BK and ending on the straight line AI, if extended to Z) the segment EZ is always equal to the segment CS parallel to the straight line AZ & ending on the curve YM. I say that the figure BKMY, comprised between the curves BK, MY, & the straight lines BY, KM, parallel to the straight line AZ, is equal to the mixtilinear figure BAIK.

Gregory gave a geometric proof, based on the use of Proposition 10, that we do not reproduce. What Proposition 11 does is to reduce the computation of the area between the curve BCK and the segment AEI to the computation of the area between the curves BCK and MSY. To express this in present-day notation, consider the particular case in which the segments BA, CE, and KI are perpendicular to AZ. Then choose coordinate axes with origin at I, the positive x-axis along IA and the positive y-axis along IK. Let the curve BCK be the graph of a differentiable and increasing function f whose inverse is denoted by g, denote the abscissa of A by a, and let the coordinates of an arbitrary point C on the curve be (x, y). Then

$$\text{area } ABKI = \int_0^a f(x)\,dx.$$

Now, the area between the curves BCK and MSY is the same as the area under a curve whose ordinate at each arbitrary point y from $f(0)$ to $f(a)$ is CS. Since

$$g'(y) = \frac{CS}{f(x)} = \frac{CS}{f(g(y))},$$

it follows that, under appropriate hypotheses,

$$\text{area } BKMY = \int_{f(0)}^{f(a)} CS \, dy = \int_{f(0)}^{f(a)} f(g(y))g'(y) \, dy.$$

Thus Gregory's proposition, that area $ABKI = $ area $BKMY$, can be rewritten as

$$\int_0^a f(x) \, dx = \int_{f(0)}^{f(a)} f(g(y))g'(y) \, dy.$$

This is integration by a change of variable.

Propositions 7 and 11 were not to remain dead ends. Gregory used both in proving the following important result [pp. 102–103], which we shall interpret after the proof:

PROP. 54. THEOREM.

Consider the parallelogram $ABIK$, and let the curve ACI be of such nature, that (having drawn any straight line [segment] DE parallel & equal to the straight line [segment] AB cutting the curve at C) the ratio AB to EC is the product of the ratio AK to AE [by itself] in the ratio of P to Q. I say that [the area of] the parallelogram $ABIK$ is to [that of] the figure $ACIK$ as $P + Q$ is to Q.

The statement about the ratio AB to EC means that

$$\frac{AB}{EC} = \left(\frac{AK}{AE}\right)^{P/Q},$$

where P and Q are just positive integers.

To prove the proposition, Gregory drew the tangent to ACI at C until it intersected the extension of AE at F. Then he drew the segment CH parallel and equal in length to EF. If this is done for every arbitrary point C on ACI, a new curve AHG results. Then Gregory wrote: "it is clear from [Proposition] 7 above that the straight line [segment] AE is to FE, or LC is to HC as P is to Q," which we shall accept with the meaning that he had figured this out,

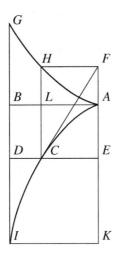

in one of his famous pieces of scrap paper, by the same method used in the example of Proposition 7. But if this ratio is valid for an arbitrary C on ACI, then

> it is evident that the figure $ACIB$ is to the figure $ACIGH$ as P is to Q; but from [Proposition] 11 above the figure $ACIGH$ is equal to the figure $ACIK$ & therefore the figure $ACIB$ is to the figure $ACIK$ as P is to Q, & *componendo* the parallelogram $ABIK$ is to the figure $ACIK$ as $P + Q$ is to Q *quod demonstrandum erat.*

The following is a transcription of this statement into more modern language:

$$\frac{\text{area } ACIB}{\text{area } ACIGH} = \frac{P}{Q},$$

from which it follows by Proposition 11 (with ACI in the role of BCK and the points F and H in place of Z and S) that area $ACIGH$ = area $ACIK$, with the result that

$$\frac{\text{area } ACIB}{\text{area } ACIK} = \frac{P}{Q}.$$

Adding 1 to both sides and simplifying, we arrive at

$$\frac{\text{area } ABIK}{\text{area } ACIK} = \frac{P + Q}{Q},$$

which was to be proved.

To interpret this proposition in current terminology, assume that the parallelogram $ABIK$ is a rectangle (it is shown as such in Gregory's own figure), and choose coordinate axes with origin at A, the positive x-axis along AK and the positive y-axis along AB. Let ACI be the graph of a differentiable function f, and let $AE = x$, $EC = y = f(x)$, $AK = a$, and $KI = AB = b$. Then, if we use lowercase letters for the integers P and Q, Gregory's hypothesis on the ratio AB/EC becomes

$$\frac{b}{f(x)} = \left(\frac{a}{x}\right)^{p/q},$$

so the equation of the curve ACI is

$$f(x) = b\left(\frac{x}{a}\right)^{p/q}.$$

Then

$$\text{area } ACIK = \int_0^a f = \frac{b}{a^{p/q}} \int_0^a x^{p/q}\, dx,$$

while area $ABIK = ab$. Gregory's proposition, rewritten as

$$\text{area } ACIK = \frac{q}{p+q}\text{area } ABIK,$$

then translates to

$$\int_0^a x^{p/q}\, dx = \frac{a^{\frac{p}{q}+1}}{\frac{p}{q}+1}$$

after some obvious simplification. Thus, what Gregory effectively did was to integrate a rational power of x.

However, Fermat had already shown this. What was new with Gregory is the statement he made in his proof about the use of Proposition 7, enclosed in quotation marks above. It can be restated as

$$\frac{x}{FE} = \frac{LC}{HC} = \frac{p}{q},$$

so that $FE = qx/p$. But $f'(x) = CE/FE = f(x)/FE$, and then

$$f'(x) = \frac{p}{qx}f(x) = \frac{p}{qx}b\left(\frac{x}{a}\right)^{p/q} = \frac{p}{q}b\left(\frac{x}{a}\right)^{\frac{p}{q}-1}\frac{1}{a},$$

which amounts to differentiating a rational power of x.

It may be a matter of wonder to us, a painful experience too, that these familiar results were obtained by the painstaking application of classical geometry. But in the absence of the now familiar concept of limit, at a time when what we know now had yet to be discovered, how else could anyone deal with problems about tangents and areas but by resorting to the well-founded Euclidean geometry? There will be an answer to this question, but not quite yet.

Without some necessary algebrization, Gregory's published results lack the agility and the power to be called a calculus. But we should remember those higher-order "derivatives" that he computed to obtain the infinite series expansions described in Chapter 4. They suggest that he may have created such a calculus, but, if so, it never saw the light through timely publication.

5.5 BARROW'S GEOMETRIC CALCULUS

In 1669 Isaac Barrow (1630–1677), the first Lucasian Professor of Mathematics at Trinity College, Cambridge, resigned his position and recommended one of his former students, the 27-year-old Isaac Newton, as his successor. With Newton's assistance, Barrow had prepared the publication of his mathematical swan song, the *Lectiones geometricæ*, which appeared the following year.[46]

To anyone perusing this volume for the first time it might appear as a collection of theorems on geometry, perhaps not a particularly interesting one at that.[47] But a closer examination reveals that quite a number of the propositions are about tangents, quadratures, and related topics. Cloaked in very thick geometric garb, practically unrecognizable, these are well-known results of the calculus. The first of these in the *Lectiones* that is significant is the following proposition from Lecture VIII [p. 64; 91]:

V. Let VEI be a straight line,[48] and let two curves YFN, ZGO be related to one another, so that if any straight line EFG is parallel to a straight line

---

[46] *Lectiones geometricæ; in quibus (præsertim) generalia curvarum linearum symptomata declarantur*, 1670; 2nd. ed. 1674. Abridged translation by Child as *The geometrical lectures of Isaac Barrow*, 1916. Page references are to the 1674 edition and to Child's translation, in this order and separated by a semicolon. The translations included here are modifications of those by Child, except for the one of *Examp*. III, which is my own.

[47] Child, the translator, in conveying his own first impression on page 28 of the Introduction, stated that the results presented by Barrow "seemed to be more or less a haphazard grouping, in which one proposition did occasionally lead to another; but certain of the more difficult constructions were apparently without any hint from the preceding propositions."

[48] Child's translation has TEI and V is not shown in his figure.

ISAAC BARROW
Portrait by Isaac Whood
Engraving by Benjamin Holl

AB [not shown] given in position, the intercepts EG, EF are always in the same ratio to one another; while the straight line TG touches one curve ZGO

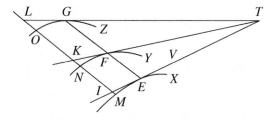

in G (and comes together with the straight line VE at T) then TF, extended, will also touch the other curve YFN.

In short, if the quotient EG/EF is constant regardless of the position of the line EFG, as long as it is parallel to AB, and if TG is tangent to ZGO at G,

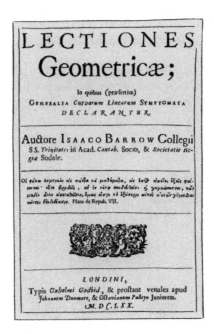

then *TF* is tangent to *YFN* at *F*. Barrow gave a simple geometric proof of this fact, but it is not the proof that is of interest to us.[49]

What matters is what this means, so rotate the figure so that the line *VEI* is horizontal and take it to be the *x*-axis with the positive direction to the left. Now consider the particular case in which the line *AB* is vertical and take it to be the *y*-axis with the positive direction upward. If the curves *ZOG* and *YFN* are denoted, in function notation, by $y = f(x)$ and $y = g(x)$, then the hypothesis that *EG/EF* equals some constant *k* means that $f = kg$, and then

$$f'(x) = \frac{EG}{TE} = k\frac{EF}{TE} = kg'(x)$$

for any *x*. Thus, in our language, Barrow has shown that if $f = kg$ then $f' = kg'$, which is one of the theorems of the calculus.[50]

---

[49] Since it depends on a previous proposition it would take a little too much space to reproduce here. But those readers who would check with the original sources should be aware of the fact that both Barrow's original and Child's translation contain typos (not the same), but they are easy to identify and correct.

[50] This is one of the "processes of differentiation in a geometrical form" that Turnbull

Here is another theorem from Lecture VIII that has the distinction of being both nontrivial and easy to explain [pp. 65; 93–94]:

IX. Let VD be a straight line, and XEM, YFN two curves so related that, if any straight line EDF is freely drawn parallel to a line given in position, the *rectangle* from DE, DF is always equal to any the same space [area];

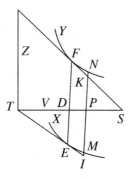

also the straight line ET touches the curve XEM at E, and concurs with VD in T; then, if we take $DS = DT$; & FS is joined; [FS] will touch the curve YFN at F.

The hypothesis is that the curves are related in such a manner that the product $DE \times DF$ is constant regardless of the location of the line EDF. In other words, if IN is any line drawn parallel to EF and if it intersects the curves at M and N, then $DE \times DF = PM \times PN$. The complete proof is as follows [pp. 65–66; 94]:

---

thought might have inspired Gregory. This is quite possible, but Child's assertion on page 31 of the Introduction to his translation of Barrow's *Lectiones*, to the effect that this proposition gives the derivative of a sum, seems to be an exaggeration. At the end of the translation of Lecture VIII there is a note by Child in which he admitted that Barrow omitted a theorem, then Child himself stated and proved that theorem in the style of Barrow [p. 99], and this is the equivalent of the differentiation of a sum. Then he expressed his belief that [p. 100] "the omission of the theorem was intentional;" stating that "Barrow may have thought it evident, ... but I prefer to think that he considered it as a corollary of the theorem of §V;" that is, the one just presented. Then Child showed how such a corollary can be obtained, but since there is no figure and it is not clear what points are represented by his letters a and b, I have been unable to understand it. In any event, it is clear that Child was passionate about Barrow's work, and eager to justify the opening assertion of his preface, that [p. xiii] "Isaac Barrow was the first inventor of the Infinitesimal Calculus."

However, after the work of Viète, Descartes, and others, mathematics had already developed as a fertile soil for the ideas of the calculus. It is not surprising that they were developed by several men of genius almost simultaneously. It is neither fair nor necessary to attempt to give all the credit to a single individual.

For [let] any [straight line] IN be drawn parallel to EF; cutting the given lines as shown. Then [51]

$$\frac{TP}{PM} > \left(\frac{TP}{PI} = \right) \frac{TD}{DE} \qquad \text{and also} \qquad \frac{SP}{PK} = \frac{SD}{DF}.$$

Therefore

$$\frac{TP \times SP}{PM \times PK} > \frac{TD \times SD}{DE \times DF} = \frac{TD \times SD}{PM \times PN}.$$

But $TD \times SD > TP \times SP$ [by the Euclidean proposition quoted here on page 241, since D is the midpoint of TS]; hence all the more,

$$\frac{TD \times SD}{PM \times PK} > \frac{TD \times SD}{PM \times PN}.$$

Therefore,

$$PM \times PK < PM \times PN; \qquad \text{or} \qquad PK < PN.$$

Thus the whole line FS lies outside the curve YFN.

In other words, what Barrow is trying to say is that if the curves are related as stated before the proof, and if ET is tangent to XEM at E, then FS is tangent to YFN at F.

What does this really mean? Is it a result that we should be excited about? To better explain it, we start by reconstructing the figure with some rectangular axes, as is the custom now, and then we shall use coordinate and function notation again. Take the line VD to be the x-axis with the positive direction to the right and let the line given in position be a vertical y-axis with the positive direction upward. Let the negative part of the y-axis represent a positive z-axis. If the curves YFN and XEM are then denoted by $y = f(x)$ and $z = g(x)$, respectively, then the fact that they are related by $DE \times DF = k$, where k is a positive constant, means that (referring to the figure) $y = k/z$. The slopes of the tangents to the graphs of g and f at E and at F are, in the arrangement shown in the figure and using the hypothesis that $TD = DS$ and current notation,

$$z' = \frac{DE}{TD} \qquad \text{and} \qquad y' = -\frac{DF}{DS} = -\frac{DF}{TD}.$$

---

[51] I am using modern fractions and symbols. Barrow wrote $TP.PM \;\sqsubset\; (TP.PI::)$ $TD.DE$ and so on.

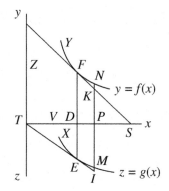

Then

$$\frac{y'}{z'} = -\frac{DF}{DE} = -\frac{DE \times DF}{DE^2} = -\frac{k}{z^2},$$

so that

$$y' = -\frac{k}{z^2}z'.$$

Thus, what Barrow essentially did is to find the derivative of $y = k/z$ geometrically, a task for which we use the chain rule today.

Immediately after the proof he considered the particular case in which the curve XEM is a straight line [p. 66; 94]:

> *Note.* If the line XEM were a straight line (and so coincident with TEI) YFN would be the ordinary *hyperbola*, whose centre is T, one *asymptote* TS, and the other TZ parallel to EF.

This statement may not be a model of clarity (in particular, TZ is not parallel to EF in Barrow's own figure), but here is what it means in today's terminology. Since no mention has been made about the location of the vertical axis, we can place it at will, as we did in the last figure, and assume that the equation of the line XEM is $z = mx$, and then $y = k/(mx)$, the equation of an ordinary hyperbola. In this case, $z' = m$ and

$$y' = -\frac{k}{z^2}z' = -\frac{k}{m^2x^2}m = -\frac{k}{mx^2}.$$

Using similar techniques, Barrow established many results in what we now call differentiation in Lectures VIII and IX.[52] However, his heavy geometric

---

[52] For the complete list see Child, *The geometrical lectures of Isaac Barrow*, pp. 30–31, but take it with a grain of salt.

approach almost hides this fact and has left many people wondering whether he really knew what he was doing. Barrow himself stated at the start of Lecture VI that the last six lectures would be partly [p. 45; 66]

> *about an investigation of tangents, without the trouble or annoyance of cal-culation ... partly about the ready determination of the dimensions of many magnitudes, with the help of designated tangents ...*

Since that is what the last six lectures actually delivered, Barrow must have known what he was doing.

But then Barrow diverged from his self-imposed method of strict geomet-ric constructions when, at the end of Lecture X [pp. 80–81; 119–120], he presented an alternative, and more algebraic

> method for finding tangents by Calculation commonly used by us. Although I hardly know, after so many well-known and well-worn methods of the kind above, whether there is any advantage in doing so. Yet I do so on the advice of a Friend;[53] and all the more willingly, because it seems to me more profitable and general than those which I have discussed. Accordingly, I proceed as follows.
>
> Let AP, PM be two straight lines given in position (of which PM cuts a given curve in M) & let MT be supposed to touch the curve at M, and to cut

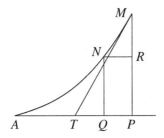

> the straight line at T; in order to find the quantity of the straight line PT; I set off an indefinitely small arc, MN, of the curve; then I draw straight lines NQ parallel to MP & NR to AP; I call $MP = m$; $PT = t$; $MR = a$; $NR = e$;

---

[53] In the draft of a letter found by Louis More in the *Portsmouth collection*, Newton stated: "A paper of mine gave occasion to Dr. Barrow [who] showed me his method of tangents before he inserted it into his 10th Geometrical Lecture. For I am that friend which he then mentioned." More, *Isaac Newton. A biography*, 1934; reprinted by Dover Publications, 1962, p. 185, n. 35.

and the other straight lines, determined by the special nature of the curve, useful for the matter in hand, I also designate by name; also I compare MR, NR (& through them MP, PT) with one another by means of an *equation* obtained by Calculation; meantime observing the following rules.

1. In the calculation, I omit all terms containing a power of a or e, or products of these (for these terms have no value).

2. After *the equation has been formed*, I reject all terms consisting of letters denoting known or determined quantities; or terms which do not contain a or e (for these terms brought over to one side of the equation, will always be equal to zero).

3. I substitute m (or MP) for a, and t (or PT) for e. Hence at length the quantity of PT is found.

Rule 3 may profit from a bit of explanation. Since the "triangles" NMR and TMP are (approximately) similar, we can write

$$\frac{m}{a} = \frac{MP}{MR} = \frac{TP}{NR} = \frac{t}{e},$$

and then whatever equation there is at the start of Rule 3 can be multiplied by $m/a = t/e$.

There is no geometric proof of this alternative method. This is an algorithm, much in the manner of Fermat's *Methodus* and not too dissimilar. Barrow was, in all likelihood, not directly acquainted with Fermat's work because his name is not mentioned in the *Lectiones*. But he mentioned those of Gregory [p. 89; 131] and Huygens [p. 94; 141], both of whom had used Fermat's method. So there is a distinct possibility of an indirect influence of Fermat on Barrow. But Barrow's method is an improvement over Fermat's *Methodus* because the use of two small quantities, a and e, instead of Fermat's E, makes Barrow's procedure better suited for applications to implicit functions. Of this alternative method Barrow gave five examples, of which we present the third, which he labeled in the margin *La Galande*, that is, the folium of Descartes. It is possible to quote it in its entirety, with some explanatory notes in brackets [pp. 82–83; not in Child]:

Examp. III.

Let AZ be a straight line given in position, & AX a magnitude [a given constant]; and also let AMO be a *curve* such that, when any straight line MP whatever is drawn normal to AZ, it will be AP *cub.* $+ PM$ *cub.* $=$

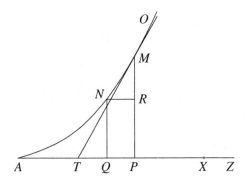

$AX \times AP \times PM$ [that is, $x^3 + y^3 = bxy$, where b is a constant].

Set $AX = b$; & $AP = f$; then $AQ = f - e$; & AQ cub. $= f^3 - 3ffe$ [omitting terms that contain powers of e, according to Rule 1]; & QN cub. $[= PR^3 = (m-a)^3] = m^3 - 3mma$ [Rule 1]. & $AQ \times QN = fm - fa - me + ae = fm - fa - me$ [Rule 1]; from which $AX \times AQ \times QN = bfm - bfa - bme$; hence the equation $f^3 - 3ffe + m^3 - 3mma = bfm - bfa - bme$; or removing terms to be rejected [by Rule 2, since $f^3 + m^3 = bfm$], $bfa - 3mma = 3ffe - bme$; substituting [Rule 3] $bfm - 3m^3 = 3fft - bmt$; or,

$$\frac{bfm - 3m^3}{3ff - bm} = t.$$

This solution can now be compared with the one given by Fermat.

5.6 FROM TANGENTS TO QUADRATURES

Barrow gave the following important result in Lecture X [pp. 78; 116–117], which, with reference to the next figure,[54] can be translated as follows:

XI. Let ZGE be any line [curve], whose axis is VD;[55] to which apply first perpendiculars (VZ, PG, DE) starting with VZ and continually increasing [in length]; also let VIF be a line such that, if any straight line EDF is drawn perpendicular to VD (cutting the *curves* at the points E, F,

---

[54] I have omitted a portion of Barrow's Figure 109. The curve $ZGEG$ branches out, at least in the 1674 edition, about midway between Z and G. One branch goes left to Z, as shown, but the other goes to V. As much as is possible to judge from a reproduction, this omitted branch looks like a later addition to the original drawing.

[55] Child's translation has A in place of V. Another translation can be found in Struik, *A source book in mathematics, 1200–1800*, pp. 255–256.

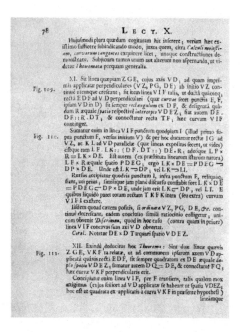

Theorem XI in the *Geometrical lectures.*

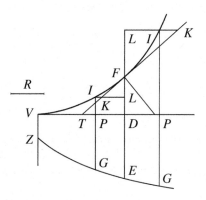

and VD in D) the *rectangle* formed by DF, & a certain designated [length] R is always equal to the *intercepted space VDEZ*; also let $DE/DF = R/DT$ [$DE.DF :: R.DT$ in Barrow's own notation]; & join the straight line TF; then it will touch the curve VIF.

As usual, this needs interpretation, which we provide in three ways. First, we split the figure in two and flip the bottom curve vertically; second, we eliminate some of the clutter in the figure and in the statement of the theorem;

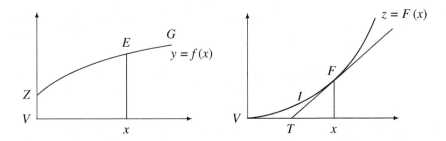

and third, we introduce some current notation as shown in the new figures. With the aid of these, in which we have taken the liberty to replace Barrow's D with x, we can restate the result as follows:

XI. *Let the curve ZGE represent an increasing function f; also let F be a function with graph VIF such that, for some constant R and at any x, the product $RF(x)$ is equal to the area under the graph of f from V to x; and select T on the Vx axis such that*

$$\frac{xE}{xF} = \frac{R}{xT}.$$

Then TF is tangent to $y = F(x)$ at $(x, F(x))$.

Of course, some obvious conditions must be imposed on the curve. Omitting the proof, which is not our concern at this point, this means that the slope of the tangent to F at x is

$$F'(x) = \frac{xF}{xT} = \frac{xE}{R} = \frac{f(x)}{R}.$$

The conclusion to be drawn from this is that, if $R = 1$, Barrow's Theorem XI is the Fundamental Theorem of Calculus in the following version: if F is a function such that

$$F(x) = \int_{V}^{x} f(t)\, dt,$$

then $F'(x) = f(x)$.

Lecture XII includes several results on quadratures and rectification, including one that is of particular interest because it may be the first of its kind
that can be expressed as integration by parts. It is a result that is part of a
collection of propositions in Appendix I, which Barrow included *"to please
a friend* who thinks them worth the trouble" [p. 110; 165].[56] It begins with
a general foreword, from which—as well as from Barrow's figure—we make
the following extract [pp. 110–111; 165–167]:

> Let *ACB* be a *Quadrant of a circle*, and let *AH*, *BG* be tangents to it; & in
> ... *AC* produced take ... *CE* ... equal to the *radius CA*; ... and let the
> hyperbola *LEO* through *E* have asymptotes *BC*, *BG*. Also let an *arbitrary
> point M* be taken in the arc *AB*, and through it draw *CMS* (meeting the
> tangent *AH* in *S*) ... and *MPL* parallel to *AC*.

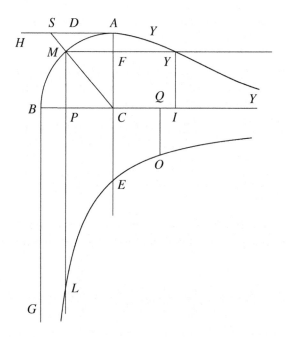

After the general introduction, Barrow gave a collection of propositions,
the fifth of which begins as follows:

---

[56] This friend was John Collins, as he revealed himself in a letter to Oldenburg of 30
September 1676.

V. Take $CQ = CP$; and draw QO parallel to CE (meeting the *hyperbola LEO* in O) ...

With this preparatory work, we are ready for the proposition that concerns us [p. 112; 167].

VIII. Let the curve AYY be such that FY is equal to AS; draw next a straight line YI parallel to AC, the *space ACIYYA* (that is, the *sum* of the *Tangents* belonging to the *arc AM*, & applied to the straight line AC, together with the *rectangle FCIY*) is equal to *half the hyperbolic space PLOQ*.

Literally, this says that the sum of the area under an arc of the graph of a tangent function (the curve AYY, whether it looks it or not) and the area of a certain rectangle equals half the area under some arc of a hyperbola. This literal interpretation may not inspire rave reviews at first reading, but a second examination using today's tools of the trade will reveal something that may be considered curious and even interesting.

For convenience, we take the radius of the quadrant ACB to be 1, denote the angle ACS by θ_0, choose C as the origin of rectangular coordinates, assume that CA is located on the positive x-axis (directed upward) and that CB is located on the positive y-axis (directed to the left). Then we have

$$CF = \cos\theta_0, \quad CA = 1, \quad \text{and} \quad FY = -\tan\theta_0.$$

Now, if D is an arbitrary point on the segment AS, denote the angle ACD by θ. Then the curve AYY is the graph of $y = -\tan\theta$, and the area of the space $AFYYA$ is

$$\int_{\cos\theta_0}^{1} [0 - (-\tan\theta)]\, dx = -\int_{1}^{\cos\theta_0} \tan\theta\, dx.$$

But we are mixing our variables here. It should be graphically clear that $x = \cos\theta$, and then the last integral becomes

$$-\int_{0}^{\theta_0} \tan\theta\, d(\cos\theta).$$

Next, the area of the rectangle $FCIY$ is

$$FY \times FC = \tan\theta_0 \cos\theta_0.$$

Finally, since $CP = \sin\theta_0$, then $CQ = -\sin\theta_0$, and if we take the equation of the hyperbola LEO to be

$$x = \frac{1}{y-1},$$

then the area of the space $PLOQ$ is

$$\int_{-\sin\theta_0}^{\sin\theta_0}\left(0 - \frac{1}{y-1}\right)dy = \int_{-\sin\theta_0}^{\sin\theta_0}\frac{1}{1-y}\,dy.$$

But $y = \sin\theta$, and then the area of the space $PLOQ$ can be expressed by the integral

$$\int_{-\theta_0}^{\theta_0}\frac{1}{1-\sin\theta}\cos\theta\,d\theta = \int_{-\theta_0}^{\theta_0}\cos\theta\,\frac{1+\sin\theta}{1-\sin^2\theta}\,d\theta$$

$$= \int_{-\theta_0}^{\theta_0}\cos\theta\,\frac{1}{\cos^2\theta}\,d\theta + \int_{-\theta_0}^{\theta_0}\frac{\sin\theta}{\cos\theta}\,d\theta.$$

The second integrand is odd and the integral vanishes. The first integrand is even and the integral equals

$$2\int_0^{\theta_0}\cos\theta\,d(\tan\theta).$$

Therefore, Barrow's result can be written as

$$-\int_0^{\theta_0}\tan\theta\,d(\cos\theta) + \tan\theta_0\cos\theta_0 = \int_0^{\theta_0}\cos\theta\,d(\tan\theta),$$

and this is integration by parts.

   Barrow's results are about tangents and quadratures. But the subject that he put together with such care and completeness was so heavily cloaked that major doses of skill and will would have been necessary to penetrate the geometric shields erected by the author. Barrow may have played the role of a moth attracted by the brilliance and elegance of the prestigious synthetic geometry of Euclid, only to get burned in the process.

   At this time Barrow left mathematics for good to devote himself to the study of theology, and it was as a Doctor of Divinity that he became Master of Trinity College in 1672.

5.7 NEWTON'S METHOD OF INFINITE SERIES

During the plague years Newton worked extensively on what would later be called the calculus, summing up his discoveries in a manuscript that has been called the October 1666 Tract, a 30-page manuscript that bears this date.[57] It was at this very early stage that he discovered the rules of differentiation and some methods of quadrature, at least for some particular kinds of what we now call functions.

For instance, in a manuscript written about 1665 he found the subnormal at a point on a curve that we would express by the equation

$$x = \frac{p(y)}{q(y)},$$

where p and q are polynomials in y. Newton called the subnormal v, and gave it in the form of a quotient whose numerator and denominator are to be computed according to the following rule [p. 323]:[58]

> *Rule* 2^d. If y is in y^e rationall denom: of x consisting of many termes [our $q(y)$], for the y^e Numerat: in y^e valor [value] of v multiply y by y^e denom of x.squared. for y^e denominat^r mu[l]tiply y^e Num: of x according to its dimensions [see below for an explanation of these words] & y^e product by y^e denom:. againe multipl y^e Denom: according to its dim: & y^e product by y^e numer^r & substract y^e less from y^e greater & divide y^e diff by y.

This says that the numerator of the expression giving v should be $y[q(y)]^2$. Next, the "Num: of x according to its dimensions" means that each term of $p(y)$ must be multiplied by the corresponding exponent of y. For instance, if $p(y) = 6y^3 - 5y^5$, to write it "according to its dimensions" means to write $18y^3 - 25y^5$. In general, the "Num: of x according to its dimensions" is what we would write as $yp'(y)$. Thus, Newton's rule can be written in the form

$$v = \frac{y[q(y)]^2}{\dfrac{yp'(y)q(y) - yq'(y)p(y)}{y}}.$$

---

[57] First published by Hall and Hall in *Unpublished scientific papers of Sir Isaac Newton. A selection from the Portsmouth Collection in the University Library, Cambridge,* 1962, pp. 15–64 = Whiteside, *The mathematical papers of Isaac Newton,* I, 1967, pp. 400–448.

[58] The portion of the manuscript that contains the work quoted here is reproduced in Whiteside, *The mathematical papers of Isaac Newton,* I, pp. 322–341. Page references are to this edition.

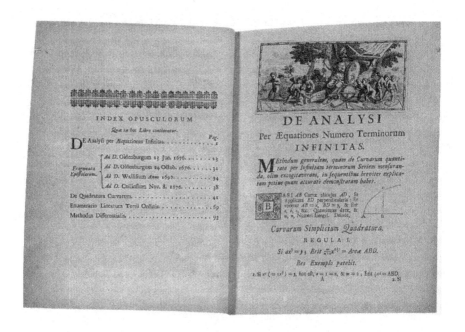

First edition of Newton's *De analysi*.
Reproduced from the virtual exhibition *El legado de las matemáticas*:
de Euclides a Newton, los genios a través de sus libros, Sevilla, 2000.

It is not difficult to draw a figure of a curve, its normal at a point, and a right triangle with legs v and y whose hypotenuse is an appropriate segment of the stated normal. Then, using current notation, it is easy to see that $v = y(dy/dx)$. It follows from the previous equation that

$$\frac{dx}{dy} = \frac{1}{\dfrac{dy}{dx}} = \frac{p'(y)q(y) - q'(y)p(y)}{[q(y)]^2}.$$

This makes it clear that Newton had discovered a quotient rule for the subnormal in the case of polynomials at a very early stage in his work on what we now call the calculus.

Later, after the appearance of Mercator's *Logarithmotechnia* and the commentaries with improvements by Wallis and Gregory, and fearing that others "would find out the rest before I could reach a ripe age for writing" (letter to Oldenburg of 24 October 1676) and would publish before him, Newton quickly put down in writing all his discoveries on this subject in a manuscript entitled *De analysi per æquationes numero terminorum infinitas*, probably his most famous mathematical work. It was circulated to a restricted circle in manuscript form in 1669, but remained unpublished until 1711. It may be fair to state that working in the setting of strict Euclidean geometry was not Newton's cup of tea at this moment. He much preferred to obtain and express his results through the use of infinite series—power series to be specific—one term at a time. To deal with the general term, and referring to the figure on the first page of the Latin edition, reproduced below (in which $AB = x$), Newton established the following Rule I (*Regula* I):[59]

If $ax^{\frac{m}{n}} = y$, it shall be $\dfrac{an}{m+n}x^{\frac{m+n}{n}} = $ Area ABD.

He saved the demonstration for the end of the paper [pp.19–20; 243–245],[60] giving it with the aid of the next figure, in which

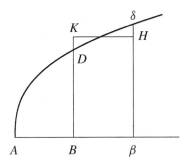

the base $AB = x$, the perpendicular ordinate $BD = y$, & area $ABD = z$, as before. Similarly let $B\beta = o$, $BK = v$, & the rectangle $B\beta HK$ (ov) equal to the space $B\beta\delta D$.

It is therefore $A\beta = x + o$, & $A\delta\beta = z + ov$.

Then he made an assumption that would be crucial in his argument:

[59] This is an elaboration of Proposition 8 of the October 1666 Tract. See Whiteside, *The mathematical papers of Isaac Newton*, I, pp. 403–404.

[60] Page references given here are to Jones' edition in *Analysis per quantitatum series, fluxiones, ac differentias: cum enumeratione linearum tertii ordinis*, 1711, and, after a semicolon, to Whiteside's translation in *The mathematical papers of Isaac Newton*, II, 1968.

If we now suppose $B\beta$ to be infinitely small, or o to be zero, v & y will be equal ...[61]

Then he gave a small example before starting the proof proper in reverse, by taking as a hypothesis the conclusion in the statement of Rule I:

Or in general if

$$\frac{n}{m+n}\, ax^{\frac{m+n}{n}} = z;$$

or, putting $\dfrac{na}{m+n} = c$ & $m + n = p$, if $cx^{\frac{p}{n}} = z$, or $c^n x^p = z^n$; then substituting $x + o$ for x & $z + ov$ (or, equivalently, $z + oy$) for z produces

$$c^n \text{ times } x^p + pox^{p-1}, \&\text{c.} = z^n + noyz^{n-1}, \&\text{c},$$

omitting of course the remaining terms which would ultimately vanish.

That is, if x is replaced by $x + o$, then the area under the curve increases from $z = ABD$ to $z + ov$ (see figure), which is approximately equal to $z + oy$ since o is very small. Thus, Newton replaced x with $x + o$ and z with $z + oy$ in $c^n x^p = z^n$, raised both sides to the nth power, expanded these powers, and announced his intention to omit the terms represented by "&c" because they contain negligible powers of o. Then he continued as follows:

Now removing the equal terms $c^n x^p$ & z^n, and dividing the rest by o, there remains

$$c^n p x^{p-1} = nyz^{n-1} \left(= \frac{nyz^n}{z} \right) = \frac{nyc^n x^p}{cx^{\frac{p}{n}}}.$$

Or, dividing by $c^n x^p$, it shall be

$$px^{-1} = \frac{ny}{cx^{\frac{p}{n}}}. \qquad \text{or} \qquad pcx^{\frac{p-n}{n}} = ny;$$

or, restoring $\dfrac{na}{m+n}$ for c & $m + n$ for p, that is, m for $p - n$ & na for pc, it shall be $ax^{\frac{m}{n}} = y$. Conversely therefore, if $ax^{\frac{m}{n}} = y$, it will make

$$\frac{n}{m+n}\, ax^{\frac{m+n}{n}} = z.$$

Q.E.D.

---

[61] In the manuscript this statement reads "Si jam supponamus $B\beta$ esse infinite parvam, sive o esse nihil, erunt v & y æquales, ... " in Whiteside, *The mathematical papers of Isaac Newton*, II, p. 242, but it is slightly different in Jones' edition, p. 20. It is not correct to assume that Newton took over the o notation from Gregory, for he had already used it in his October 1666 tract on fluxions. See Whiteside, *The mathematical papers of Isaac Newton*, I, pp. 414–415. Note, in particular, his statement: "And those terms are infinitely little in wch o is."

If we were to interpret this in modern terms, we would write $f(x) = c^n x^p = z^n$, which after incrementing becomes

$$f(x + o) = f(x) + noyz^{n-1} \text{ \&c.}$$

Subtracting $f(x)$ from both sides and dividing by o, we obtain

$$\frac{f(x + o) - f(x)}{o} = nyz^{n-1} \text{ \&c,}$$

and omitting the terms represented by "&c" because they would ultimately vanish is equivalent to taking the limit as $o \to 0$. In short,

$$y = \frac{z}{nz^n} \lim_{o \to 0} \frac{f(x + o) - f(x)}{o} = \frac{cx^{\frac{p}{n}}}{nc^n x^p} c^n px^{p-1} = \frac{c}{n} px^{\frac{p}{n}-1} = ax^{\frac{m}{n}}.$$

Newton gave six examples of Rule I before stating Rule II [p. 2; 209], to the effect that if y is the sum of terms of the kind stated in Rule I, even if they are infinite in number, then the area is the sum of the areas corresponding to the individual terms. At that time, Newton must have thought that this was evident and in no need of proof. After several examples, he stated the third and last Rule [pp. 5; 211–213], basically asserting that if y "be more compounded than the foregoing [more than the sum of terms of the given kind], it is to be reduced to simpler terms." He considered several possibilities and dealt with them by example. In this way, he found the area under $y = aa/(b+x)$ by long division, obtaining an infinite series for the right-hand side and performing the quadrature term by term; for the curve $y = \sqrt{aa + xx}$ the binomial theorem gives a series expansion for the right-hand side; and in the more complex example

$$y = \frac{\sqrt{1 + ax^2}}{\sqrt{1 - bx^2}}$$

both the binomial theorem and long division would bring home the bacon (the division is simpler if both numerator and denominator are multiplied first by the denominator) [pp. 5–8; 213–219].

To pinpoint Newton's contribution to the subject at this stage, it is clear that the method of using an infinitely small quantity o to perform what we recognize today as a differentiation was not new with Newton. In fact, we have seen the work of Fermat, which is quite similar. The sequence of steps by each of these two authors is as follows: Fermat—who used the letter E for the increment of the variable—adequates, removes common terms, removes higher powers of E, divides by E, and gets the solution; Newton equates, removes higher powers of o, removes equal terms, divides by o, and gets the solution. Gregory did much the same thing, other than adding the twist of introducing rectangles in his procedure. This priority was acknowledged by Newton, who admitted having obtained the idea for the method from Fermat and also mentioned the work of Gregory and Barrow.[62]

Newton's originality at this point consists in transforming a procedure known to work in isolated cases into a calculus of greater generality. While previous authors limited themselves to finding the tangent to one particular curve at a time, Newton was interested in a general method. And because of his use of infinite series, his method applies to functions with radicals, and that was new. Furthermore, a most important aspect of Newton's originality consists in the interplay between what we now call differentiation and integration, as we have seen in his proof of Rule I. Note how he found the function by differentiating the area under its graph from the origin to a variable point. That is, Newton was using what eventually became the fundamental theorem of calculus "and by applying it to abstract equations, directly and invertedly," he made it widely applicable. Gregory's and Barrow's books also contain many results that, if properly interpreted, turn out to be theorems of analysis, including the fundamental theorem in both works (which Gregory applied "directly and invertedly" in the proof of his Proposition 54). But the heavy geometric flavor of their presentation—almost incomprehensible today in Barrow's case—greatly limited the usefulness and applicability of their results. In Newton's hands, this new subject became a practical calculus.

5.8 NEWTON'S METHOD OF FLUXIONS

Newton would reelaborate and amplify his view of the calculus in subsequent works, reinterpreting it first as the fluxional calculus, in a manuscript possibly entitled (the first folium of the autograph is missing) *Tractatus de methodis*

---

[62] In the draft of a letter, now in the University Library, Cambridge (Add. 3968.30, 441^r), Newton stated: "I had the hint of this method from Fermat's way of drawing tangents and by applying it to abstract equations, directly and invertedly, I made it general. Mr. Gregory and Dr. Barrow used and improved the same method in drawing tangents." See More, *Isaac Newton. A biography*, p. 185, n. 35. However, Whiteside, in *The mathematical papers of Isaac Newton*, I, p. 149, argues with good reason that "Newton had no direct knowledge of Fermat's work but knew [only] the outlines of his tangent-method." It is difficult to say, today, what Newton knew or did not know, but he claimed only to have had the hint of the method from Fermat and, as the French say, *à bon entendeur, demi mot.*

serierum et fluxionum (Treatise on the method of fluxions and infinite series).
Newton tried to publish it beginning in 1671, but was unsuccessful. After a
number of pages of preliminaries, in which Newton elaborated on such topics
as long division of polynomials, he got around to considering equations in
which some quantities are constant "to be looked on as known and determined
and are designated by the initial letters *a*, *b*, *c* and so on," [p. 73] [63] and others
are variable:

> I will hereafter call them fluents and designate them by the final letters *v*, *x*,
> *y* and *z*. And the speeds with which they each flow and are increased by their
> generating motion (which I might readily call fluxions or simply speeds) I
> will designate by the letters *l*, *m*, *n* and *r*. Namely, for the speed of the
> quantity *v* I shall put *l*, and so for the speeds of the other quantities I shall
> put *m*, *n* and *r* respectively. [64]

It would be difficult to remember that the speeds (fluxions) at which *v*, *x*, *y*,
and *z* flow are designated by the letters *l*, *m*, *n*, and *r*. Newton would change
the notation twenty years later when, in December 1691, he denoted those
fluxions by $\dot{v}$, $\dot{x}$, $\dot{y}$, and $\dot{z}$. [65] This is particularly appealing if we think of the
fluents as varying with time and of the dot as indicating derivative with respect
to time, and we shall use this notation in our presentation even if it was 20
years before its time in 1671.

   Then Newton posed the first of twelve problems to be discussed in this
new tract: [66] *Given the relation of the flowing quantities to one another, to
determine the relation of the fluxions.* Keep in mind while reading his method
that it refers to an equation of the form now written as $F(x, y) = 0$ or a
higher-dimensional equivalent [p. 75]:

> Arrange the equation by which the given relation is expressed according to
> the dimensions of some flowing quantity, say *x* [that is, rearrange $F(x, y)$ in

---

[63] Page references are to and translations from Whiteside, *The mathematical papers of
Isaac Newton*, III, 1969, pp. 32–353.

[64] Three centuries before, Richard Suiseth had already used the words *fluxus* and *fluens*
in this connection in his *Liber calculationum*, f. 9^r, col. 2 and f. 75^v, col. 1.

[65] This notation appeared first in print in the portion of a new tract, *De quadratura curvarum*, that was included as an appendix to Wallis' *De algebra tractatus; historicus &
practicus* of 1693 (see below, pages 312 and 315, for details). Jones adopted these "dotted"
fluxions in his 1711 printing of *De quadratura curvarum*, as did Colson in his translation of
the fluxonian tract, so the practice of replacing Newton's original fluxions with their dotted
equivalents dates back to the early eighteenth century.

[66] Its notable length is due to the presentation of many examples, the subdivision of some
of the problems into a number of subcases, and the consideration of related questions.

order of decreasing powers of x], and multiply its terms by any arithmetical progression [not any; he multiplied the terms by 0, 1, 2, 3, ... from right to left; that is, he multiplied each by the exponent of x] and then by $\dot{x}/x$. Carry out this operation separately for each of the fluent quantities and then put the sum of all the products equal to nothing, and you have the desired equation.[67]

To show how this is done he provided five examples, of which we select two. The first and simplest is as follows:

If the relation of the quantities x and y be $x^3 - ax^2 + axy - y^3 = 0$, I multiply the terms arranged first according to x and then to y in this way.

Multiply	$x^3 - ax^2 + axy - y^3$			Mult.	$-y^3 + axy - ax^2 + x^3$		
by	$\dfrac{3\dot{x}}{x}$.	$\dfrac{2\dot{x}}{x}$.	$\dfrac{\dot{x}}{x}$. 0	by	$\dfrac{3\dot{y}}{y}$.	$\dfrac{\dot{y}}{y}$.	0
there comes	$3\dot{x}x^2 - 2\dot{x}ax + \dot{x}ay$		*.	comes	$-3\dot{y}y^2 + a\dot{y}x$		*.

And the sum of the products is $3\dot{x}x^2 - 2a\dot{x}x + a\dot{x}y - 3\dot{y}y^2 + a\dot{y}x = 0$, an equation which gives the relation between the fluxions $\dot{x}$ and $\dot{y}$. Precisely, should you assume x arbitrarily the equation $x^3 - ax^2 + axy - y^3 = 0$ will give y [not necessarily; it may not be solvable for y], and with these determined it will be

$$\frac{\dot{x}}{\dot{y}} = \frac{3y^2 - ax}{3x^2 - 2ax + ay}.$$

This procedure resembles what we now call implicit differentiation. More complicated cases involving products, quotients, and roots can be solved by additionally using substitutions [p. 77]:

Example 4. If $x^3 - ay^2 + \dfrac{by^3}{a+y} - x^2\sqrt{ay + x^2}$ expresses the relation between x and y, I put z for $by^3/(a+y)$ and v for $x^2\sqrt{ay + x^2}$ and thence obtain the three equations

$$x^3 - ay^2 + z - v = 0, \quad az + yz - by^3 = 0, \quad \& \quad ax^4y + x^6 - v^2 = 0.$$

[67] This is an elaboration of Proposition 7 of the October 1666 Tract. See Whiteside, *The mathematical papers of Isaac Newton*, I, p. 402.

The first gives $3\dot{x}x^2 - 2a\dot{y}y + \dot{z} - \dot{v} = 0$,[68] the second $a\dot{z} + \dot{z}y + \dot{y}z$
$- 3b\dot{y}y^2 = 0$ and the third $4a\dot{x}x^3y + 6\dot{x}x^5 + a\dot{y}x^4 - 2\dot{v}v = 0$ for the
relations of the speeds $\dot{v}, \dot{x}, \dot{y}$ and $\dot{z}$.

After this, all he had to do was to solve the last equation for $\dot{v}$, the preceding
one for $\dot{z}$, and take the results to $3\dot{x}x^2 - 2a\dot{y}y + \dot{z} - \dot{v} = 0$. Then, if the values
of z and v are restored,

there comes the equation sought

$$3\dot{x}x^2 - 2a\dot{y}y + \frac{3ab\dot{y}y^2 + 2b\dot{y}y^3}{a^2 + 2ay + y^2} - \frac{4a\dot{x}xy + 6\dot{x}x^3 + a\dot{y}x^2}{2\sqrt{ay + x^2}} = 0$$

by which the relation of the speeds $\dot{x}$ and $\dot{y}$ is designated.

Newton took this example and the idea of substitution from the October 1666
Tract.[69] This combination of the original method and substitution, which in
this context is equivalent to an application of the chain rule, is quite powerful, and permits what we now call implicit differentiation of very involved
equations.

After his five examples Newton provided a demonstration of his method.
He chose the letter o to denote an infinitely small amount of time, in which x
increases by $\dot{x}o$ and y increases by $\dot{y}o$ (Newton called these increments "moments"), and then he reasoned as follows [p. 80]:

Given therefore any equation $x^3 - ax^2 + axy - y^3 = 0$, and substitute $x + \dot{x}o$
for x and $y + \dot{y}o$ for y, and there will emerge

$$\left.\begin{array}{l} x^3 + 3\dot{x}ox^2 + 3\dot{x}^2o^2x + \dot{x}^3o^3 \\ - ax^2 - 2a\dot{x}ox - a\dot{x}^2o^2 \\ + axy + a\dot{x}oy + a\dot{y}ox + a\dot{x}\dot{y}o^2 \\ - y^3 - 3\dot{y}oy^2 - 3\dot{y}^2o^2y - \dot{y}^3o^3 \end{array}\right\} = 0.$$

---

[68] As in the first example; that is,

$$\begin{array}{l|l|l|l} x^3 - ay^2 + z - v & -ay^2 + x^3 + z - v & -v + x^3 - ay^2 + z & z + x^3 - ay^2 - v \\ \dfrac{3\dot{x}}{x}. \quad 0 & \dfrac{2\dot{y}}{y}. \quad 0 & \dfrac{\dot{v}}{v}. \quad 0 & \dfrac{\dot{z}}{z}. \quad 0 \\ 3\dot{x}x^2 \quad *. & -2a\dot{y}y^2 \quad *. & -\dot{v} \quad *. & \dot{z} \quad *. \end{array}$$

and so on for the others.

[69] Whiteside, *The mathematical papers of Isaac Newton*, I, pp. 411–412.

Now by hypothesis $x^3 - ax^2 + axy - y^3 = 0$, and when these terms are deleted and the rest divided by o there will remain

$$3\dot{x}x^2 + 3\dot{x}^2ox + \dot{x}^3o^2 - 2a\dot{x}x - a\dot{x}^2o + a\dot{x}y$$
$$+ a\dot{y}x + a\dot{x}\dot{y}o - 3\dot{y}y^2 - 3\dot{y}^2oy - \dot{y}^3o^2 = 0.$$

And since above o is supposed to be infinitely small, so that it may express the moments of quantities, the terms that are multiplied by it will be worth nothing in respect to the others. I therefore reject them and there remains $3\dot{x}x^2 - 2a\dot{x}x + a\dot{x}y + a\dot{y}x - 3\dot{y}y^2 = 0$, as in Example 1 above.[70]

With this already settled (although present-day readers may have trouble accepting this as a demonstration), Newton was ready for Problem 3: *To determine maxima and minima*, which he solved as follows [p. 117]:

When a quantity is greatest or least, at that moment its flow neither increases nor decreases. For if it increases, that proves that it was less and will at once be greater than it now is, and conversely so if it decreases. Therefore seek its fluxion by Problem 1 and set it equal to nothing.

EXAMPLE 1. If the greatest value of x in the equation

$$x^3 - ax^2 + axy - y^3 = 0$$

be desired, seek the fluxions of the quantities x and y and there will come $3\dot{x}x^2 - 2a\dot{x}x + a\dot{x}y - 3\dot{y}y^2 + a\dot{y}x = 0$. Then when $\dot{x}$ is set equal to zero, there will remain $-3\dot{y}y^2 + a\dot{y}x = 0$ or $3y^2 = ax$. With the help of this you might eliminate one or the other of x and y in the primary equation, and by the resulting equation determine the other, and then both by $-3y^2 + ax = 0$.

Problem 4: *To draw tangents to curves.* This is more complicated because there was no single way of describing curves in Newton's times. Or, as he put it, "Tangents are drawn in various ways according to the various relationships of curves to straight lines" [p. 121]. Thus, he considered nine modes and gave various examples of each. Mode 1 is presented by means of the following figure, in which $AB = x$ and the ordinate $BD = y$ moves "through an indefinitely small space to the position bd ... " [p. 123].

---

[70] This is the same method already used in the October 1666 Tract to prove Proposition 7 using the example $x^3 - abx + a^3 - dyy = 0$. See Whiteside, *The mathematical papers of Isaac Newton*, I, p. 414. Thus, we see that this tract contained the germ of Newton's future work on the calculus.

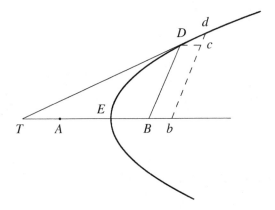

Now let Dd be extended till it meets AB in T: this will then touch the curve in D or d and the triangles dcD, DBT will be similar, so that

$$\frac{TB}{BD} = \frac{Dc}{cd}.$$

BD is known from the equation of the curve, and the fraction on the right is the ratio of the moment $\dot{x}o$ of x over the moment $\dot{y}o$ of y, which is equal to the ratio of their fluxions. Then the preceding equation is solved for TB. As a first example Newton returned to the curve $x^3 - ax^2 + axy - y^3 = 0$, for which he had already computed the fluxional ratio

$$\frac{\dot{x}}{\dot{y}} = \frac{3y^2 - ax}{3x^2 - 2ax + ay},$$

and then

$$BT = \frac{3y^3 - axy}{3x^2 - 2ax + ay}.$$

Problems 5 and 6 of this tract on fluxions are devoted to curvature questions, Problems 7 to 9 are on quadratures, and Problems 10 to 12 refer to rectification. From all these we select two examples on quadratures. Newton began his discussion by referring to a figure like the one below, in which the perpendicular AC is 1, and denoted AB by x and the area ADB under the curve by z [pp. 195–197]. If the segment BED moves a little to the right while remaining perpendicular to AB, the increments in the areas $AB \times BE = x$ and $ABD = z$ are a narrow rectangle and a narrow quasi-rectangle with heights BE and BD. Therefore, these increments are (approximately) in the ratio BE/BD.

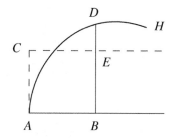

Since the increments in x and z in a small amount of time are $\dot{x}o$ and $\dot{z}o$, we obtain the basic formula

$$\frac{\dot{x}}{\dot{z}} = \frac{BE}{BD} = \frac{1}{BD},$$

"so that when we set $\dot{x}$ equal to unity, $BD = \dot{z}$." Thus, if BD is given by the equation of the curve we have $\dot{z}$, and then it is possible, at least in principle, to find z from $\dot{z}$.

Newton found an interesting use for this formula in Problem 8, where the following task is posed. Consider two curves FDH and GEI, as shown in the next figure, and establish the following notation: $AB = x$, $BD = v$, area

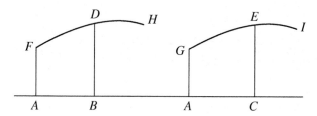

$AFDB = s$, $AC = z$, $CE = y$, and area $AGEC = t$. If the curve FDH is given and a relationship between z and x is also given, find a second curve GEI such that the areas s and t are equal.

To solve the problem we use the previous equation, which in the notation for the two new curves becomes

$$\frac{\dot{x}}{\dot{s}} = \frac{1}{v} \qquad \text{and} \qquad \frac{\dot{z}}{\dot{t}} = \frac{1}{y}.$$

Newton said [p. 199]:

Hence if, as above, we assume $\dot{x} = 1$, and $v = \dot{s}$, it will be $\dot{z}y = \dot{t}$ and thence
$$\frac{\dot{t}}{\dot{z}} = y.$$

A little while later he continued:

By hypothesis therefore $s = t$ and thence $\dot{s} = \dot{t} = v$, and

$$y = \left(\frac{\dot{t}}{\dot{z}} = \right) \frac{v}{\dot{z}}.$$

His first example was the curve $ax - xx = vv$ (from now on we write squares in the current fashion) and the relationship between the variables was given by $ax = z^2$. From Problem 1, $a = 2\dot{z}z$, and then

$$y = \frac{v}{\dot{z}} = \frac{2vz}{a}.$$

Since substituting $x = z^2/a$ in $ax - x^2 = v^2$ and simplifying gives

$$v = \frac{z}{a}\sqrt{a^2 - z^2},$$

we obtain

$$\frac{2z^2}{a^2}\sqrt{a^2 - z^2} = y.$$

This is the "equation of the curve whose area equals the area of the circle."

It looks like the solution to a beautiful game, but there is a bit more to it than that. If those two curves enclose the same area, if A is the origin, and if we select $b > 0$, what we have in today's notation is

$$\int_0^b \sqrt{ax - x^2}\, dx = \int_0^{\sqrt{ab}} \frac{2z^2}{a^2}\sqrt{a^2 - z^2}\, dz.$$

More generally, if we write $f(x) = \sqrt{ax - x^2}$ and $x = g(z) = z^2/a$, then, under appropriate hypotheses on g,

$$\int_0^b f(x)\, dx = \int_0^{g^{-1}(b)} f(g(z))g'(z)\, dz.$$

This is integration by substitution.

As a first example of Problem 9, *To determine the area of any proposed curve*, Newton considered the case of the hyperbola $BD = x^2/a$, where a is a constant [p. 211]. Then, from the basic formula obtained before Problem 8,

$$\dot{z} = BD\dot{x} = \frac{\dot{x}x^2}{a},$$

"and there will (by Prob 2) emerge $x^3/3a = z$." For us this is just integration. For Newton, since Problem 2, *From the relation of the fluxions determine the relation of the fluents*, is the converse of Problem 1, "it ought to be resolved the contrary way" [p. 83], so let the following serve as an example of his contrary way:

I divide	$\dfrac{\dot{x}x^2}{a}\,[\,+0\dot{x}x + 0\dot{x}\,]$	I divide	$\dot{z}$
by $\dfrac{\dot{x}}{x}$ making	$\dfrac{x^3}{a}$	by $\dfrac{\dot{z}}{z}$ making	z
Then I divide by	3. 2. 1.	Then by	1.
making	$\dfrac{x^3}{3a}$	making	z.

The total $x^3/3a - z = 0$ will be the desired relationship of x and z.

This procedure requires what we now call indefinite integration, which is easy for polynomials. But Newton was well aware, way before we were, of the fact that this problem "cannot always be resolved by this practice" [p. 85].

After this, Newton assumed the validity of Rules 1 and 2 already presented in *De analysi* and relied on the method of infinite series to perform what we would call term-by-term integration in numerous examples.

It should be said that our selections from the tract on series and fluxions do not do justice to the power of Newton's methods. Newton's paper contains a wealth of modes, cases, and examples that show the full range of his methods, plus an extensive table of areas (table of integrals) [pp. 236–241 and 244–255] in the part of the *Tractatus* devoted to quadratures.

It is clear now that Newton's work represents an essential change, having replaced the geometric approach of his predecessors with an algebraic one. This represents a totally new general approach and supersedes the consideration of particular cases.

5.9 WAS NEWTON'S TANGENT METHOD ORIGINAL?

When Newton provided his demonstration of the rule for "implicit differentiation" as applied to the equation $x^3 - ax^2 + axy - y^3 = 0$ (page 297) he did not give any credit to Barrow, but this is exactly the method of tangents that Barrow showed Newton "before he inserted it into his 10th Geometrical Lecture." Referring back to this method and to the figure on page 281, if the coordinates of N are x and y, then the coordinates of M are $x + e$ and $y + a$.

Since the point M is also on the given curve, substituting $x + e$ for x and $y + a$ for y, "there will emerge" the same equation that Newton obtained but with e in place of $\dot{x}o$ and a in place of $\dot{y}o$. By Barrow's Rule 1, all terms containing powers of a or e are to be omitted, which is equivalent to rejecting the terms that still contain o after division by o. By Barrow's Rule 2, terms that do not contain a or e are to be rejected, which is equivalent to deleting the terms $x^3 - ax^2 + axy - y^3$. Thus Barrow's method and Newton's basic method are one and the same.

On this, Newton had the following comment in a letter to Collins of 10 December 1672:[71]

> I remember I once occasionally told Dr. Barrow when he was about to publish his Lectures that I had such a method of drawing Tangents but some divertisement or other hindered me from describing it to him.

But we don't have to take this statement as proof that Newton already "had such a method of drawing Tangents" when he assisted Barrow in the publication of the *Lectiones*. In a manuscript written in September 1664,[72] after moving all the terms of an equation "to one side soe y^t it be $= 0$," he wrote the following explanation of his method:

> Multiply each terme of y^e equat: by so many units as x hath dimensions in y^t terme [that is, multiply by the exponent of x], divide it by x & multiply it by y for a Numerator. Againe multiply each terme of y^e equation by soe many units as y hath dimensions in each terme & divide it by $-y$ for a denom: in the valor of v.[73]

He gave three examples of this method, the third of which is this:

And if $x^4 - yyxx + aayx - y^4 = 0$. then $\dfrac{4yx^3 - 2y^3x + aayy}{4y^3 - aax + 2yxx} = v.$

It is easy to carry out his instructions to obtain this result and to verify it using today's implicit differentiation.

---

[71] Turnbull, *The correspondence of Isaac Newton*, I, 1959, p. 248.

[72] Whiteside, *The mathematical papers of Isaac Newton*, I, pp. 236–238.

[73] By v, which Newton defined via a figure that is unnecessary to us, he meant what we now would write as yy'. Thus, the last statement in the quotation means that the quotient so constructed equals yy'.

a of the method of tangents for what we call "implicit functions"
ɔ the mind of many in the second half of the seventeenth century.
Newton were not the first nor would they be the last. The Dutch
mathematician Johann Hudde had already described such a "General Rule" to
Frans van Schooten, in a letter in Dutch of 21 November 1659:[74]

Arrange all the terms of the equation that expresses the nature of the curve,
in such a manner that they are = 0 and remove from this equation all the
fractions that have x or y in their denominators. Multiply the terms of the
highest order in y by an arbitrarily chosen number, or even by 0, and multiply
the term of the next highest order in y, by the same number minus one unit, and
continue in the same manner with all other terms of the equation. Similarly
multiply by an arbitrarily chosen number or by 0 the term of the highest order
in x: the term of the next highest order in x, must be multiplied by the same
number minus one unit, and the same with the others. When the first of these
products is divided by the second, the quotient multiplied by $-x$ is AC.[75]

Hudde's only example was this (the asterisk is just a place holder for a nonex-
istent term in y):

Let the equation that expresses the nature of the curve be

$$ay^3 + xy^3 + b^2y^2 - x^2y^2 - \frac{x^3}{2a}y^2 * + 2x^4 - 4ab^3 = 0$$

1. Multip. by 1. $+1$.   0.   0.   0. -1. -2. -2.
2. Multip. by 0. $+1$.   0. $+2$. $+3$. $+4$.   0.

1. Yields $ay^3 + xy^3$ $-4x^4 + 8ab^3$

2. Yields $+xy^3$ $-2x^2y^2 - \frac{3x^2}{2a}y^2 + 8x^4$

consequently, $AC = \dfrac{ay^3 + xy^3 - 4x^4 + 8ab^3}{xy^3 - 2x^2y^2 - \frac{3x^2}{2a}y^2 + 8x^4}$ times $-x$.[76]

---

[74] It was later translated into French as "Extrait d'une lettre du feu M. Hudde à M. van
Schooten, Professeur en Mathématiques à Leyde. Du 21. Novembre 1659," 1713. My
English translations are from Gerhardt's reprint.

[75] As in Newton's case, Hudde included a figure, and the segment AC is what we would
recognize today as y/y'.

[76] This is the result as given by Hudde, but the industrious reader who wants to use

It is easy to observe, and Hudde was quite clear on this point and naturally proud of it, that the progression of multipliers used in his method is completely general in the selection of the starting number (on the right of the steps labeled "Multip." above he wrote "or by any other Arith. progr."). This makes the method simple to use because one can arrange matters so that the multiplier of the power with the largest number of terms is zero:

> Which has been seen in the preceding example, in which 0 has been placed first under y^2 and, second, under the terms that do not contain x.

But it was René François Walther de Sluse (1622–1685), from Liège, who developed this method first, according to the following statement by John Collins in a letter to Newton of 18 June 1673:[77]

> As to *Slusius Method of Tangents* consider this: it was well understood by him when he published his book *de Mesolabio*.[78] But he did not want to make it public then because he was unwilling to prevent his Friend Angelo Riccio.[79]

By 1671 Sluse had been advised that Ricci was otherwise occupied and no further mathematical output was to be expected from him. Moreover, Sluse himself had "fallen upon a very easy method of proving things" regarding the method of tangents. Both of these statements are contained in a letter from Sluse to Oldenburg of 17 December 1671. He announced then his intention to publish it the next year, and in another letter of 17 January 1673 he sent it

---

Newton's method or today's implicit differentiation to verify that they lead to the same conclusion may get in a spot of trouble. To avoid it, it must be observed that the equation of the curve gives

$$8ab^3 - 4x^4 = 2ay^3 + 2xy^3 + 2b^2y^2 - 2x^2y^2 - \frac{x^3}{a}y^3,$$

so that the last numerator becomes

$$3ay^3 + 3xy^3 + 2b^2y^2 - 2x^2y^2 - \frac{x^3}{a}y^3.$$

Then the three methods lead to the same result.

[77] Turnbull, *The correspondence of Isaac Newton*, I, p. 288.

[78] *Mesolabum et problemata solida*, 1659.

[79] Michelangelo Ricci was a friend and pupil of Torricelli, who published 19 pages of mathematics on maxima and minima as *Michaelis Angeli Riccii geometrica exercitatio*, 1666.

to Oldenburg for submission to the Royal Society. In this letter Sluse, who wrote y for x and v for y, gave the following general rule:

1. Drop from the equation those parts [terms], which do not contain either y or v, placing on one side all the terms containing y, & on the other those that contain v, with their signs $+$ or $-$. For the sake of simplicity, we call this the right-, and that the left-hand side.

2. On the right-hand side, set in front of each term the exponent to which v is raised, or, what is the same, extend the terms in that way.

3. Do the same on the left-hand side, but preceding each term with the Exponent of the power of y. But & this is done throughout: One of the y's in each term is changed into an a [the notation he used for the subtangent, or v/v'].

On the next page Sluse gave several examples of the application of this rule, from which we select that of the curve $y^5 + by^4 = 2qqv^3 - yyv^3$, which fully shows the procedure. Step 1 leads to the equation $y^5 + by^4 + v^3yy = 2qqv^3 - yyv^3$. Note that the term yyv^3 must be included in both sides because it contains both y and v. Multiplying each term by the exponent of y on the left-hand side and by the exponent of v on the right-hand side, the equation becomes $5y^5 + 4by^4 + 2v^3yy = 6qqv^3 - 3yyv^3$. Finally, changing one y in each term on the left-hand side into a we obtain $5y^4a + 4by^3a + 2v^3ya = 6qqv^3 - 3yyv^3$, and

$$a = \frac{6qqv^3 - 3yyv^3}{5y^4 + 4by^3 + 2v^3y}.$$

Sluse got this far, but if we remember that $a = v/v'$ we can write

$$v' = \frac{5y^4 + 4by^3 + 2v^3y}{6q^2v^2 - 3y^2v^2}.$$

The same result is obtained by any of the other methods.

In conclusion, Barrow, Newton, Hudde, and Sluse independently developed what is essentially the same method of tangents. But Newton's method was the most powerful because, by his use of substitutions, he could deal with radicals.

5.10 NEWTON'S FIRST AND LAST RATIOS

Newton's creativity with respect to the calculus had probably reached its peak with the preceding works, but he had published nothing. Then his attention turned to matters of natural philosophy (which we call physics today), and in 1687 he published his immortal book, the one that made Newton the figure that he is: *Philosophiæ naturalis principia mathematica.*[80]

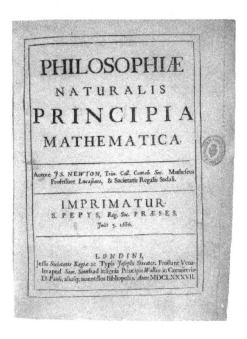

Reproduced from the virtual exhibition *El legado de las matemáticas*: *de Euclides a Newton, los genios a través de sus libros*, Sevilla, 2000.

While the main thrust of this book is about physical matters, it touches upon some of the concepts of the calculus, but in a very different language from the one he had used before. The first publication of Newton's calculus may have been a disappointment to those who already knew his work, if they understood at all that he was talking about that subject. There are, in his new presentation, no maxima and minima, no implicit differentiation, no infinite series, no rectification, little of what his close colleagues had come to expect and admire.

---

[80] Translated into English by Andrew Motte as *The Mathematical principles of natural philosophy*, 1729. Page references are to this translation, of which there are recent reprints.

The new approach hits the reader with the force of unexplained language, for the title of Book I, Section I is [p. 41] *Of the method of first and last ratio's of quantities, by the help whereof we demonstrate the propositions that follow.* Of these we present a selection. The first one is preliminary and quite general [pp. 41–42]:

LEMMA I.

Quantities, and the ratio's of quantities, which in any finite time converge continually to equality, and before the end of that time approach nearer the one to the other than by any given difference, become ultimately equal.

The second proposition is about quadratures and we include the proof [pp. 42–43]:

LEMMA II.

If in any figure AacE terminated by the right lines Aa, AE, and the curve acE,

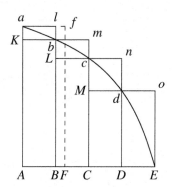

there be inscrib'd any number of parallelograms Ab, Bc, Cd, &c. comprehended under equal bases AB, BC, CD, &c. and the sides Bb, Cc, Dd, &c. parallel to one side Aa of the figure; and the parallelograms aKbl, bLcm, cMdn, &c. are completed. Then if the breadth of those parallelograms be suppos'd to be diminished, and their number to be augmented in infinitum: *I say that the ultimate ratio's which the inscrib'd figure AKbLcMdD, the circumscrib'd figure AalbmcndoE, and curvilinear figure AabcdE, will have to one another, are ratio's of equality.*

For the difference of the inscrib'd and circumscrib'd figures is the sum

of the parallelograms *Kl*, *Lm*, *Mn*, *Do*, that is, (from the equality of all
their bases) the rectangle under one of their bases *Kb* and the sum of their
altitudes *Aa*, that is, the rectangle *ABla*. But this rectangle, because its
breadth *AB* is suppos'd diminished *in infinitum*, becomes less than any given
space. And therefore (by Lem. I) the figures inscribed and circumscribed become ultimately equal one to the other; and much more will the intermediate
curvilinear figure be ultimately equal to either. *Q.E.D.*

The rectangles in Lemma II are assumed to be "comprehended under equal
bases," but this restriction was lifted in the following [p. 43]:

LEMMA III.

*The same ultimate ratio's are also ratio's of equality, when the breadths AB,
BC, CD, &c. of the parallelograms are unequal, and all are diminished, in
infinitum.*

For suppose *AF* equal to the greatest breadth, and compleat the parallelogram *FAaf*. This parallelogram will be greater than the difference
of the inscrib'd and circumscribed figures; but, because its breadth *AF* is
diminished *in infinitum*, it will become less than any given rectangle. *Q.E.D.*

From this result, Newton was able to obtain four corollaries.

Cor. I. Hence the ultimate sum of those evanescent parallelograms will
in all parts coincide with the curvilinear figure.

Cor. 2. Much more will the rectilinear figure, comprehended under the
chords of the evanescent arcs *ab*, *bc*, *cd*, &c. ultimately coincide with the
curvilinear figure.

Cor. 3. And also the circumscrib'd curvilinear figure comprehended under the tangents of the same arcs.

Cor. 4. And therefore these ultimate figures (as to their perimeters *acE*,)
are not rectilinear, but curvilinear limits of rectilinear figures.

The remaining selections are on tangents [pp. 45–47] and use the following
figure.[81]

---

[81] Newton's own figure contains additional letters, segments, and one arc that are of
interest in the proof of Lemma VIII, a result not discussed here.

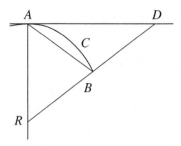

LEMMA VI.

If any arc ACB given in position is subtended by its chord AB, and in any point A in the middle of the continued curvature, is touch'd by a right line AD, produced both ways; then if the points A and B approach one another and meet,[82] *I say the angle BAD, contained between the chord and the tangent, will be diminished* in infinitum, *and ultimately will vanish.*

LEMMA VII.

The same things being supposed; I say, that the ultimate ratio of the arc, chord, and tangent [the segment AD], *any one to any other, is the ratio of equality.*

We could think of a rectangular coordinate system in which the coordinates of A are (x, y) and those of B are $(x + \Delta x, y + \Delta y)$. The slope of the chord AB is $\Delta y / \Delta x$ and, since the ultimate ratio of the segments AB and AD is the ratio of equality, the slope of the tangent AD is the last ratio of the quantities Δy and Δx.

At the end of Section I there is a Scholium in which Newton states why he chose to give geometric proofs, which was basically to conform to tradition and avoid criticism, although he did not put it that way. Then he anticipated an objection that undoubtedly would arise, as it did [pp. 54–55]:

Perhaps it may be objected, that there is no ultimate proportion of evanescent quantities; because the proportion, before the quantities have vanished, is not the ultimate, and when they are vanished, is none.

and explained that [p. 55]

---

[82] Imagine the segment RBD decreasing in length and approaching RA as R is kept fixed and B and D approach A. In this approach, B remains on the fixed arc AC.

by the ultimate ratio of evanescent quantities is to be understood the ratio of the quantities, not before they vanish, nor afterwards, but with which they vanish.

To clinch it, near the end of the Scholium he added [p. 56]

For those ultimate ratio's with which quantities vanish, are not truly the ratio's of ultimate quantities, but limits towards which the ratio's of quantities, decreasing without limit, do always converge; ...

No, he did not have a definition of limit, but he clearly had the idea and his language in this statement is the one we use today. In this respect he was closer than his predecessors and contemporaries, and closer than most of his successors for over a century, to the modern definition of derivative in terms of limits.

Then in Section II of Book II, *On the motion of bodies*, he inserted the following [vol. II, p. 17]:

Lemma II.

The moment of any Genitum *is equal to the moments of each of the generating sides drawn into the indices of the powers of those sides, and into their coefficients continually.*

This is far from clear. What is a Genitum, to start with? Newton explained [pp. 17–18]:

Quantities of this kind are products, quotients, roots, rectangles, squares, cubes, square and cubic sides, and the like.

Indeed, that is what the lemma proper does, to evaluate what we call the derivatives [moments] of products, powers, roots, quotients, and the like.

Wherefore the sense of the Lemma is, that if the moments of any quantities A, B, C; &c. increasing or decreasing by a perpetual flux, or the velocities of the mutations which are proportional to them, be called a, b, c, &c. the moment or mutation of the generated rectangle AB will be $aB + BA$; ... in general, that the moment of any power $A^{\frac{n}{m}}$ will be $\frac{n}{m}aA^{\frac{n-m}{m}}$.

5.11 NEWTON'S LAST VERSION OF THE CALCULUS

In 1691, about to turn 49, Newton was ready to collect his thoughts on the calculus and give the world a mature presentation of his work on this subject. In November he started writing a tract entitled *De quadratura curvarum* that he left unfinished. In December he wrote and finished a revised and augmented version, which he left untitled but is usually known by the same name as the unfinished tract.[83] It begins by recalling the 1676 letter to Leibniz about the binomial theorem, and just before stating his first theorem he wrote [p. 51]:

> On this basis I have tried also to render the method of squaring curves simpler, and have attained certain general theorems.

The first of these is to perform the quadrature of the curve whose general ordinate is $dz^\theta (e + f z^\eta)^\lambda$ [p. 51], where z is the variable, d, e, and f are constants, and θ, η, and λ are just exponents. The general ordinate is first expanded using the binomial theorem and then integrated term by term as in *De analysi*. In general, we should say that there is not much in the treatise that conceptually advances the calculus per se. In Proposition IV, Problem I, "Given an equation involving any number of quantities, to find the fluxions; and vice versa" he repeated the method already used in the *Tractatus fluxionum*, but there is a novelty: the use of "pricked" letters for the fluxions [p. 65]:

> Let a, b, c, d, e be determinate, unalterable quantities; v, x, y, z fluent quantities, that is indeterminate ones increasing or decreasing by a perpetual motion; $\dot{v}$, $\dot{x}$, $\dot{y}$, $\dot{z}$, their fluxions, namely, their speeds of increasing or decreasing ...[84]

It is Newton's proficiency in performing quadratures that shows brilliantly in this tract. This is shown by the following imposing result, which will need

---

[83] Both of these versions, with English translations, were first published by Whiteside in *The mathematical papers of Isaac Newton*, VII, 1976, pp. 24–48 and 48–129. Quotations are from and page references to this work.

[84] In some preparatory drafts for *De quadratura*, Newton played around with some condensed notation for higher-order fluxions. For instance, he wrote

$$-2b\overset{7}{z}\overset{1}{z} - \overset{6}{1}\overset{2}{2} - \overset{5}{3}\overset{3}{0} - \overset{4}{2}\overset{4}{0},$$

where z with an integer n on top means the nth fluxion of z, and the last three terms are just short (excessively short) for $-12b\overset{6}{z}\overset{2}{z} - 30b\overset{5}{z}\overset{3}{z} - 20b\overset{4}{z}\overset{4}{z}$. See Whiteside, *The mathematical papers of Isaac Newton*, VII, pp. 162–163.

some explanation.[85]

PROP. V. THEOR. III

If the Curve's abscissa AB be z, & for $e + fz^\eta + gz^{2\eta} + hz^{3\eta} + \&c.$ write R: also let the Ordinate be $z^{\theta-1}R^{\lambda-1} \times \overline{a + bz^\eta + cz^{2\eta} + dz^{3\eta} + \&c.}$ & put $\dfrac{\theta}{\eta} = r$, $r + \lambda = s, s + \lambda = t, t + \lambda = v, \&c.$

The Area will be $= z^\theta R^\lambda$ *times* $+ \dfrac{\frac{1}{\eta}a}{re}$

$$+ \frac{\frac{1}{\eta}b - sfA}{r+1 \times e}z^\eta$$

$$+ \frac{\frac{1}{\eta}c - \overline{s+1} \times fB - tgA}{\overline{r+2} \times e}z^{2\eta}$$

$$+ \frac{\frac{1}{\eta}d - \overline{s+2} \times fC - \overline{t+1} \times gB - vhA}{\overline{r+3} \times e}z^{3\eta}$$

$$+ \frac{[\ ] - \overline{s+3} \times fD - \overline{t+2} \times gC - \overline{v+1} \times hB}{\overline{r+4} \times e}z^{4\eta}$$

$+ \&c.$[86]

Here $A, B, C, D, \&c.$ denote all the given coefficients of the separate terms in the series with their $+$ & $-$ signs, namely

A the coefficient $\dfrac{\frac{1}{\eta}a}{re}$ *of the first term*

B the coefficient $\dfrac{\frac{1}{\eta}b - sfA}{r+1 \times e}$ *of the second term*

C the coefficient $\dfrac{\frac{1}{\eta}c - \overline{s+1} \times fB - tgA}{\overline{r+2} \times e}$ *of the third term*

And so on hereafter.

---

[85] I have reproduced this proposition and the example following it from pages 49 to 51 of Jones' edition of *De quadratura* of 1711. See the bibliography for the full reference.

[86] There is a missing term in the coefficient of $z^{4\eta}$—replaced here with [ ]—in the 1711 statement of the theorem. The reason is that Newton had originally stated this term to be $\frac{1}{\eta}e$ but then realized that the letter e, which is naturally expected after d, had already been used in the expression for R. For some reason, he deleted the originally written term rather than correct it, and it went to the printer in this manner.

We skip the demonstration that Newton provided. The explanation consists in defining first the expressions

$$R = e + fz^{\eta} + gz^{2\eta} + hz^{3\eta} + \&c.$$

and

$$S = a + bz^{\eta} + cz^{2\eta} + dz^{3\eta} + \&c.$$

(Newton had used the letter S for a similar purpose in the previous proposition, so we use it here), where η is just a constant used to construct the exponents. So are θ and λ, which in Newton's examples were either integers or rational numbers. Then define new numbers

$$r = \frac{\theta}{\eta}, \quad s = r + \lambda, \quad t = s + \lambda, \quad v = t + \lambda, \quad \&c.$$

With the notation so established, the proposition states that

$$\int z^{\theta-1} R^{\lambda-1} S \, dz = z^{\theta} R^{\lambda} \left[\frac{\frac{1}{\eta}a}{re} + \frac{\frac{1}{\eta}b - sfA}{(r+1) \times e}z^{\eta} + \frac{\frac{1}{\eta}c - (s+1)fB - tgA}{(r+2)e}z^{2\eta} \right.$$

$$\left. + \frac{\frac{1}{\eta}d - (s+2)fC - (t+1)gB - vhA}{(r+3)e}z^{3\eta} + \&c. \right],$$

and this is an integration formula (not included in today's calculus textbooks).

Newton's first example, which is necessary for a full understanding of what is going on, was to perform the quadrature of

$$\frac{3k - lzz}{zz\sqrt{kz - lz^3 + mz^4}},$$

where k, l, and m are constants. Using present-day notation, this can be rewritten as

$$z^{-2}(kz - lz^3 + mz^4)^{-\frac{1}{2}}(3k - lz^2) = z^{-\frac{3}{2}-1}(k - lz^2 + mz^3)^{\frac{1}{2}-1}(3k - lz^2).$$

Then, in the notation of the proposition,

$$a = 3k, b = 0, c = -l, e = k, f = 0, g = -l, h = m, \lambda = \tfrac{1}{2}, \eta = 1,$$
$$\theta - 1 = -\tfrac{5}{2}, \theta = -\tfrac{3}{2} = r, s = -1, t = -\tfrac{1}{2}, v = 0.$$

Then $A = -2$, $B = 0$ because $b = f = 0$, $C = 0$ because $c - tgA = -l + \tfrac{1}{2}(-l)(-2) = 0$, and, in the same manner, all the terms after the first vanish to infinity & the area of the Curve comes forth

$$-2\sqrt{\frac{k - lz^2 + mz^3}{z^3}}.$$

After a few more equally impressive results on quadratures, Proposition XI is about recovering the fluents from their fluxions, involving quite a number of cases and examples. Proposition XII and its corollaries give the Taylor series expansion, as already discussed in Chapter 4, and then Newton went on to solve problems on motion and centers of gravity.

Although Newton took the almost completed manuscript of the quadratura on a three-week visit to London in the winter of 1691/1692, and there he expressed an interest in publishing it soon, the fact is that he lost interest in this matter, and the manuscript remained unpublished in this form until 1976, three centuries after he wrote the letter to Leibniz with which he began the tract. However, Newton sent two extracts from it (now lost) to Wallis in the summer of 1692, and Wallis published them in his very extended Latin version of his *Treatise of algebra, both historical and practical* of 1685.[87] This was the first printed appearance of Newton's pricked letters, a notation rapidly adopted by large sections of the mathematics world.

But Wallis could not, during his lifetime, persuade Newton to publish the entire *De quadratura* tract. Instead, he wrote a severely truncated version of it in 1693 (without the Taylor or Maclaurin expansions), which, retitled *The rational quadrature of curves*, became Book 2 of his projected, but never published, treatise on Geometry. But this manuscript was not destined, as so many others, for oblivion. In 1703, Newton took it out of whatever drawer or shelf in which it was gathering dust, wrote an Introduction and a Scholium for it, and set out to publish it. Why? What was Newton's motivation for such zeal in publishing one of his mathematical works while allowing others to remain dormant? Very simply, he had been stung by a plagiarist. Newton was about to publish his *Opticks* when the incident took place, and decided to append to it the abridged *De quadratura* as the second of *Two treatises of the species and magnitude of curvilinear figures*. The following is a part of what he had to say at the end of the "Advertisement" to the volume:[88]

And some Years ago I lent out a Manuscript containing such Theorems, and having since met with some Things copied out of it, I have on this

[87] The extracts can be seen in Wallis, *Opera mathematica*, **2**, 1693, pp. 390–396, and the pricked letters appear on page 392. See the bibliography for reprints.

[88] From Whiteside, *The mathematical papers of Isaac Newton*, VIII, 1981, p. 92.

Occasion made it publick, prefixing to it an *Introduction* and subjoyning a *Scholium* concerning the Method.[89] And I have joined with it another small Tract concerning the Curvilinear Figures of the Second Kind, which was also written many Years ago ...

It is in the new Introduction that we find Newton's newest point of view on the calculus, starting with the assertion [p. 41; 123]:

Mathematical Quantities I here consider not as consisting of least possible parts ["indivisibles" in an earlier draft], but as described by a continuous motion.

Then he made a statement that may be viewed as a glimpse into the concept of limit [pp. 42; 123–125]:

Fluxions are very closely near as the augments [increments] of their Fluents begotten in the very smallest equal particles of time, &, to speak accurately, they are in the first ratio of the nascent augments ...

which, if the fluents are denoted by x and y, we could explain by writing

$$\lim_{\Delta x \to 0} \frac{\Delta y}{\Delta x}.$$

Newton explained it as follows, referring to the next figure (we have inserted arc symbols in the text for clarity) [p. 42; 125]:

Let the ordinate BC advance from its place BC into a new place bc. Complete the parallelogram $BCEb$, and draw the tangent line VTH to touch the Curve at C and meet bc & BA extended in T & V : & the augments of the Abscissa AB, Ordinate BC & Curved Line ACc now begotten will be Bb, Ec & $\overset{\frown}{Cc}$; & in the first ratio of these nascent augments are the sides of the triangle CET, consequently the fluxions of AB, BC & $\overset{\frown}{AC}$ are as the sides CE, ET & CT of that triangle CET, & can be expressed by means of the same sides, or what is the same by the sides of the triangle VBC similar to it.

It comes to the same if the fluxions be taken in the last ratio of the vanishing parts. Draw the straight line Cc & produce it to K. Let the Ordinate bc

---

[89] The new introduction and scholium can be seen in Whiteside *The mathematical papers of Isaac Newton*, VIII, pp. 123–159, while the revisions to the text proper are in pp. 92–105. Double page references in the rest of this section are to Jones' Latin edition first, from which the quotations are taken, and then to Whiteside's translation.

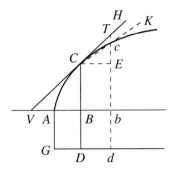

go back into its former place BC, & as the points C & c come together, the straight line CK will coincide with the tangent CH, & so the vanishing triangle CEc will as it attains its last form end up akin to the triangle CET, & its vanishing sides CE, Ec & $\overset{\frown}{Cc}$ will ultimately be to one another as are the sides CE, ET & CT of the other triangle CET, & in this proportion in consequence are the fluxions of the lines AB, BC & $\overset{\frown}{AC}$. If the points C & c are at any small distance apart from each other the straight line CK will be a small distance away from the tangent CH. In order that the line CK shall coincide with the tangent CH & so the last ratios of the lines CE, Ec & $\overset{\frown}{Cc}$ be discovered, the points C & c must come together & entirely coincide. The most minute errors are not in Mathematical matters to be scorned.

In this we can see the evolution of Newton's thought regarding the problem of tangents. In *De analysi*, at the age of 26, he did not hesitate to accept infinitely small quantities, while in the published version of *De quadratura*, at 61, he had come to realize that they are not acceptable. Since the writing of the *Principia* he used the notion of first and last ratios instead of infinitesimals. This new concept is not properly defined, and, of course, the finite ratios (for instance, Ec/CE) do not approach an ultimate ratio, but just a number. He knew that, as we have seen above.

It will be interesting to show how he dealt in *De quadratura* with a case already presented long ago in *De analysi*, that of a simple power [pp. 43–44; 127–129]:

Let the quantity x flow uniformly & the fluxion of the quantity x^n need to be found. In the time that the quantity x comes in its flux to be $x + o$, the quantity x^n will come to be $(x + o)^n$, that is [when expanded] by the method of infinite series, $x^n + nox^{n-1} + \dfrac{nn - n}{2} oox^{n-2} +$ &c. *And so the augments o*

&

$$nox^{n-1} + \frac{nn - n}{2}oox^{n-2} + \&c.$$

are one to the other as 1 & $nx^{n-1} + \dfrac{nn - n}{2}ox^{n-2} + $ &c. Now let those augments come to vanish, & their last ratio will be 1 to nx^{n-1} : consequently the fluxion of the quantity x is to the fluxion of the quantity x^n as 1 to nx^{n-1}.

In the next paragraph he added [p. 44; 129]:

> By similar arguments there can by means of the method of first & last ratios be gathered the fluxions of lines whether straight or curved in any cases whatever, ... & to investigate the first and last ratios of nascent or vanishing finites, is in harmony with the Geometry of the Ancients : & I wanted to show that in the Method of Fluxions there should be no need to introduce infinitely small figures [quantities] into Geometry.

5.12 LEIBNIZ' CALCULUS: 1673–1675

Starting in 1673, Leibniz investigated problems on tangents and quadratures in some manuscript papers that remained unpublished for almost two centuries. They are in the nature of notes for his own use, possibly made as he read the works of Descartes, Sluse, St. Vincent, Barrow, and James Gregory among others. In this way he developed his own ideas, discovered some interesting formulas, and introduced his very famous notations d and $\int$. These appeared first in a manuscript of 29 October 1675, which is part of a collection of three manuscripts under the common title *Analysis tetragonistica ex centrobarycis* (Analysis of quadratures by means of centers of gravity).[90] These manuscripts

---

[90] Copies of Leibniz' mathematical manuscripts are available from the *Gottfried Wilhelm Leibniz Bibliothek / Niedersächsische Landesbibliothek* (Lower Saxony State Library), Hannover. A complete catalog (not completely up to date) can be found in Bodemann, *Die Leibniz-Handschriften der Königlichen Öffentlichen Bibliothek zu Hannover*, 1895, and the bibliography contains references to this catalog for Leibniz' manuscripts. Many of these manuscripts have been subsequently printed in *Die Entdeckung der differentialrechnung durch Leibniz*, 1848; *Die Geschichte der höheren Analysis, erste Abtheilung, Die Entdeckung der höheren Analysis*, 1855; *Der Briefwechsel von Gottfried Wilhelm Leibniz mit Mathematikern*, 1899 (all edited by Gerhardt); and in *Sämtliche Schriften und Briefe*, Ser. VII, **5** (edited by Probst, Mayer and Sefrin-Weis). English translations of many of these manuscripts can be found in Child, *The early mathematical manuscripts of Leibniz; translated from the Latin texts published by Carl Immanuel Gerhardt with critical and historical notes*, 1920. References given in the text for Leibniz' manuscripts are to the manuscript folio first, and then by page to the printed versions listed in the bibliography, in the order listed there.

GOTTFRIED WILHELM LEIBNIZ
Stipple engraving by Benjamin Holl, 1834.
Lithograph from the author's personal collection.

were not meant for publication, and exposing their contents to the public view may fall under the heading of indiscretion, to say the least. Nevertheless, curiosity killed the cat.

The first of these manuscripts is dated October 25 and 26, and in the second part we find a significant result. Before presenting it, it is convenient to explain some of Leibniz' notation at that time. First, he used the symbol ⊓ as an equal sign and then he chose the abbreviation omn. of the Latin word *omnia* [*omnes* in Leibniz' usage] to mean "all," although we may prefer to read it as "the sum of all." Thus, if ω is a variable that assumes several values in a certain context, he denoted the sum of all the values of ω's by omn. ω. When summing terms that consist of more than one letter, he found it convenient to cover them with an overline to denote the extent of the terms to be added, as in omn. $\overline{x\omega}$ to sum the products $x\omega$. Finally, he used the abbreviation ult., from the Latin *ultima*, to mean "last."

It must be said, before quoting from his work, that Leibniz managed to construct a practical calculus that could solve real problems, but was never a model of clarity. Even his contemporaries, more used to the language of his time than we could ever be, found it heavy going to read and understand Leibniz. Nevertheless, at the risk of doing injustice to the author, we must plunge in. The reader has been forewarned.

At a certain point in the manuscript of October 26 he changed the topic he was discussing and inserted the following [f. 1ᵛ; pp. 120–121; 150–151; 268; 70–71]:

Another [thing]: The moments of the differences about a [straight line] perpendicular to the axis, are equal to the complement of the sum of the terms

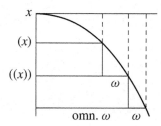

or: The Moments of the Terms are equal to the complement of the sum of the sums.

Or

$$\text{omn } \overline{x\omega} \sqcap \text{ ult.} x, \ \overline{\text{omn. } \omega,}, \ - \text{omn. } \overline{\text{omn. } \omega}$$

Let $x\omega \sqcap az$. becomes: $\omega \sqcap \dfrac{az}{x}$. it makes

$$\text{omn. } \overline{az} \sqcap \text{ ult.} x \text{ omn } \overline{\dfrac{az}{x}} - \text{omn. omn } \overline{\dfrac{az}{x}}$$

Therefore,

. . .

And

$$\text{omn } a \sqcap \text{ult } x \cdot \text{omn } \overline{\dfrac{a}{x}} - \text{omn. omn. } \overline{\dfrac{a}{x}}$$

Which last theorem exhibits the sum of logarithms in terms of the given quadrature of the Hyperbola.[91]

---

[91] I am trying to reproduce the manuscript formulas as closely as possible, including Leibniz' punctuation (some of his commas look like periods) or lack thereof and his switch from omn. to omn without a period. However, the copies that I have used are not totally

To explain this result we can start by drawing a more detailed figure, representing the series of ordinates y (located horizontally) of a curve OC, and

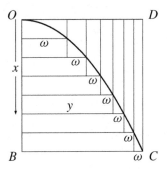

Leibniz assumed that the distance between any two consecutive abscissas is an infinitesimally small unit, which we shall denote by z. The reader who wonders how this distance can be a unit and at the same time infinitesimally small is not alone. Referring then to the figure, in which the difference of any two successive ordinates has been labeled ω, if x denotes the abscissa corresponding to an arbitrary ordinate y, then the area of the curvilinear figure OCD is the sum of all the products $x\omega$ (which Leibniz called moments by analogy with physics, in which a moment is the product of a weight times the distance to the axis; the roles played here by ω and x), that is, omn. $\overline{x\omega}$, giving the left-hand side of his first equation. But this area is that of the rectangle $OBCD$ minus that of the curvilinear figure OCB. The area of the

---

clear (see the sample reproduced on page 326) and I cannot make impossible claims about my readings. The reader should be aware of the fact that the previously printed versions of Leibniz' formulas do not always coincide with the manuscript forms. For example, the last equation appears as

$$\text{omn. } a \sqcap \text{ult. } x \text{ omn. } \overline{\frac{a}{x}} - \text{omn. omn. } \overline{\frac{a}{x}}$$

in *Die Geschichte*, as

$$\text{omn. } a \sqcap \text{ult. } x. \text{ omn. } \overline{\frac{a}{x}} - \text{omn. omn. } \frac{a}{x}$$

in *Der Briefwechsel*, and as

$$\text{omn. } a \sqcap \text{ult. } x \text{ omn. } \frac{a}{x} - \text{omn. omn. } \frac{a}{x}$$

in Child's translation. The differences shown here may not be very significant, but it gets worse, and there are instances in which Leibniz' meaning has been completely altered.

rectangle is the product of the last x and the segment BC, that is, the product ult. x $\overline{\text{omn.}\ \omega}$. The area of the curvilinear figure OBC is the sum of the areas of the rectangles formed by the unit and the ordinates, and each ordinate is the sum of all the ω's up to that particular spot, or omn. ω, and then the area of OBC is omn. $\overline{z\,\overline{\text{omn.}}\ \omega}$. In conclusion,

$$\text{omn.}\ \overline{x\omega}\ \sqcap\ \text{ult.}\ x\ \overline{\text{omn.}\ \omega}\ -\ \text{omn.}\ \overline{z\,\overline{\text{omn.}}\ \omega},\ ^{92}$$

where we have eliminated the unnecessary commas of the original.[93] Leibniz did not write the unnecessary unit z. This result will be interpreted below in today's terminology.[94]

Since the curve in this discussion is completely general, we can consider particular cases. For instance, if $x\omega = az$, where a is a constant and $z = 1$,[95] we obtain

$$\text{omn.}\ \overline{az}\ \sqcap\ \text{ult.}\ x\ \text{omn.}\ \overline{\frac{az}{x}}\ -\ \text{omn.}\ \text{omn.}\ \overline{\frac{az}{x}}.$$

Of course, x should not come near zero in this equation, and we can assume that the abscissa of the ordinate OD is $x = 1$ in this case. Note then that omn. $\overline{az/x}$ is the sum of the areas of the rectangles formed by the unit z and the ordinates a/x, and therefore equals the area under the hyperbola $y = a/x$ starting at $x = 1$, which is a logarithm. Then

$$\text{omn.}\ \text{omn.}\ \overline{\frac{az}{x}}$$

is a sum of logarithms. Thus, Leibniz has obtained "a sum of logarithms in terms of the known quadrature of the hyperbola."

On 29 October 1675, in *Analyseos tetragonisticæ pars 2da.* (Second part of analytical quadrature), Leibniz considered a problem in inverse tangents that shows the use of the so-called characteristic triangle, shown in the next figure as GWL (perhaps Leibniz was so proud of this feature that he labeled

---

[92] Leibniz has used omn. ω with two meanings on this line. First it means the last y, to be multiplied by ult. x, and then it denotes an arbitrary y to be multiplied by z and added up.

[93] They had a meaning for Leibniz: ult. x was to be multiplied by everything between the next two commas, then everything after the third comma was to be subtracted from this product.

[94] For those who are impatient or prefer to figure it out by themselves, this is a formula for integration by parts.

[95] Now Leibniz wrote the unnecessary unit z. Since the left-hand side of the equation is an area, the right-hand side must also be an area, and, to preserve dimensionality in those geometry-dominated times, Leibniz inserted a length $z = 1$. It is surprising that he did not do so in his first result, where it would have served the same purpose.

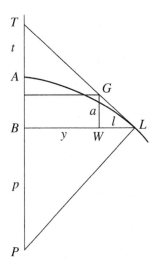

the triangle with his own initials). Here TL is the tangent to the curve AL at L, PL is the normal at L, l takes the place of ω in the previous figure and represents the difference of consecutive ordinates, while a replaces z as the infinitesimally small unit that separates these ordinates, so we shall be able to take $a = 1$. Leibniz did not place any of these lowercase letters in his figure (while he drew some additional segments that are not shown here because they are not needed in our presentation), but established the notation in this manner [f. 2ʳ; p. 123; 153; 290; 79]:

$$BL \sqcap y. \quad WL \sqcap l. \quad BP \sqcap p. \quad TB \sqcap t. \quad AB \sqcap x. \quad GW \sqcap a. \quad y \sqcap \text{omn. } l.$$

Then he continued as follows (except that we display equations on their own lines as is the habit today) [ff. 2ʳ–2ᵛ; p. 124; 154; 291–292; 80]:

Furthermore

$$\frac{l}{a} \sqcap \frac{p}{\text{omn.}\, l \sqcap y}.$$

Therefore

$$p \sqcap \frac{\overline{\text{omn } l}}{a} \, l.$$

$$\cdots$$

Therefore

$$\text{omn } p \sqcap \text{omn}, \frac{\overline{\text{omn } l}}{a}, l$$

but I have elsewhere proved that

$$\text{omn. } p \sqcap. \frac{y^2}{2}, \quad \text{or} \quad \sqcap \frac{\overline{\text{omn } l \boxed{2}}}{2}.\ {}^{96}$$

Therefore we have a theorem that I regard as admirable, and [that will be] of great service to this new calculus in the future (*Ergo habemus theorema quod mihi videtur admirabile, et novo huic calculo magni adjumenti loco futurum*), namely

$$\frac{\overline{\text{omn } l \boxed{2}}}{2} \sqcap \text{omn } \overline{\text{omn } l \frac{l}{a}},$$

whatever *l* may be. That is if all the *l*, are multiplied by the last one, and another [set of] all *l*, in turn by their last, and so on as often as it can be done[,] the sum of all these [products] will be equal to half the sums of the squares, whose sides are the sums of the *l*, or all the *l*.[97] this is a very beautiful and not at all obvious theorem. Likewise so is the Theorem:

$$\text{omn } \overline{xl} \sqcap x \overline{\text{omn } l} - \text{omn } \overline{\text{omn } l}.\ {}^{98}$$

putting ...

at which point he explained the notation in the last equation.

We shall interpret the meaning of these results shortly. But first we shall establish the left-hand side in the admirable theorem by evaluating

$$\text{omn. } p = \text{omn. } \frac{\overline{yl}}{a} = \frac{1}{a} \text{ omn. } \overline{yl}.$$

Leibniz would provide a geometric explanation later in his life, when he wrote a recollection of his discoveries on this subject as *Historia et origo calculi differentialis* in 1714 [pp. 9; 400; 40–41; 11]. Referring to a figure sharing many common elements with the previous one, he stated:

---

[96] The boxed 2 in this equation is an exponent, and the infinitesimal unit *a* need not be written explicitly. I do not know whether Leibniz actually proved this formula in a previous manuscript, although he made the statement several times, for instance, in the manuscript *De triangulo curvarum characteristico* of January 1675. On the basis of a figure like the one included here, but writing *D* instead of *B* and *M* instead of *P*, Leibniz obtained the first equation in this quotation and then simply stated [f. 1ʳ; p. 184] "Ergo summa omnium reductarum ipsius *DL* semiquadrato;" that is, that the sum of all the *p*'s is $\frac{1}{2}y^2$.

[97] Leibniz' use of plurals is confusing. This means "equal to half the square whose side is the sum of the *l*'s, or all the *l*'s."

[98] This is his first theorem in the previous quotation, but replacing ω with *l* according to the new notation. Notice that Leibniz has deleted the abbreviation ult., thereby adding to the confusion of using omn *l* with two different meanings.

But straight Lines which increase from nothing [the y's of an increasing function that passes through the origin] each multiplied by its corresponding Element [the l for each y] make a triangle.

This triangle is shown in the following figure. On the vertical axis we have placed the successive increments l, and on the horizontal each ordinate y is

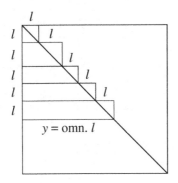

the sum of all the preceding l's: $y = \text{omn.}\, l$. The sum of the infinitesimal rectangles yl is approximately equal to that of the triangle, which, if we are given a "last" y, is $y^2/2$. That is,

$$\text{omn.}\, \overline{yl} = \frac{y^2}{2} = \frac{(\text{omn.}\, l)^2}{2}.$$

In Leibniz' notation, and in view of the fact that $p \sqcap yl/a$,

$$\text{omn.}\, p \sqcap \frac{\overline{\text{omn.}\, l\;\boxed{2}}}{2a},$$

and we see that Leibniz had simply dropped the unit a to write the left-hand side of his admirable theorem.

At this point in the *Analysis tetragonistica* Leibniz decided to improve his notation, stating: "It will be useful to write $\int$. for omn. so that $\int l$ [stands] for omn. l. that is the sum of all the l" [f. 2^{v}; p. 125; 154; 292; 80]. He chose the symbol $\int$ as an elongated s that was used at that time, and it was meant as the initial of the Latin word "summa" for sum. With this new symbol $\int$, the admirable theorem becomes

$$\frac{\overline{\int l}^{\,2}}{2a} \sqcap \int \overline{\int l\,\frac{l}{a}}$$

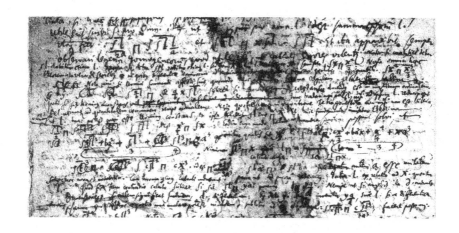

The portion of folio 2^v of the "Analysis Tetragonistica" manuscript containing Leibniz'
famous $\int$ and d notations (upper left on the second line from the top,
and lower right on the third line from the bottom).
Courtesy of the *Gottfried Wilhelm Leibniz Bibliothek*, Hanover.

(note that Leibniz reinserted the unit a), and the equation

$$\text{omn } \overline{xl} \sqcap x \, \overline{\text{omn } l} - \text{omn } \overline{\text{omn } l}$$

becomes

$$\int \overline{xl} \sqcap x \overline{\int l} - \int \int l \, .$$

They appear in the manuscript in the following form, and these are not the

$$\frac{\underline{\int}^2}{2a} \sqcap \int \int \underline{\overline{l}}_a \quad et \quad \int \overline{xl} \sqcap x \overline{\int l} - \int \int l \, ,$$

same as some of the formulas made available later in printed form.[99]

    These equations are still foreign-looking today, but then Leibniz introduced
the letter d to help rewrite each difference l of y's in a more palatable way.

---

[99] These equations have been electronically cleaned from a scan of the image at the top
of this page in order to make them more readable, at the risk of throwing away the baby
with the bath water. There is an old spill on this page which makes it difficult to read the
equation on the right. The printed versions appear in Gerhardt's *Die Geschichte* as [p. 125]

$$\frac{\int \overline{l^2}}{2} \sqcap \int \overline{\int \overline{l} \frac{l}{a}} \quad et \quad \int \overline{xl} \sqcap x \int \overline{l} - \int \int \overline{l} \, ,$$

However, his first attempt may be shocking. In his own words, he introduced this d operator as follows [f. 2ᵛ; pp. 126; 155; 293–294; 82] (once again, most of his commas look like periods):

> Given l. [its] relation to x. search for $\int l$. Which is done now from the contrary calculus namely if $\int l \sqcap ya$, we put $l \sqcap \dfrac{ya}{d}$. Then as $\int$. will increase so d. will diminish the dimensions. But $\int$. means a sum, d. a difference: From the given y we can find $\dfrac{ya}{d}$, or l. or the difference of the y.

Not very clear and not what we might have expected. Even accepting that a is the infinitesimal unit, what is it doing here, what is this talk about dimensions, and why is d in the denominator? The equation $\int l = ya$ and the statement "$\int$ will increase the dimensions" suggest that Leibniz was thinking of $\int l$ as an area,[100] and then it cannot equal the length y. It is necessary to multiply y by some other length a (which we must take to be unity) to make the equation homogeneous. The same kind of argument justifies that ya/d is a length.

However, homogeneity cannot be a sufficient explanation for Leibniz' decision to put d in the denominator, especially in conjunction with his statement that d means a difference. What difference? What does it mean to say that ya/d is l, the difference of the y's? The differences between consecutive y's could very well be all different, so which is the one that d produces from their sum $\int l$? A simpler explanation, which comes with a clarification, can be sought in the figure on page 323. The similarity of the triangles TBL and

---

in *Der Briefwechsel* as [p. 154]

$$\frac{\int \overline{l}^2}{2} \sqcap \int \int \overline{l}\frac{l}{a} \qquad \text{et} \qquad \int \overline{xl} \sqcap x \int \overline{l} - \int \overline{\int l},$$

and in Child's translation in the form [p. 80]

$$\frac{\int \overline{l^2}}{2} = \int \int \overline{l}\frac{l}{a}, \qquad \text{and} \qquad \int \overline{xl} = x \int \overline{l} - \int \int l.$$

The first left-hand side is now meaningless: a sum of squares has replaced what previously was the square of a sum. Whether the changes from manuscript to printed form to translation are due to transcriber and translator or to problems at the printer's, this sample should give an idea of some of the difficulties faced by a modern reader trying to understand Leibniz from Child's translation. However, Child pointed out on several occasions that he had been unable to see the original manuscripts, and could only translate from *Die Geschichte*.

[100] Recall that in the first quotation in this section Leibniz had used the notation omn. ω with two meanings, and one of them was as a sum of rectangular areas.

GWL immediately gives

$$l = \frac{ya}{t}.$$

This is what Leibniz might have had in mind, and this is a real quotient. And now we know which *l* is produced by this process: the one corresponding to the particular subtangent *t* at a given *y*. But *l* is a difference of *y*'s and Leibniz may have wanted to emphasize this point by the use of the letter *d* instead of *t*.

However, this is reading Leibniz' mind in retrospect. If we just read his words, he did not even make it easy for us to learn by example, for this is what he had to say in that respect [f. 2ᵛ; p. 126; 155; 294; 82]:

> Hence one equation may be transformed into the other, just as from the equation:

$$\int c \, \overline{\int l}^{\,2} \sqcap \frac{c\overline{\int l}^{\,3}}{3a^3}$$

we can obtain:

$$c \, \overline{\int l}^{\,2} \sqcap \frac{c\overline{\int l}^{\,3}}{3a^3 d}. \ 101$$

What we get from this, if we are not too inquisitive, is that $\int$ and d are inverse operators. What $\int$ does, d (in the denominator) undoes. If we also want to know what the equations mean, note that $\int l = y$, and then, discarding the common constant c, recalling that $a = 1$, and assuming that $\int l^{\,3}$ means $(\int l)^3$, the second equation reads

$$y^2 = \frac{y^3}{3d}.$$

Thus, $1/d$ does here what the derivative does today. The first equation is a little more problematic and needs interpretation. If Leibniz meant

$$\int c \, \overline{\int l}^{\,2} \sqcap \frac{c\overline{\int l}^{\,3}}{3a^3},$$

then, discarding the c and recalling that $a = 1$, this is equivalent to

$$\int y^2 = \frac{y^3}{3},$$

---

101 This is my reading of the equations in the manuscript. Both Gerhardt and Child have $\int \overline{l^2}$ on their left-hand sides. Child also has $\int \overline{l^3}$ on both right-hand sides, as Gerhardt has in *Die Geschichte*.

which we regard as correct. Of course Leibniz' notation at this point is horrendous and inconvenient, and we are all happy that it did not last long. On November 11, in a new manuscript entitled *Methodi tangentium inversæ exempla* (Examples of the inverse method of tangents), he carried on with the d in the denominator for a while and then, without explanation or transition he wrote (this is out of context and its precise meaning is not in question now) [f. 1^r; p. 34; 134; 162; 323–324; 95]

$$\frac{x^2}{2} + \frac{y^2}{2} = \int \overline{\frac{a^2}{y}} \quad \text{or} \quad d\,\overline{x^2 + y^2} = \frac{2a^2}{y}.$$

From this moment on, after a little wavering, the d is always upstairs. Accepting this new position of d, as we must, we are now ready to rewrite some of his previous equations in this new notation. From $\int l = y$ we obtain $l = dy$. Then Leibniz' admirable result, which previously interpreted with the symbol $\int$ (page 325) and putting $a = 1$ was

$$\frac{\overline{\int l}^2}{2} \sqcap \int\int\overline{l}\,l,$$

now becomes

$$\frac{y^2}{2} = \int y\,dy.$$

Not much to write home about, since much more than this had already been obtained by Fermat and Gregory. Also, the equation $\int \overline{xl} \sqcap x\int\overline{l} - \int\int\overline{l}$ becomes

$$\int x\,dy = xy - \int y\,dx,$$

but where did the dx come from? It is just the infinitesimal unit separating ordinates, so inserting it does no harm.

Eventually, Leibniz would go on to develop the rules of the operator d, but on November 11, 1675, he just toyed with the idea that $d(xy)$ and $d(x/y)$ might be equal to $dx\,dy$ and dx/dy [f. 2^r, p. 38; 137; 165; 328; 100], respectively, but discarded it. Keep in mind, before passing judgment, that Leibniz never meant any of these musings to be published. He was just making up his own mind about these matters.

5.13 LEIBNIZ' CALCULUS: 1676–1680

Very little happened in Leibniz' work that was of any consequence to the calculus for one year, except that he kept using his new notation and getting enough consistency in writing the necessary differentials in all his "integrals." But then, already on his way back to Germany, he stopped in London for about a week in October 1676. He probably wanted to check on what the English had been up to, either because he had heard something about it or out of simple curiosity.

He did find something and put it to good use. Collins was very open with him and showed him a copy of Newton's *De analysi*, of which Leibniz made an ample extract.[102] But it is clear from what he copied and what he did not that this did not advance his own research in any manner. He also made an extract of a letter of 10 December 1672 from Newton to Collins. The extract is as follows:[103]

Letter of Newton, 1672: ABC is any angle, $AB \sqcap x$, $BC \sqcap y$. Take for

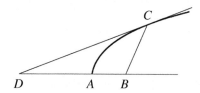

example, the equation

$$\overset{0}{\underset{3}{x^3}} - \overset{1}{\underset{2}{2x^2y}} + \overset{0}{\underset{2}{bx^2}} - \overset{0}{\underset{1}{b^2x}} - \overset{2}{\underset{0}{by^2}} - \overset{3}{\underset{0}{y^3}} \sqcap 0.$$

Multiply the equation by an arithmetical progression, both for the second dimension y and for x; the first product will be the numerator, and the other divided by x will be the denominator of a fraction which will express BD, thus,

$$BD \sqcap \frac{-2x^2y + 2by^2 - 3y^3}{3x^2 - 4xy + 2bx - b^2}.$$

Moreover, that this is only a corollary or a case of a general method for both mechanical and geometrical lines, whether the curve is referred to a single straight line, or to another curve, without the trouble of calculation, and other abstruse problems about curves, etc. This method differs from that of Hudde and also from that of Sluse, in that it is not necessary to eliminate irrationals.

---

[102] It can be seen in its entirety, in Latin only, in Whiteside, *The mathematical papers of Isaac Newton*, II, pp. 248–259.

[103] It appears in Gerhardt, "Leibniz in London" (1891) 257–176. English translation in Child, *The early mathematical manuscripts of Leibniz*. This quotation is from page 194 of the translation.

Or, as Newton himself put it,[104]

> Suppose CB applyed to AB in any given angle be terminated at any Curve line AC, and calling AB x & BC y let the relation between x & y be expressed by any æquation as
>
> $$x^3 - 2x^2y + bx^2 - b^2x - by^2 - y^3 = 0$$
>
> whereby the curve is determined. To draw the tangent CD the Rule is this.

Here he repeated the procedure already outlined in Section 5.8 to solve Problem 4 and obtained BD as copied by Leibniz. He also stated that his method is not "limited to equations wch are free from surd [irrational] quantities."

In short, what Leibniz got out of this is that Newton had a method (the one shown in Example 4 on page 296) to find tangents that was similar to that of Sluse but in which it was not necessary to eliminate irrationals. Leibniz must have worked quickly to justify this rule using his new differentials because he included such a justification in a manuscript entitled *Calculus tangentium differentialis, adjecta sub finem machina construendi æquationes per logarithmicam* of November 1676. Here we find first some rules for finding differences and sums [f. 3^r; pp. 56; 140; 229–230; 614–615; 124]:

$$d\overline{x} \sqcap 1 \qquad d\overline{x^2} \sqcap 2x \qquad d\overline{x^3} \sqcap 3x^2 \qquad \&c$$

$$d\frac{1}{x} \sqcap -\frac{1}{x^2} \qquad d\frac{1}{x^2} \sqcap -\frac{2}{x^3} \quad \& \quad d\frac{1}{x^3} \sqcap -\frac{3}{x^2} \qquad \&c^{105}$$

$$d\sqrt{x} \sqcap \frac{1}{\sqrt{x}} \qquad \&c^{106}$$

From these the following general rules may be derived for the differences and sums of simple powers

$$d\overline{x^e} \sqcap e, x^{e-1} \quad \text{and conversely} \quad \int \overline{x^e} \sqcap \frac{x^{e+1}}{e+1}.$$

---

[104] Turnbull, *The correspondence of Isaac Newton*, I, pp 247–248.

[105] Gerhardt corrected the last power of x to x^4 in *Der Briefwechsel*, as did the *Sämtliche* (except for full disclosure in a footnote), but that is not what Leibniz wrote, and both Gerhardt, in *Die Entdeckung* and in *Die Geschichte*, and Child wrote the right-hand side in the second equation on this line as $-2/x^2$. The manuscript has $-\frac{2}{x^3}$ in which the exponent of x looks like a broken 3. I have read it as a 3, as did Gerhardt in *Der Briefwechsel* and the *Sämtliche*.

[106] Gerhardt, in *Der Briefwechsel*, corrected the right-hand side of this equation to $1/2\sqrt{x}$.

Needless to say, the e used here stands for an arbitrary exponent and not for Euler's number e. This gives the derivative and integral of a power of x, but there is no explanation about how these rules were obtained beyond following the pattern shown by a few particular cases. As for these, Leibniz had already obtained $d\overline{x^3} = 3x^2$ as his first example after the introduction of d, and $dx^2 = 2x$ follows by the application of d to his admirable theorem. Then, from the first of these general rules he obtained again some of the preceding particular cases:

Hence $d\dfrac{\overline{1}}{x^2}$ or $d,\overline{x^{-2}}$ will be $-2x^{-3}$ or $-\dfrac{2}{x^3}$ and $d\overline{\sqrt{x}}$ or $d,\overline{x^{1/2}}$ will

be $\left(\boxed{\frac{1}{2}-1}\right) -\frac{1}{2}x^{-1/2}$ or $-\dfrac{1}{2}\sqrt{\dfrac{1}{x}}$.

Note that in all of Leibniz' work to this point dx is just the infinitesimal unit between successive values of x. But at this point he generalized this as follows [f. 3^r; pp. 56–57; 140–141; 230; 615; 124–125]:

Let $y \sqcap x^2$. and it will be $d\overline{y} \sqcap 2x\, d\overline{x}$ [107] therefore $\dfrac{dy}{dx} \sqcap 2x$. Such reasoning is general, and it does not depend on what the progression for the x may be. In the same manner this general rule therefore holds:

$$\frac{dx^e}{dx} \sqcap e, x^{e-1} \quad \text{and in turn} \quad \int \overline{x^e\, dx} \sqcap \frac{x^{e+1}}{e+1}.$$

From now on the infinitesimal distances between consecutive ordinates can be anything besides the unit.

Next Leibniz showed us how to deal with arbitrary equations, even in implicit form [f. 3^r; p. 57; 141; 230; 615–616; 125]:

Let there be any equation whatever, for instance

$$ay^2 + byx + cx^2 + f^2x + g^2y + h^3 \sqcap 0$$

and writing $y + d\overline{y}$ for y, and similarly $x + d\overline{x}$ for x, we have by omitting [terms] which should be omitted another equation:

$$
\begin{array}{l}
\begin{array}{llllll}
ay^2 & +\,byx & +\,cx^2 & +\,f^2x & +\,g^2y & +\,h^3 \\
\hline
a2d\overline{y}y & +\,byd\overline{x} & +\,2cxd\overline{x} & +\,f^2d\overline{x} & +\,g^2d\overline{y} & \\
& +\,bxdy & & & &
\end{array} \Big\} \sqcap 0 \\[2pt]
\hline
ad\overline{y}^2 \;+\,bd\overline{x}d\overline{y} + cd\overline{x}^2 \Big| \sqcap \boxed{0}
\end{array}
$$

[107] In the previous quoted line the circled difference $\frac{1}{2}-1$ is for Leibniz' own reckoning and clearly to be discarded from the final statement. This equation also contains several circled terms, and I have discarded them in compliance with Leibniz' intention.

This is the origin of the rule published by Sluse.[108]

Before moving ahead we must interpret the last equation. All the terms to the left of the brace are those that result from replacing x and y by $x + dx$ and $y + dy$ and then expanding. The terms in a half box on the top line add up to zero because of the original equation, which is the equivalent of Fermat's "remove the common terms," Barrow's "reject all ... terms which do not contain a or e," or Newton's "removing equal terms." The terms in another half box on the bottom line add up to approximately zero because they are too small compared to the rest. From those that remain, the central group, $d\overline{y}/d\overline{x}$, can be found.

Accompanying Leibniz on his discovery trip up to this point may lead some readers to think that his main contribution to the calculus was notation, notation, notation.[109] However, in this paper of November 1676 we find the following important contribution to the calculus [f. 3^v; p. 58; 142; 231; 616; 126]:

Furthermore it will be worthwhile to adjust this kind of work to irrationals and compound fractions.

$$d\sqrt[2]{a + bz + cz^2}, \quad \text{put} \quad a + bz + cz^2 \sqcap x$$

---

[108] Leibniz used $d\overline{x}$ and $d\overline{y}$ (except that he forgot to overline the y in one instance) where Newton had used $\dot{x}o$ and $\dot{y}o$ and Barrow before him—in a similar Example 1 in Lecture X—had used a and e. Child thought [p. 125] that the similarity of Leibniz' words *omissis omittendis* for "omitting which should be omitted" to Barrow's *rejectis rejiciendis* is indicative of Leibniz' debt to Barrow. In the letter to Collins of 10 December 1672, from which Leibniz made an extract while in London, Newton stated: "I am heartily glad at the acceptance wch our Reverend friend Dr Barrow's Lectures finds with forreign Mathematicians, and it pleased me not a little to understand that they are falln into the same method of drawing Tangents with me." Then, he proceeded to quote an example of what these foreign mathematicians had done. It is not inconceivable that, on reading Newton's letter, Leibniz rushed to his copy of the *Lectiones geometricæ* and used Barrow's method after replacing e with $d\overline{x}$ and a with $d\overline{y}$. But he had already considerable practice with his differentials, so he may have produced this proof on his own and then checked with Barrow and picked up some similar language in his explanation. Or this similarity may be coincidental.

[109] Such readers would not be alone, for Newton himself once said: "But certainly no man would call this Notation a new method of Analysis ... " Quoted from "Counter observations on Leibniz' review [of W. Jones, *Analysis per quantitatum series, fluxiones, ac differentias: cum enumeratione linearum tertii ordinis*]," in Whiteside, *The mathematical papers of Isaac Newton*, II, 1968, p. 273. Ironically, Leibniz himself seems to have contributed to this view when, referring to himself in the third person, he wrote:

Furthermore it is now to be explained how Our [friend] arrived gradually to a new Notation, that he called the differential calculus.

In *Historia et origo calculi differentialis* [pp. 13–14; 404; 49; 16].

and

$$\frac{d\,\overline{\sqrt[2]{x}}}{d\overline{x}} \sqcap -\frac{1}{2\sqrt{x}}, \quad \text{now} \quad \frac{d\overline{x}}{d\overline{z}} \sqcap b + 2cz$$

therefore

$$d\,\overline{\sqrt[2]{a + bz + cz^2}} \sqcap -\frac{b + 2cz}{2d\overline{z},\sqrt{a + bz + cz^2}}. \,110$$

What we have here is an example of the chain rule in its present form, although a similar device had already been used by Gregory and Newton.

It was at this time, in an untitled letter of 21 June 1677, that Leibniz sent to Oldenburg, to be forwarded to Newton, that he first made "public" his recent discoveries on the calculus. It was a reply to Newton's of 24 October 1676, containing a lengthy explanation of his discovery of the binomial theorem but very little on his new calculus, that Leibniz had just received with a great delay due to his moving from Paris to Hanover. Leibniz' reply was full of praise for Newton and included all of his knowledge about the calculus at that point.

He started with a definition of his differentials [pp. 154; 241; 168; 213; 219–220], calling

$d\overline{y}$ ... the difference of two close [values of] y ... and $d\overline{x}$ ... the difference of two close [values of] x ...

and continued with an example of "implicit differentiation," very similar to the one described above, stating that this method "is useful at a time, when irrational [quantities] intervene" [p. 155; 242; 170; 214; 220]. Next he included the evaluation of the differential of $\sqrt[3]{a + by + cy^2}$ by the method of the chain rule [pp. 155–156; 243; 170–171; 214; 221], defined a differential equation as one in which x is to be obtained from $d\overline{x}$ and provided an example [p. 156; 243; 172; 215; 221], and described the method of inverse tangents [pp. 158–159; 244–246; 175–178; 216–217; 223–224], claiming that it was generally more powerful than the method of infinite series. In one of the included figures he used subscripts in labeling all its points [Fig. 29 at the end of the volume; p. 244; 174; 216; 219].

But it was not until July 11, 1677, in a manuscript on a "General method to obtain the tangents to curves without calculation, and without reducing

---

[110] Gerhardt's version in *Die Entdeckung* and in *Die Geschichte*, Child's translation, and the *Sämtliche* follow the original manuscript, which contains two wrong signs and the dz in the denominator. However, Gerhardt's version in *Der Briefwechsel* has corrected these errors and introduced two new ones: $d\overline{z}/d\overline{x}$ instead of $d\overline{x}/d\overline{z}$ and no dz in the last quotient.

irrational or fractional quantities," written in French instead of Latin,[111] that he gave the complete set of rules of his differential calculus. By that time Leibniz could "differentiate" sums, differences, products, quotients, powers, and roots. After giving a long example on "implicit differentiation," much as he had given in the previous paper, he stated [f. 3^v; p. 61; 144; 130]:

Now let the formula or equation, or magnitude ω, be equal to $\dfrac{\lambda}{\mu}$, I say that $d\omega$ will be equal to

$$\frac{\mu\,d\overline{\lambda} - \lambda\,d\overline{\mu}}{\mu^2}.$$

This will be sufficient to deal with fractions.

Finally let ω be equal to [replaced by] $\sqrt[z]{\omega}$. I say that $d\overline{\omega}$ will be equal to

$$\frac{d\overline{\omega}}{z^{\frac{z-1}{z}}\sqrt{\overline{\omega}}} \quad 112$$

which will suffice for the proper treatment of irrational magnitudes.[113]

It can be said, and it has been said, that Leibniz offered these and similar formulas without proof. This may be technically correct, but proofs are not necessary because they can be easily supplied by the reader using the same procedure that Leibniz had used to justify the tangent method. To wit, in the case of a quotient, if $\omega = \lambda/\mu$, then $\mu\omega = \lambda$ and we would have $(\mu + d\mu)(\omega + d\omega) = \lambda + d\lambda$. Now, omitting what can be omitted, as Leibniz would have put it, or removing common terms and rejecting products of differentials, this boils down to $\mu\,d\omega + \omega\,d\mu = d\lambda$, which is the product rule,[114] and then

$$d\omega = \frac{d\lambda - \omega\,d\mu}{\mu} = \frac{\mu\,d\lambda - \lambda\,d\mu}{\mu^2},$$

---

[111] *Méthode générale pour mener les touchantes des Lignes Courbes sans calcul, et sans réduction des quantités irrationelles et rompues.*

[112] This differential is in error. The root in the denominator should be $z^{\frac{z}{z-1}}\sqrt{\overline{\omega}}$.

[113] The overlines in Child's translation are different throughout this quotation. In *Die Entdeckung* and in *Die Geschichte* the letter ω in the square roots is not overlined, but it is in the manuscript.

[114] Newton had already given a form of the product rule in an Addendum to the fluxional treatise, probably written in the winter of 1671–1672. After stating it in verbal form as Theorem 1 of the Addendum, he stated: "Let there be $A : B = C : D$ [where A to D are fluents], and then $A \times \mathrm{fl}(D) + D \times \mathrm{fl}(A) = B \times \mathrm{fl}(C) + C \times \mathrm{fl}(B)$ [where fl denotes fluxion]. This can be demonstrated in the same manner as the solution of Problem 1, or, alternatively, this way." At which point he gave a geometric demonstration. See Whiteside, *The mathematical papers of Isaac Newton*, III, pp. 330–333.

which is the quotient rule. The rule for the root can be easily obtained from his newly developed chain rule.

In this manuscript Leibniz went on to give an "Algorithm of the new analysis for maxima and minima, and for tangents" in which, surprisingly, no mention is made of either maxima or minima. Then he revised the first draft of this manuscript, possibly with a view to publication or to be included in some correspondence. In the revision, retitled *Methode nouvelle des Tangentes, ou de Maximis et Minimis. ita ut non sit opus tollere irrationalitates*, he gave credit to Fermat, Descartes, Hudde, and Sluse, but did not refer to Barrow at all [f. 1ᵛ; p. 62; 145; 131].

There is another manuscript written before Leibniz' first publication that never made it to print in his time. One remarkable thing about it is the manner in which he expressed himself on these matters, using subscripts for successive points on a curve or axis, providing proofs of some of the formulas of the differential calculus, and feeling sure of himself and of his new calculus at that time (the manuscript is undated but it may have been written about 1680). This shows in the grandiloquent title of the manuscript: "The elements of the new calculus for differences and sums, tangents and quadratures, maxima and minima, dimensions of lines [rectification], surfaces, solids, and for other things that transcend common means of calculation."[115] Some excerpts will show the polish of his ideas at this time. Leibniz was getting ready to publish.

On tangents he had the following to say [f. 1ʳ; pp. 149–150; 137]:

Let CC be a [curved] line whose axis is AB, let BC be ordinates normal to this axis, to be called y, the abscissæ on the axis AB to be called x. now the Differences CD of the abscissæ will be called dx such are $_1C_1D$, $_2C_2D$, $_3C_3D$ etc.[116] On the other hand the straight lines $_1D_2C$, $_2D_3C$, $_3D_4C$ (the differences of the ordinates BD) will be called dy. If now these dx and dy are taken to be infinitely small, ... then it is clear that the straight line joining these two points, such as $_2C_1C$, ... when produced to meet the axis at $_1T$, will be tangent to the curve, and $_1T_1B$ ... will be to the ordinate $_1B_1C$, as $_1C_1D$ is to $_1D_2C$ or if $_1T_1B$, or $_2T_2B$, etc. in general called t

---

[115] *Elementa calculi novi pro differentiis et summis, tangentibus et quadraturis, maximis et minimis, dimensionibus linearum, superficierum, solidorum, aliisque communem calculum transcendentibus.*

[116] I am using subscripts for convenience, but Leibniz just wrote (mostly) smaller numbers with the same baseline as the letters, as in $1C1d$, $2C2d$, $3C3d$, actually using here a lower-case d. The manuscript figure also shows an additional curve from A to the line $_4B_4C$, which serves no purpose. It was not shown in *Die Geschichte* and it is not shown here.

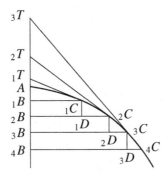

then $t : y :: [=] dx : dy$. And thus to find the differences of series is to find tangents.

In other words, to find the slope of the tangent at a point, y/t, one must find the quotient dy/dx, so that it is imperative to learn how to find dy at a given point of what we now call a function of x.

Turning to quadratures, he wrote (if I interpret Leibniz' punctuation correctly) [f. 1^r; p. 150; 138]:

> Furthermore differences are the opposites of sums. therefore the ordinates are the sums of their differences, so $_4B_4C$ [117] is the sum of all the differences, such as $_3D_4C$, $_2D_3C$, etc. up to A. even if they are infinite in number, which I describe as $\int dy$ equ. to y. Also the [area] of a figure, I describe by the sum of all the rectangles [obtained] from the ordinates times the differences of the abscissæ, such as $_1B_1D + _2B_2D + _3B_3D$. etc. On the other hand the small triangles $_1C_1D_2C$, $_2C_2D_3C$, etc. since they are infinitely small compared with the said rectangles, may be omitted with impunity, thus I denote in my calculus the area of the figure by $\int y \, dx$. or the sum of the rectangles, formed by each y, times the corresponding dx.

Then, "soaring high" (*altius assurgentes*), he obtained the area between a positive curve and the x-axis as follows [f. 1^r; pp. 150–151; 138]:

> ... let the given curve of the figure to be squared be EE,[118] of which the ordinates are EB [119] which we call $\underline{e}$. They are proportional to the differences

---

[117] The manuscript has $_4B_4B$.

[118] No figure was provided by Leibniz, but the one included here is drawn in his own style, with the origin at A and the direction of the positive x-axis down. For clarity, there are two implied positive y-axes, one directed to the left for the curve EE and one to the right for the curve CC.

[119] The manuscript has ED.

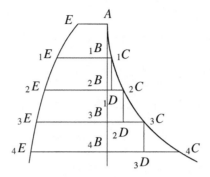

of the ordinates BC, or to dy, or let $._1B_1E : _2B_2E :: [=] _1D_2C : _2D_3C$, and so on, or let as A_1B is to $_1B_1C$, or, as $_1C_1D$ is to $_1D_2C$, or as dx is to dy be as a constant or always permanent straight line [segment] a is to $_1B_1E$ or e. then we have $dx : dy :: a : e$ or $e\,dx$ equ. $a\,dy$ Therefore $\int e\,dx$ equ $\int a\,dy$

To interpret this result, we split the figure in two and rearrange the parts as shown next, eliminating some of the clutter. We have also taken the liberty to replace Leibniz' B with x. Then we can restate the result in present-day

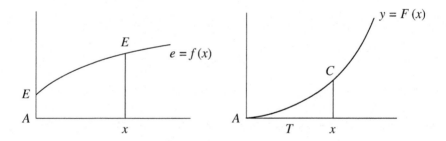

terms as follows. Let the curve EE represent a positive function, which we shall denote by $e = f(x)$, and let $y = F(x)$ be a function with graph AC such that for some constant a and any x,

$$\frac{dy}{dx} = \frac{e}{a}.$$

Then the area under the graph of e from A to x is

$$\int_A^x e\,dx = ay = aF(x).$$

If we take $a = 1$ and note that $F(A) = 0$, which Leibniz did not state explicitly, the result is that if $F'(x) = f(x)$ then

$$\int_A^x f(x)\, dx = F(x) - F(A),$$

the fundamental theorem of calculus. Of course, this is nothing new at this time. This result had been frequently used by Newton, who had not published anything on the calculus yet, but it had been proved in print by both Gregory and Barrow. In fact, Leibniz' presentation is quite similar to Barrow's except for the differential and integral notation. Perhaps this is precisely what Leibniz intended: to show how a known geometric result can be expressed in his new calculus.

Further ahead in the same manuscript, he discussed some of the other topics promised in the title [f. 1^v; pp. 151; 139–140]:

> ... for $_1C_2C$ is equ. to $\sqrt{dx \cdot dx + dy \cdot dy}$. From this we have at once a method for finding the length of the curve by means of some quadrature,
> ...

and then [ff. 1^v–2^r; p. 152; 141]:

> On the other hand if it is required to find the centers of lines, and the surfaces generated by their rotation, ex. gr. the surface [generated] by the rotation of the line AC. about AB., we must only find
>
> $$\int y\sqrt{dx\,dx + dy\,dy} \quad \ldots$$
>
> Therefore the [whole] thing is at once reduced to the quadrature of some plane figure, if we substitute for y and dy the values [obtained] from the nature of the ordinates and the tangents to the curve.

As for the differential calculus, he mostly repeated what he had already written in the July 1677 manuscript, but this time he included proofs. For instance, he found that [f. 2^v; p. 154; 143]

> $\overline{dxy}$ equ $\overline{x + dx}$ times $\overline{y + dy} - xy$ or $x\,dy + y\,dx + dx\,dy$ and the omission of the quantity $dx\,dy$, which is infinitely small with respect to the rest, given that dx and dy are infinitely small, will produce $x\,dy + y\,dx$ ...

For the quotient rule he computed

$$d\frac{y}{x} \quad \text{equ} \quad \frac{y+dy}{x+dx} - \frac{y}{x} \quad \text{or} \quad \frac{x\,dy - y\,dx}{xx + x\,dx} \text{ 120}$$

whereby writing xx for $xx + x\,dx$ since $x\,dx$ can be omitted as being infinitely small with respect to xx makes:

$$\frac{x\,dy - y\,dx}{xx} \cdots$$

Then he found the differentials of xyv (what is known today as Leibniz' rule) and $y/(vz)$, and concluded with the differentials of powers and roots.

5.14 THE ARITHMETICAL QUADRATURE

The longest mathematical manuscript of Leibniz, containing 51 propositions, did not even mention the calculus or differentials. But at the core of the manuscript, he proved rigorously that the area under a curve can be approximated as closely as desired by what we now call "Riemann sums." It was written in 1675–1676, shortly after a time in which he thought of dx as an infinitesimal unit. However, in *De quadratura arithmetica circuli ellipseos et hyperbolæ cujus corollarium est trigonometria sine tabulis* (The arithmetical quadrature of the circle, the ellipse, and the hyperbola whose corollary is trigonometry without tables) he showed tremendous maturity and progress, and for this reason it seems that its placement in this overview of his work should follow that of the 1680 work discussed in the previous section. Leibniz left the manuscript in Paris before his departure for Germany in October, hoping that its publication could be used to secure him a position at the Academy of Sciences. He left the manuscript, together with some drafts for corrections and extensions, in the hands of his friend Soudry, who would be in charge of preparing the final copy.

Soudry, unfortunately, died suddenly in 1678, and the manuscript remained in possession of Friedrich Adolf Hansen, who had discussed its publication with Soudry. He then put it in the hands of Christoph Brousseau, from Hanover but a resident of Paris. Brousseau, in turn, passed it to Isaac Arontz, who was planning to travel to Hanover, for delivery to Leibniz. This happened in 1680, while Soudry's copy became and remains lost.[121]

---

[120] The word "or" and the numerator of the last fraction are crossed out in the manuscript.

[121] These facts about the whereabouts of the *De quadratura arithmetica* manuscript are from Knobloch, "Leibniz et son manuscrit inédite sur la quadrature des sections coniques,"

After unsuccessfully attempting to publish the original manuscript, Leibniz gave up, and later he refused to publish it because it had become antiquated. Instead, he wrote an abbreviated version, *Compendium quadraturæ arithmeticæ*, using differential and integral notation, but it contains only a few sketches of proofs.

Before stating Proposition VI, which is at the heart of the manuscript, Leibniz set the scene with some necessary preliminaries based on a very detailed figure. The most interesting part of this figure for us includes the corner near the origin at A, which is too small a portion of Leibniz' larger figure to comfortably distinguish points, letters, and segments. For this reason, we have taken the liberty to greatly enlarge and somewhat distort a portion of the figure containing this corner, as shown on the next page, and we have kept only those elements that are relevant in the following discussion. The positive x-axis is located vertically down from A and the positive y-axis horizontally to the right of A. As we well know by now, Leibniz placed his subscripts on the left of their letters. However, the propositions and proofs that we are about to present are somewhat lengthy, and such an arrangement may be tiring for the present-day reader. Thus, in the enclosed figure and in the quotations and explanations that follow, we write the subscripts on the right, as is today's fashion. Then, if "$C_1 C_2$ etc. C_4" is a given curve (we would talk about a continuous function that has a tangent at each point), he constructed the tangents CT to the curve, and, from each T, the perpendicular to the corresponding ordinate BC generates the point D. These points form a new curve "$\underline{D_1 D_2}$ etc. D_4."[122] Next, the chords "$C_1 C_2$ etc. up to $C_3 C_4$" (here and in what follows, "up to" means to the end of the curve, however many C's are placed on it, and it was just convenience for Leibniz to end at $C_3 C_4$) cut the right angle TAB at [f. 8^v; p. 28; 49; 8][123]

> points M. that fall between the points T, as M_1 is between T_1 and T_2, and M_3 is between T_3 and T_4. and similarly from these points of intersection M drop other perpendiculars $M_1 N_1 P_1$, up to $M_3 N_3 P_3 \ldots$

1986. This paper also contains proofs of all these assertions. Knobloch's analysis can also be seen in the introduction to his 1993 edition of *De quadratura arithmetica*.

[122] Leibniz used a great deal of underlining in the *De quadratura arithmetica* and, in particular, in his presentation of Propositions VI and VII, of which this is an example. I have decided to omit it from this point on, but the interested reader can see it in Knobloch's edition.

[123] Page references are to manuscript folio, to Knobloch's edition of the *De quadratura arithmetica*, 1993; to Parmentier's French translation *Quadrature arithmétique*, 2004; and to Hamborg's German translation *Über die arithmetische Quadratur*, 2007.

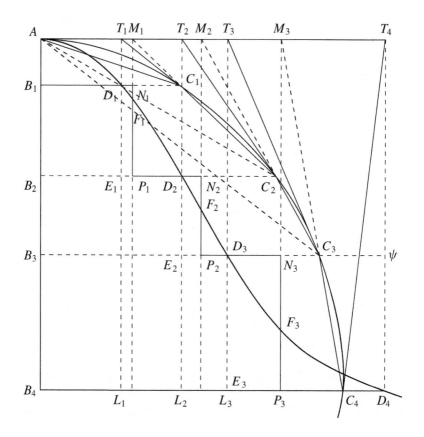

to the ordinates $B_1 N_1$, up to $B_4 N_4$.

Then the text of Leibniz' Proposition VI is as follows [f. 8^v; p. 29; 51; 8–9]:

Thus these rules and preparations [concluded], *I say that we can consider in the* [two] *curves points C between C_1 and C_4 and points D between D_1 and D_4 in such number and so close to each other, that the rectilinear gradiform space $N_1 B_1 B_4 P_3 N_3 P_2 N_2 P_1 N_1$ composed of the rectangles $N_1 B_1 B_2 P_1$, and others up to $N_3 B_3 B_4 P_3$ which are comprised between the ordinates $N_1 B_1$ or if needed their prolongations, and others up to $N_3 B_3$, and by the intervals $B_1 B_2$, and others, up to $B_3 B_4$.; differs by a quantity smaller than any given* [quantity] *whatever from the quadrilinear space $D_1 B_1 B_4 D_4 D_3$ etc. D_1 (delimited by the axis $B_1 B_4$, the outside ordinates $B_1 D_1$, $B_4 D_4$ and the new curve $D_1 D_2$ etc. D_4).*

And the same demonstration holds for any other mixtilinear and gradiform
space formed by the continuous application of straight lines on a certain axis.
One can take it as proved that the areas of figures can be obtained from the
sum of straight lines, as addressed by the method of indivisibles. It is required
that the curves or at least the parts in which they are split, have their concavity
on the same side, and lack reversion points.

Reversion points (*Puncta reversionum*) were immediately defined by Leibniz as those at which the ordinate is tangent to the curve. After this, he provided
a proof in eight stages, which he numbered in parentheses and later referred
to as articles.

In (1), the fact that the arc $D_1 D_2$ has no reversion points allowed Leibniz
to construct the point F_1 as the intersection of the segment $N_1 P_1$, parallel to
$T_1 D_1$, with the arc $D_1 D_2$. Then he added [f. 9^r; p. 30; 53; 10]: "In the same
manner the other portion $D_2 D_3$ will cut the straight line $N_2 P_2$ at F_2. etc."

In (2) he extended each segment $T D$ until it intersects the next ordinate at E.
He called each rectangle $N_1 B_1 B_2 P_1$, $N_2 B_2 B_3 P_2$, ... an Elementary Rectangle,
each rectangle $D_1 E_1 D_2$, $D_2 E_2 D_3$, ... the complementary rectangle (he left
it to the reader to imagine the remaining vertex of each of these rectangles).
Leibniz asserted that the [absolute value of the] difference between the partial
Quadrilinear $D_1 B_1 B_2 D_2 D_1$ and its elementary rectangle is smaller than the
corresponding complementary rectangle, which in current notation we write
as

$$|D_1 B_1 B_2 D_2 D_1 - N_1 B_1 B_2 P_1| < D_1 E_1 D_2.$$

He proved this by noting that the partial quadrilinear and the elementary rectangle have in common the Quintilinear space $D_1 B_1 B_2 P_1 F_1 D_1$. Subtracting it
from each leaves the trilinear spaces $D_2 P_1 F_1 D_2$ and $D_1 N_1 F_1 D_1$, respectively.
Thus, in today's notation,

$$|D_1 B_1 B_2 D_2 D_1 - N_1 B_1 B_2 P_1| < |D_2 P_1 F_1 D_2 - D_1 N_1 F_1 D_1|.$$

In (3) [f. 9^r; p. 31; 55; 11] Leibniz stated:

Then it will suffice to show that the difference between these two trilinears
is smaller than the complementary rectangle $D_1 E_1 D_2$, which is clear ...

and then explained how this is clear.

In (4) [f. 9; p. 31; 57; 11–12] he stated that the same is true for the difference
between the next partial Quadrilinear $D_2 B_2 B_3 D_3 D_2$ and the corresponding
elementary rectangle $N_2 B_2 B_3 [P_2]$, and for the others. Then the difference

between the total Quadrilinear $D_1 B_1 B_4 D_4 D_3$ etc. and the rectilinear gradiform space (*spatio rectilineo gradiformi*) $N_1 B_1 B_4 P_3 N_3 P_2 N_2 P_1 N_1$ is smaller than the sum of all the complementary rectangles. In current notation, and if there are n quadrilinears, this means that [124]

$$\left| \sum_{i=1}^{n} \left[D_i B_i B_{i+1} D_{i+1} D_i - N_i B_i B_{i+1} P_i \right] \right| < \sum_{i=1}^{n} D_i E_i D_{i+1}.$$

In (5) Leibniz observed that the rectangles in the last sum have bases $D_4 E_3 = D_4 L_3$, $D_3 E_2 = L_3 L_2$, up to $D_2 E_1 = L_2 L_1$, and heights $D_1 E_1 = B_1 B_2$, $D_2 E_2 = B_2 B_3$, ... If the heights $B_1 B_2$, $B_2 B_3$, ... are all equal, then the sum of the areas of these rectangles equals the product of the sum of the bases, $D_4 L_1$, and $B_3 B_4$. Otherwise, the sum of these areas is smaller than the product of the sum of the bases and the largest of the heights. If we assume that the largest height is the last, $B_3 B_4 = \psi D_4$, then the sum of the areas of all the complementary rectangles is less than or equal to that of the rectangle $\psi D_4 L_1$.

In (6) Leibniz simply stated that it follows from articles 4 and 5 that the difference between the total Quadrilinear $[D_1 B_1 B_4 D_4 D_3 D_2 D_1]$ and the gradiform space $[N_1 B_1 B_4 P_3 N_3 P_2 N_2 P_1 N_1]$ is smaller than the sum of the complementary rectangles, and therefore smaller than the rectangle $\psi D_4 L_1$.

In (7) [f. 9^v; pp. 31–32; 59; 12–13] he noted that the height of the rectangle $\psi D_4 L_1$ can be assumed to be smaller than any given quantity, since the D points can be taken arbitrarily close to each other, and in any number whatever. Thus, the area of the rectangle $\psi D_4 L_1$ can be made arbitrarily small.

Finally, in (8), Leibniz concluded that it follows from articles 6 and 7 that the difference between the total Quadrilinear $[D_1 B_1 B_4 D_4 D_3 D_2 D_1]$ and the gradiform space $[N_1 B_1 B_4 P_3 N_3 P_2 N_2 P_1 N_1]$ can be made smaller than any given quantity.

This is the modernity of Leibniz' thought and methods. The "infinitesimal unit" of his earlier tentative manuscripts is nowhere present in this treatment. Instead, this is a modern proof by *reductio ad absurdum* of the integrability of a large class of functions, and Leibniz was justifiably proud when he stated that [f. 10^v; p. 35; 71; 18]

I believe none is simpler and more natural, and closer to the direct demonstration, than that which not only simply shows, that there is no difference

---

[124] This makes implicit use of the triangle inequality. Leibniz had already established a result that amounts to the triangle inequality as Proposition V of this tract.

between two quantities, so that they are equal, (which elsewhere is usually proved [by showing] that neither is greater nor smaller than the other by a double reasoning) but one that [uses] only one middle term, namely either inscribed or circumscribed, not both simultaneously[.]

We could not have deflated Leibniz' joy by pointing out that there is some unfinished business in this work, because he had already anticipated us and finished that business when he wrote this assessment. The apparently unfinished business is this: we wanted to approximate the area "under" C_1C_2 etc. C_4, but we have only approximated the area "under" D_1D_2 etc. D_4. Leibniz took care of this in Proposition VII.[125] If we agree to call C_1C_2 etc. C_4 the first curve and D_1D_2 etc. D_4 the second curve, the conclusion of this proposition reads [f. 10^r; p. 33; 65; 15]:

... *the space comprised between the axis (on which the ordinates are drawn,) the two outside ordinates, and the second comprising curve, is double the space comprised between the first curve and the two straight lines whose ends are joined to the given right angle.*

Leibniz explained that these two right lines are the segments AC_1 and AC_3 [f. 10^r; p. 34; 67; 15]. Thus, the proposition states that the area of the quadrilinear figure $D_1B_1B_3D_3D_2D_1$ is twice that of the trilinear figure $C_1AC_3C_2C_1$ (in either case, $D_3D_2D_1$ and $C_3C_2C_1$ are arcs of curve). The proof, which could have used any number of ordinates rather than just three, was done in five stages.

In (1) Leibniz assumed that the conclusion does not hold, and denoted by Z the difference between twice the Trilinear area and the Quadrilinear one. Then constructed the polygons AC_1C_2A, $AC_1C_2C_3A$, and so on, reminded us of the construction of the points M on the Y-axis and of the perpendiculars MN, and then denoted their intersections with the ordinates B_2C_2, B_3C_3, etc. by S_1, S_2, etc. These were previously called P, so a new figure is not necessary.

In (2) [f. 10^v; p. 34; 67; 15] he assumed that the construction of the inscribed polygons is such that

the difference between the polygon $AC_1C_2C_3A$ and the Trilinear $C_1AC_3C_2C_1$ [recall that $C_3C_2C_1$ is an arc of curve here], as well as the difference between the rectilinear Gradiform space $B_1N_1S_1N_2S_2B_3B_1$, and the Quadrilinear $D_1B_1B_3D_3D_2D_1$ [same for $D_3D_2D_1$], each one separately, is smaller,

---

[125] Leibniz had already found this proposition in May 1673. See *Sämtliche Schriften und Briefe*, Ser. III, **1**, p. 115.

than the fourth part of Z. We can in fact continue the very same [process] until [each] is smaller than any given quantity.

In (3) Leibniz used Proposition 1 of this tract to show that [f. 10^v; p. 35; 69; 17]

twice the triangle AC_1C_2 is the rectangle $[N_1B_1B_2]$ and twice the triangle AC_2C_3 is the rectangle $[N_2B_2B_3]$, and so of the others ... [126]

Therefore, the sum of all the rectangles of this type, which is the Gradiform space, will be twice the sum of all the corresponding triangles, or the inscribed polygon.

In (4) he denoted the Quadrilinear $[D_1B_1B_3D_3D_2D_1]$ by Q, the inscribed polygon $[AC_1C_2C_3A]$ by P, and the Trilinear $[C_1AC_3C_2C_1$, where $C_3C_2C_1$ is an arc of curve] by T. The difference between Q and the gradiform space, whose area equals $2P$ by Article 3, is smaller than $Z/4$ by article 2. Also by article 2, the difference between P and T is smaller than $Z/4$. Leibniz did not use absolute values, but if we do, all this means that

$$|Q - 2T| \leq |Q - 2P| + |2P - 2T| < \tfrac{1}{4}Z + \tfrac{2}{4}Z = \tfrac{3}{4}Z.$$

Finally, Leibniz remarked in (5) that comparing this inequality with the definition of Z in article 1 shows that $|Q - 2T|$ is smaller than itself, which is absurd. Thus, the original assumption cannot hold, and $Q = 2T$.

What this result does is allow us to compute the area $AB_3C_3C_2C_1A$, between the C curve and the x-axis, as the sum of $T = \tfrac{1}{2}Q$ and that of the triangle AB_3C_3A, and the method is applicable if Q is easier to compute than the original area. Appropriately, in his *Historia et origo calculi differentialis*, Leibniz referred to this procedure as the transmutation method (*Methodo transmutationum*) [p. 11; 402; 44; 13].

This method and much additional work allowed Leibniz to perform the general quadrature of ellipses and hyperbolas in Proposition XLIII of *De*

---

[126] We can use his argument while keeping the notation now in use as follows. Prolong $C_2C_1M_1$ in this direction and drop a perpendicular AH to this prolongation (the reader is in charge of drawing the figure). Now prolong $B_1N_1C_1$ to the right and drop a perpendicular C_2G to this prolongation. The triangles AM_1H and C_2C_1G are similar because $\angle AM_1H = \angle C_2C_1G$. Then,

$$\frac{AH}{AM_1} = \frac{C_2G}{C_1C_2} \quad \text{or} \quad \frac{AH}{B_1N_1} = \frac{B_1B_2}{C_1C_2},$$

so that $B_1N_1 \times B_1B_2 = AH \times C_1C_2$. The right-hand side is twice the area of the triangle AC_1C_2 that has base C_1C_2 and height AH.

quadratura. We have neither the space nor the desire to embark on such a long trek. Instead, to give an example of the method of transmutation in the particular case of the circle, we follow a shorter route, as Leibniz himself did at the end of 1675, in an untitled and unsent letter in French to Jean-Paul de La Roque, editor of the *Journal des sçavans.*[127] Leibniz provided essentially the same figure as in *De quadratura* but different notation, but it is simpler if we use the same notation as above. He assumed that the curve $AC_1C_2C_3$ is an arc of the circle of radius a centered at $x = a$ whose length is no larger than that of a quadrant, and stated that (recall that we are replacing his new notation with the present one) [p. 91; 349]

> Since the curve $AC_1C_2C_3$ is an arc of circle, the curve of intercepts, namely $AD_1D_2D_3$, can be related to the right angle BAT, by this equation
>
> $$\frac{2az^2}{a^2 + z^2} \sqcap x,$$

where z represents the ordinate BD. Leibniz did not explain how he obtained this equation, but some explanation is preferable. Note that we would write the equation of the circle as $(x - a)^2 + y^2 = a^2$, which gives $y^2 = 2ax - x^2$. Then, according to Leibniz' own rules of the calculus, $2y\,dy = 2a\,dx - 2x\,dx$, or

$$\frac{dy}{dx} = \frac{a - x}{y}.$$

If we now imagine a small characteristic triangle at C_3, with legs dy and dx, it is clearly similar to the triangle $C_3D_3T_3$, and then

$$\frac{dy}{dx} = \frac{y - z}{x}.$$

Therefore, solving for z, using the quotient previously obtained for dy/dx, and the expression given above for y^2 yields

$$z = y - x\frac{dy}{dx} = y - x\frac{a - x}{y} = \frac{y^2 - ax + x^2}{y} = \frac{ax}{y} = \sqrt{\frac{a^2x}{2a - x}}.$$

Squaring and solving for x provides the equation given by Leibniz.

---

[127] Leibniz had first communicated the quadrature of the circle, but without proof, to Henry Oldenburg in the summer of 1674. It also appears as Proposition X in a communication to Huygens of October 1674. See *Sämtliche Schriften und Briefe*, Ser. III, **1**, pp. 117 and 165.

Direct use of the transmutation theorem calls for the evaluation of the area $Q = AB_3 D_3 D_2 D_1 A$, but Leibniz did this indirectly, computing first the area

$$AT_3 D_3 D_2 D_1 A = \int x \, dz = \int \frac{2az^2}{a^2 + z^2} \, dz$$

(these equations are for our convenience, because Leibniz did not use the integral sign in this letter or in *De quadratura*). To do this he employed [pp. 91–92; 349–350]

> the beautiful method of Nicolaus Mercator [long division] according to which since a is unity, and $\dfrac{x}{2}$ equal to $\dfrac{z^2}{1+z^2}$ the same $x[/2]$ will be equal, to $z^2 - z^4 + z^6 - z^8$ etc to infinity. And the sum of all the $x[/2]$, equal to the sum of all the $z^2 - z^4$, etc. now ... the sum of all the z^2 [from $z = 0$ to $z = b$] will be $\dfrac{b^2}{3}$ and the sum of all the z^4 will be $\dfrac{b^5}{5}$ etc. (by the quadrature of parabolas) hence the sum of all the x, or the space $AT_3 D_3 [D_2 D_1]A$, or the difference between the rectangle $T_3 AB_3[D_3]$, and twice the segment of the circle, $A[C_1 C_2]C_3 A$, will be $\dfrac{b^3}{3} - \dfrac{b^5}{5} + \dfrac{b^7}{7} - \dfrac{b^9}{9}$ etc.

Accepting that the radius of the circle is $a = 1$ and writing z instead of Leibniz' b for an arbitrary ordinate BD, these statements can be rendered in current notation as follows. First, the "sum" of all the quotients $x/2$ is

$$\int \frac{x}{2} \, dz = \int \frac{z^2}{1+z^2} \, dz = \frac{z^3}{3} - \frac{z^5}{5} + \frac{z^7}{7} - \frac{z^9}{9} + \cdots,$$

and then, noting that the area of the rectangle $T_3 AB_3 D_3$ is xz and that the circular segment $AC_1 C_2 C_3 A$ was denoted by T, Leibniz observed that

$$\int x \, dz = xz - 2T.$$

It follows from Lemma VII of *De quadratura*, a result included as the main theorem in Leibniz' letter, that

$$\tfrac{1}{2} Q = T = \frac{xz}{2} - \int \frac{x}{2} \, dz = \frac{xz}{2} - \frac{z^3}{3} - \frac{z^5}{5} + \frac{z^7}{7} - \frac{z^9}{9} + \cdots,$$

and adding to this the area of the triangle $AB_3 C_3 A = \tfrac{1}{2}xy$ gives the desired quadrature

$$\int y \, dx = \frac{xz + xy}{2} - \frac{z^3}{3} + \frac{z^5}{5} - \frac{z^7}{7} + \cdots.$$

Although it is possible to rewrite the first fraction in terms of z only, it would not add to the meaning of the result. Just observe that at the point $(1, 1)$ of this circle its tangent is vertical, so that the corresponding value of z is 1. Putting $x = y = z = 1$ in the preceding equation gives the area of a quadrant of the circle, which is $\pi/4$; that is,

$$\frac{\pi}{4} = 1 - \frac{1}{3} + \frac{1}{5} - \frac{1}{7} + \cdots .$$

Leibniz obtained this identity in *De quadratura* in the following form [f. 24^r; p. 79; 219; 82]:

PROPOSITION XXXII

The Circle is to the circumscribed Square, or the arc of the Quadrant is to the Diameter, as

$$\frac{1}{1} - \frac{1}{3} + \frac{1}{5} - \frac{1}{7} + \frac{1}{9} - \frac{1}{11} \text{ etc.}$$

is to unity.

Clearly, Leibniz meant the areas of the circle and the square and the lengths of the quadrant and the diameter. This expansion is known as the Leibniz formula, and also sometimes as the Gregory–Leibniz formula.[128]

5.15 LEIBNIZ' PUBLICATIONS

Leibniz may have been ready to publish by 1680, but he also seemed somewhat reluctant, at least judging by the amount of time he let pass before he finally did so, perhaps fearing that his infinitesimal quantities may not be well received.

---

[128] In a letter of 15 February 1671 to John Collins, Gregory wrote: "Sit radius $= r$, arcus $= a$, tangens $= t$, secans $= s$, erit

$$a = t - \frac{t^3}{3r^2} + \frac{t^5}{5r^4} - \frac{t^7}{7r^6} + \frac{t^9}{9r^8},"$$

(quoted from Turnbull, *James Gregory tercentenary memorial volume*, p. 170 and *The Correspondence of Isaac Newton*, I, 1959, p. 62). Putting $r = 1$ and $t = 1$ for $a = \pi/4$ gives the Leibniz formula, but Gregory never wrote it.

However, the Leibniz and Gregory series had already appeared in Jyesthadeva's *Yuktibhasa* and in the anonymous *Tantrasangrahavyakhya*, and are very likely the work of Madhava. For details see Rajagopal and Rangachari, "On an untapped source of medieval Keralese mathematics," 1978, pp. 91–96.

But he had essentially completed his investigations and he may have been concerned that his friend and colleague Ehrenfried Walter von Tschirnhaus (1651–1708) might get ahead of him and publish his own thoughts on the subject. When Leibniz helped some friends found the journal *Acta Eruditorum*

The first page of Leibniz' publication on the differential calculus
From Struik, *A source book in mathematics, 1200–1800.*

in 1682, he was invited to submit a paper, and Leibniz chose to write on the quadrature of the circle, giving the formula stated above but without a proof.[129]

Finally, two years later he unveiled his differential calculus, in the October

---

[129] "De vera proportione circuli ad quadratum circumscriptum in numeris rationalibus, à G. G. Leibnitio expressa," 1682.

issue of the same journal, with a paper entitled "A new method for maxima and minima, as well as tangents, not hampered by fractional nor irrational quantities, and a singular calculus created for them, by G. W. L."[130] It begins as follows [p. 467; 168: 220]: [131]

Let an axis AX, and several curves, such as VV, WW, YY, ZZ, whose

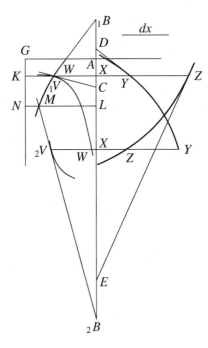

ordinates, normal to the axis, VX, WX, YX, ZX, are called respectively, v, w, y, z; & this [portion] AX cut off from the axis, is called x. Let the tangents VB,[132] WC, YD, ZE intersect the axis respectively at points B, C, D, E. Now some arbitrarily selected straight line is called dx, & the line that is to dx, as v (or w, or y, or z) is to XB (or XC, or XD, or XE)[133] is

---

[130] "Nova methodvs pro maximis et minimis, itemque tengentibus, quæ nec fractas, nec irrationales quantitates moratur, & singulare pro illis calculi genus, per G. G. L."

[131] Page references are to the *Acta*, to *Opera mathematica* and to *Leibnizens Mathematische Schriften* in this order. I follow the *Acta*'s text except for obvious errors.

[132] Two possible arrangements are shown in the figure, labeled $_1V_1B$ and $_2V_2B$.

[133] The last four segments are incorrectly given in the *Acta* and in the *Opera* as VB, WC, YD, and ZE, but the mistake was corrected in the *Schriften*. The mistake can be attributed to Leibniz himself, since the printed version is, on this point, identical to a manuscript draft that

called dv (or dw, or dy or dz) or the difference of these v (or these w, or y, or z) This posed the rules of calculus are such:

If a constant quantity a is given, then da equals 0, & $d\,\overline{ax}$ will be equ. $a\,dx$: if y is equ. v (or if any ordinate whatever of a curve YY, is equal to the any ordinate whatever of a corresponding curve VV) then dy will be equ. dv. Now *Addition & Subtraction*: if $z - y + w + x$, is equ. v, it shall be $d\,\overline{z - y + w + x}$ [134] or dv, equ. $dz - dy + dw + dx$. *Multiplication,* $d\,\overline{xv}$ equ. $x\,dv + v\,dx$, or given that y equ. xv, dy becomes equ. $x\,dv + v\,dx$. In fact it is a matter of choice whether we employ a formula, such as xv, or shorten it to a letter, such as y. Note that & x & dx are treated in this calculus, in the same manner, as y & dy, or any other indeterminate letter with its difference. Note also that it is not always possible to go back from a differential Equation, except with some caution, of which [we will talk] elsewhere. On to *Division,*

$$d\frac{v}{y} \text{ or } \left(\text{given that } z \text{ equ. to } \frac{v}{y} \right) dz \text{ equ. } \frac{\pm v\,dy \mp y\,dv}{yy}.$$

It is not necessary to continue in detail. In the next paragraph he remarked that [p. 468; 168; 221]

at the very moment when v neither increases nor decreases, but is flat, dv is exactly equ. 0, ... at this very same place v, namely the ordinate LM, is *maximum* (or ... *Minimum*), ...

A few lines below he defined point of inflection (*punctum flexus contrarii*), characterized it in terms of changing concavity, and (carelessly) stated the following:

Therefore a point of inflection takes place, when neither v nor dv are 0, and yet $d\,dv$ is 0.

After dealing (not as clearly as one might wish) with the matter of the ambiguous signs that have just shown up in the case of division, Leibniz went on to deal with powers and roots, very much as in the unpublished manuscript

---

is still extant (probably not the one used by the printer, since there are significant differences between the draft and the printed version) with the same title as the paper (Bodemann XXXV, vol. V / 25, f. 3^r). In this translation I use the corrected version.

[134] The overbar is too long in the *Acta* and it covers the d. This was corrected in the *Opera* and in the *Schriften*. The same for the product rule below. There are no such overbars in the manuscript draft mentioned in the previous footnote.

discussed in Section 5.13 [p. 469; 169; 222], specifically stating the equations "$dx^a, = a.x^{a-1} dx$" and

$$d \sqrt[b]{x^a} = \frac{a}{b} dx \sqrt[b]{x^{a-b}}.^{135}$$

Then he named this calculus the "differential" calculus, which is the name that we still use today, and then he tooted his own horn for a little while, asserting, in particular, that [p. 470; 170; 223]

> it is also clear that our method extends to transcendental curves, which cannot be reduced by Algebraic calculation, or do not have a fixed degree, ...

But he did not include such an extension in this paper. Leibniz gave next (after defining *tangent* as a line passing through two points of a curve at an infinitely small distance) three examples of the differential calculus. He stated the first example after denoting division by [pp. 470; 170; 223–224]

> $x : y$ which is the same as x divid. by y or $\frac{x}{y}$. Let the *first* or given equation be,
>
> $$x : y + \overline{a + bx} \ \overline{c - xx} : \text{squar.} \ \overline{ex + fxx}$$
> $$+ ax\sqrt{gg + yy} + yy : \sqrt{hh + lx + mxx} \quad \text{equ. } 0.^{136}$$
>
> expressing the relation between x & y, ..., assuming that $a, b, c, e, g, h, l,$ m are given [constants]; it is desired

to find the quotient $dx : dy$, to make a long story shorter.

The stated equation is somewhat complicated and, after solving the problem, Leibniz explained that this was done on purpose to show the power of the calculus rules. But since we are already familiar with that power, we deal

---

[135] The *Acta* used capital X in these equations, except for dx instead of dX. The *Opera* used X throughout, but the *Schriften* used only x.

[136] The *Acta* and the *Opera* reproduced this equation as in the manuscript draft, except for using the abbreviation "quadrat." (translated here as "squar.") instead of "quadr." However, The *Schriften* did not place a bar over $ex + fxx$ but placed one over $\overline{a + bx} \ \overline{c - xx}$. We would prefer to write this equation as

$$\frac{x}{y} + \frac{(a + bx)(c - x^2)}{(ex + fx^2)^2} + ax\sqrt{g^2 + y^2} + \frac{y^2}{\sqrt{h^2 + lx + mx^2}} = 0.$$

In one of a large number of (disorganized) enclosures that accompany the manuscript draft (on a page numbered 32 in a more recent hand), Leibniz himself wrote this equation using fraction bars but no parentheses.

with the third term only as an example, thereby avoiding the use of the quotient rule. After denoting $g^2 + y^2$ by r, Leibniz found that [p. 471; 170; 224] "$d, ax\sqrt{r}$ is $+ ax\, dr : 2\sqrt{r} + a\, dx\sqrt{r}$," [137] which we would prefer to read as

$$d(ax\sqrt{r}) = \frac{ax}{2\sqrt{r}}\, dr + a\sqrt{r}\, dx.$$

A few lines below he noticed that "dr is $2y\, dy$" and, therefore,

$$d(ax\sqrt{r}) = \frac{axy}{\sqrt{r}}\, dy + a\sqrt{r}\, dx$$

(an equation that Leibniz did not write explicitly). This is an example of the chain rule, although not as involved as the one in his November 1676 paper.

By the new calculus rules and the use of substitutions, Leibniz found the differentials of the remaining three terms, expressed them in terms of dx and dy only, added the results to the one already found here, equated the sum to zero, and solved for $dx : dy$.

In the second example Leibniz obtained the refraction law [pp. 471–472; 171–172; 224–225], and the third, which is the crowning point of this paper, is the solution of a difficult, long-standing problem, introducing something new in what we now call mathematical analysis. The problem had been posed by Florimond de Beaune (1601–1652) to Descartes, who could not solve it. This is the problem in Leibniz' words and with reference to his figure [p. 473; 172; 226]:

> To find a curve WW of such a nature that, its tangent WC extended to the axis, XC is always equal to the same constant straight line, a.

Leibniz had first attempted the solution in a manuscript of July 1676, *Methodus tangentium inversa*, but was not completely successful. Since the slope of the tangent is $XW/XC = w/a$, we have to solve the equation

$$\frac{dw}{dx} = \frac{w}{a},$$

which, having set dx equal to a constant b, Leibniz wrote as "w æqu. $\frac{a}{b}\, dw$." However it is expressed, what we have here is a differential equation, possibly the first one ever in print.[138]

---

[137] This is the *Schriften*'s corrected formula. The *Acta* and the *Opera* have $d, ax\sqrt{r}$ is $- ax\, dr : 2r\sqrt{} + a\, dx\sqrt{r}$. The manuscript draft has $d, xa\sqrt{r}$ is $- \overline{xa\, dr : 2\sqrt{r}} + dx\, a\sqrt{r}$.

[138] Leibniz had already defined differential equation (*æquationem differentialem*) in his letter to Oldenburg of 21 June 1677 [p. 156; 243; 172; 215; 221].

From a present-day course in differential equations we know that the solutions are

$$w = Ce^{x/a}, \qquad \text{or} \qquad x = a\,l\!\left(\frac{w}{C}\right),$$

where C is an arbitrary constant and l denotes the natural logarithm, but there was no such course when Leibniz wrote this equation, so he improvised. To write his solution, first in current terminology, consider an arithmetic progression of abscissas: $x, x+dx, x+2dx, \ldots$, and if w is the ordinate corresponding to x, the one corresponding to $x + dx$ is

$$w + dw = w + \frac{w}{a}\,dx = w + \frac{b}{a}w = w\!\left(1 + \frac{b}{a}\right).$$

This is how each ordinate is obtained from the previous one, so that they form a geometric progression with ratio $1 + b/a$. Recalling now our discussions in Chapter 2 on geometric and arithmetic progressions, we conclude that the relationship between the given abscissas and their corresponding ordinates is logarithmic, which is more simply expressed by the equation $x = a\,l(w/C)$.

Leibniz' own solution is much briefer, for, without writing a single additional equation, he simply expressed himself, and concluded his paper, as follows:

> ... if the x are in arithmetic progression, the w will be in Geometric progression, or if the w are numbers, the x will be logarithms: therefore the curve WW is logarithmic.

In this manner, the solution of the de Beaune problem was a feather in the cap of the new differential calculus.

Now let us backtrack to the start of this paper. Let us use our imagination and try to place ourselves in the position of one of its readers who was a contemporary of Leibniz. Such a reader would not have studied the calculus that we have studied and would not have read the manuscripts that we have previously discussed at length. What would such a reader make of all this? The description of the curves and their tangents seems to be clear, but what is all that about dx and dv and the rules of the calculus? The calculus of what? In fact, an eminent reader who was a contemporary of Leibniz, Johann Bernoulli, a younger brother of Jakob who would become professor of mathematics first at Groningen and then at Basel, wrote down his opinion for the record in his autobiography: this calculus of Leibniz was "an enigma rather than an explication ..." [139]

But let us not pursue this line of thought at this moment. Since Newton had not published a line on this subject yet, this is how the differential calculus was presented to the world. Two years later, Leibniz gave us a little glimpse at his integral calculus, but at that time without its present name, [140] although he called it "summatorium, aut tetragonisticum" [p. 193; 233].

There is too much talk in this paper, but eventually Leibniz got around to proving that the quadratures of the circle and the hyperbola, that is, what we now call the integrals

$$\int \sqrt{a^2 \pm x^2}\, dx,$$

are transcendental, because if they were algebraic [pp. 190; 228–229]

> it would follow with their help that the angle, or proportion or logarithm can
> be cut in a given proportion from straight line to straight line ...

(nobody ever said that he was a model of clarity).

It was in this paper that the integral sign appeared in print for the first time. There is a grand total of six integral formulas in it, of which the first three are variants of $\frac{1}{2}xx = \int x\,dx$, and the fourth is given as follows [p. 192; 231]:

> Let a be an arc, x its versed sine [that is, $x = 1 - \cos a$], it will be $a = \int dx : \sqrt{2x - xx}$. [141]

The next one is about the cycloid, a curve about which we are not as emotional today as they used to be back then. The last one, a repeat of the fourth but with a misprint [p. 193; 233], is there just to exhort the reader not to forget to write the dx.

Leibniz continued working, writing, and publishing on his new calculus, including a proof of the fundamental theorem of calculus that is strongly

---

[139] Fragments from Bernoulli's autobiography are contained in Wolf, "Erinnerungen an Johann I Bernoulli aus Basel," in Grunert, ed., *Archiv der Mathematik und Physik*, 1849. This purposely incomplete quotation (to be continued) is from page 693.

[140] "De geometria recondita et analysi indivisibilium atque infinitorum" (Of abstruse geometry and the analysis of indivisibles and infinites), 1686. Page references are to *Opera Mathematica* and *Leibnizens Mathematische Schriften*, in this order.

[141] The overbar covers the integral sign in the *Opera* but appears correctly in the *Schriften*.

reminiscent of Barrow's.[142] Two years later he finally incorporated some transcendental functions in his version of the calculus. Specifically, he differentiated the general exponential function, following a suggestion by Johann Bernoulli, as follows [p. 314; 330; 324]:[143]

Let $x^v = y$ then $v \log x = \log y$; now

$$\log x = \int \frac{dx}{x} \quad \text{and} \quad \log y = \int \frac{dy}{y}.$$

Therefore,

$$v \int \frac{dx}{x} = \frac{dy}{y} :$$

which differentiating, becomes

$$v \frac{dx}{x} + dv \log x = \frac{dy}{y}. \text{[144]}$$

From this equation Leibniz should have concluded (and possibly did) that

$$dy = d(x^v) = vx^{v-1}dx + x^v dv \, \log x,$$

but, instead, his published equation contains a misprint. However, Leibniz' paper contains the correct formula for the particular case in which v is a constant e, which is $d(x^e) = ex^{e-1} \, dx$ [p. 314; 331; 325].

In the same paper Leibniz showed some insight into the nature of second-order differentials [p. 314; 331; 325]. Although without a reference to the de Beaune problem that he had solved in 1684, he considered the same differential equation, which, replacing w with x and x with y, is

$$\frac{dx}{dy} = \frac{x}{a}.$$

---

[142] This is, essentially, the proof in the manuscript discussed at the end of Section 5.13. It appeared in "Supplementum geometriæ dimensoriæ, seu generalissima omnium tetragonismorum effectio per motum: similiterque multiplex constructio lineæ ex data tangentium conditione" (Supplement on the geometry of measurement; or most generally on all quadratures effected by motion: similarly, multiple constructions of curves from data on the tangent condition), 1693.

[143] "Responsio ad nonnullas difficultates, a Dn. Bernardo Niewentiit circa methodum differentialem seu infinitesimalem motas" (Response to several difficulties of Dn. Bernardo Niewentiit about the differential method or motivated by infinitesimals), 1695. Page references are to the three works mentioned in the bibliography.

[144] I have modernized Leibniz' notation. Here is his statement in the *Opera*: "Nempe sit $x.^v = y$ fiet $v.\log.x = \log.y$; jam $\log.x = \int, dx : x$ & $\log.y = \int, dy : y$. Ergo $v. \int, dx : x = \int, dy : y$: quam differentiando, fit $v \, dx : x + dv. \log.x = dy : y$."

Or $dx = xdy:a$. Therefore $ddx = dxdy:a$. Whence taking $dy:a$ from the prior equation becomes $xddx = dxdx$: whence it is clear that x is to dx, as dx is to ddx.

Today we would write Leibniz' proportion as

$$\frac{x}{dx} = \frac{dx}{ddx},$$

which shows in this example that dx is the mean proportional between x and ddx.[145]

5.16 THE AFTERMATH

For reasons that may seem incomprehensible today, a dispute flared up between the partisans of Newton's calculus and the partisans of Leibniz' calculus. There were even allegations of plagiarism raised against Leibniz, for he had been to London, where he might have seen a copy of *De analysi* and stolen Newton's ideas. Such allegations are pure nonsense, because Newton's calculus is as similar to Leibniz' calculus as an egg is to a chestnut, and the latter's efforts to arrive at an understanding through notation are palpable.[146] However, Leibniz had purchased a copy of Barrow's *Lectiones geometricæ* while in London, and, doubtlessly, he had studied it and may have found inspiration from it to view quadratures and differentiation as inverse processes. There is also a similarity between Barrow's geometric constructions and those of Leibniz, but also a very profound difference. For, whereas Barrow seemed to be content in that geometric milieu, Leibniz felt the need to leave it through the choice of a notation that would allow him to turn such procedures into an algorithmic calculus. Thus, his work represents a leap into the future from where Barrow stood.[147]

There is also a substantial difference in approach between Newton and Leibniz. Newton was basically a practical man, interested in differentiating

---

[145] For additional information on Leibniz' development of the calculus consult Parmentier's *La naissance du calcul différentiel*, 2000. This is a collection of 26 papers by Leibniz, originally published in the *Acta Eruditorum*, translated into French, with notes and an introduction.

[146] For a full account of the dispute see Hall, *Philosophers at war: The quarrel between Newton and Leibniz*, 1980. A popular account can be found in Bardi, *The calculus wars. Newton, Leibniz, and the greatest mathematical clash of all time*, 2006.

[147] For a possible influence of the work of Barrow, Newton, and Leibniz on each other see Feingold, "Newton, Leibniz, and Barrow too. An attempt at a reinterpretation," 1993.

and integrating everything in sight, with applications to practical problems, and from the very beginning provided us with what we now call tables of integrals. Leibniz was basically interested in finding the working algorithms of the calculus, but did not neglect applications in his early work. We have seen how he provided the answer to the de Beaune problem via the solution of the first ever differential equation, and this was a triumph for Leibniz' calculus.

And then the inevitable happened. Newton and Leibniz had given to the world a calculus that worked but whose foundations were shaky and insufficiently understood. Someone was bound to pick on this, and someone did. The first was the Dutch physician and mathematician Bernard Nieuwentijt (1654–1718).[148] He could not understand a number of points from either Barrow, or Newton, or Leibniz, from quantities that now are positive but then are zero to how infinitesimals could add up to something finite. Leibniz printed a reply,[149] but the fact is that Leibniz, or anyone at that time, did not understand these fine points either.

This meant that the problem could not go away until it was resolved. The leading critic in the British Isles was George Berkeley (1685–1753), one of the most prominent empiricist philosophers, a prolific writer and critic of science, philosophy, mathematics, and politics. In 1734, the same year he became the Anglican bishop of Cloyne (County Cork, Ireland), he published *The analyst; or a discourse addressed to an infidel mathematician*,[150] his only work written in England. Berkeley found this infidel guilty of [I]

> the misleading of unwary Persons in matters of the highest Concernment [religion], and whereof your mathematical Knowledge can by no means qualify you to be a competent Judge.

Then, it was only fair for Berkeley to [II]

> take the Liberty to inquire into the Object, Principles, and Method of Demonstration admitted by the Mathematicians of the present Age, ...

---

[148] *Considerationes circa analyseos ad quantitates infinite parvas applicatæ principia, & calculi differentialis usum in resolvendis problematibus geometricis*, 1694; and *Analysis infinitorum, seu curvilineorum proprietates ex polygonorum natura deductæ*, 1695.

[149] "Responsio ad nonnullas difficultates a Dn. Bernardo Niewentiit circa methodum differentialem seu infinitesimalen motas."

[150] Since this booklet is divided into fifty sections, references will be given by section numbers in roman numerals. It is generally believed that the infidel mathematician of the title was the Astronomer Royal Edmund Halley (see the Editor's Introduction to vol. 4 of *The works of George Berkeley Bishop of Cloyne*, pp. 56–57). The relevant portions of *The analyst* are reproduced in Smith, *A source book in mathematics*, 1929, pp. 627–634.

In sections III to VIII Berkeley criticized the method of fluxions but just in
verbal form, remarking on how difficult it is for the imagination to comprehend
"those Increments of the flowing Quantities in *statu nascenti*" [IV], or to
"digest a second or third Fluxion . . ." [VII]. But then he became more serious
in criticizing some statements made by Newton after that of Lemma II in
Book II of the *Principia* [Motte's translation, p. 18]. Newton had designated
the "momentary increments or decrements [of flowing quantities] by the name
of Moments;" and then stated that if the sides of a rectangle, A and B, are
"increasing or decreasing by a continual flux" with velocities a and b, then
"the moment or mutation of the generated rectangle AB will be $aB + Ba$;"
and then provided a very dubious proof of this fact [p. 19]. In Berkeley's
view, however, if you subtract AB from the product of $A + a$ and $B + b$ what
remains is $aB + Ba + ab$, and this gave him occasion to turn Newton's own
words against him [IX]:

> And this holds universally be the Quantities a and b what they will, big or
> little, Finite or Infinitesimal, Increments, Moments, or Velocities. Nor will
> it avail to say that ab is a Quantity exceedingly small: Since we are told that
> *in rebus mathematicis errores quàm minimi non sunt contemnendi.*[151]

As for Newton's way of finding the fluxion of x^n, reproduced on page 317, he
had the following to say [XIII]:

> For when it is said, let the Increments vanish,[152] *i. e.* let the Increments
> be nothing, or let there be no Increments, the former Supposition that the
> Increments were something, or that there were Increments, is destroyed, and
> yet the Consequence of that Supposition, *i. e.* an Expression got by virtue
> thereof, is retained.

This is enough to get the idea of Berkeley's objections, which he extended
to Leibniz' calculus too. His problem was not with the results, which he
accepted as valid, but rather with the faulty logic used to arrive at them. He
concluded section XXXV with some questions that are frequently quoted (it
is irresistible):

> And what are these fluxions? The Velocities of evanescent Increments? And
> what are these same evanescent Increments? They are neither finite Quan-
> tities nor Quantities infinitely small, nor yet nothing. May we not call them
> the Ghosts of departed Quantities?

---

[151] Whiteside's translation of these Latin words can be found in the last sentence of the
quotation from Newton's *De quadratura* on page 317.

[152] They are called "augments" in Whiteside's translation.

The calculus was in trouble and mathematicians knew it. They rose to its defense and provided counterarguments to Berkeley's own, but the bishop held his own because he was on the right side. Explanations were necessary. They ware attempted in a big way by Colin Maclaurin (1698–1746), a professor first at Aberdeen and later at Edinburgh. In response to Berkeley's criticisms he wrote a lengthy treatise, seeking to justify the calculus by means of Euclidean and Archimedean geometry.[153] This impossible task may have been instrumental in turning mathematicians in the British Isles away from infinitesimals.

The continental mathematicians were not so squeamish, and they plunged right in. We are in debt to Jakob and Johann Bernoulli, who were the first to make sense of Leibniz' calculus. Johann had called this calculus an enigma, as we saw above, but that was an incomplete quotation. Leinbiz' calculus was [154]

> an enigma rather than an explication; but that was enough for us, to master all its secrets in a few days, as witnessed by the quantity of papers that we published next on the subject of infinitesimals.

The result was that Leibniz and the Bernoulli brothers created most of the elementary differential calculus before the end of the seventeenth century.

Johann Bernoulli was the first to write a differential calculus textbook, the *Lectiones de calculo differentialium*, from lectures of 1691 and 1692, but it was not published until 1922. The reason is that Bernoulli had sold the rights to his work to Guillaume François Antoine de l'Hospital (1661–1701), marquis de Saint-Mesme, count of Autremont, Lord of Oucques, etc.[155] L'Hospital, a mathematician himself, had been tutored by Bernoulli on the new differential calculus during the latter's stay in Paris (1691–1692) and had acquired the exclusive rights to those teachings. In 1696 l'Hospital published the first differential calculus book to appear in print: *Analyse des inifiniment petits*

---

[153] *Treatise on fluxions*, 1742. In Articles 858 to 861 of Book II (based on Articles 255 and 261 of Book I), Maclaurin gave, proved, and illustrated by example what we now call the second derivative test for maxima and minima [Art. 858]: "When the first fluxion of the ordinate vanishes, if at the same time its second fluxion is positive, the ordinate is then a *minimum*, but it is a *maximum* if its second fluxion is negative . . ." A portion of the *Treatise* containing this topic is reproduced in Struik, *A source book in mathematics, 1200–1800*, pp. 338–341; reprinted in Calinger, *Classics of Mathematics*, pp. 476–478.

[154] Translated from Wolf, "Erinnerungen an Johann I Bernoulli aus Basel," p. 693.

[155] But usually referred to as the Marquis de l'Hôpital (using the modern French spelling), today as well as then, at least in writings that I have seen by Berkeley, Johann Bernoulli, and Leibniz.

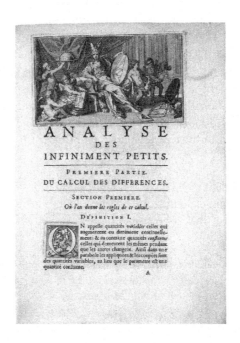

Reproduced from the virtual exhibition *El legado de las matemáticas*:
de Euclides a Newton, los genios a través de sus libros, Sevilla, 2000.

pour l'intelligence des lignes courbes, and acknowledged that he had made
free use of the discoveries of Leibniz and the Bernoullis. Besides marking the
first appearance in print of a complete presentation of the new calculus, this
book is distinguished by the use of the word *intégrale* to replace *summatorium*,
as proposed by the Bernoulli brothers,[156] and a new theorem known and used
today as l'Hôpital's rule. It appeared in Part I, Section IX, page 145.

   Mixing current terminology with l'Hospital's notation, the function represented by the curve AMD is the quotient of the function represented by ANB
over that represented by COB (the ordinates AC and PO are positive, but they
are drawn below the x-axis to avoid clutter). These last two functions vanish at
the point $x = a$, represented by the letter B. L'Hospital stated his proposition

---

[156] This word was introduced by Jakob Bernoulli in "Analysis problematis antehac
propositi," 1690, p. 218 = *Jacobi Bernoulli, Basileensis, Opera*, Tomus primus, 1744,
p. 423.

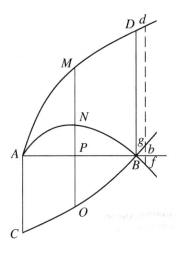

as follows:

PROPOSITION I.

Problem.

163. *Let AMD be a curve (AP = x, PM = y, AB = a) such that the value of the ordinate y is expressed by a fraction, the numerator & denominator of which, each of them becomes zero when x = a, that is when the point P falls over* [coincides with] *the given point B. It is asked what must then be the value of the ordinate BD.*

Then l'Hospital gave a procedure for this task (which we would be very happy to revise today) using "an ordinate *bd* infinitely close to *BD*," in which he revealed what to do:

> to find the ratio of *bg* to *bf*. ... & therefore if we take the differential of the numerator, & divide it by that of the denominator, after having made $x = a = Ab$ or AB, we'll have the value that is sought of the ordinate *bd* or *BD*.

However, shortly after l'Hospital's death, Bernoulli felt free to claim this rule as his own discovery.[157] This claim was validated much later when his corre-

---

[157] In "Perfectio regulæ suæ edita in libro Gall. Analyse des infiniment petits art. 163. pro determinando valore fractionis, cujus Numerator & Denominator certo casu evanescunt," 1704, p. 376 = *Johannis Bernoulli, Opera Omnia*, 1742, vol. I, N°. LXXI, p. 401.

spondence was published, for in a letter of 22 July 1694, Bernoulli had indeed communicated it to l'Hospital along with the first example in the *Analyse*.[158] Bernoulli (not Leibniz) had also developed the integral calculus, which he wrote under the title *Lectiones mathematicæ de methodo integralium, aliisque conscriptæ in usum Ill. Marchionis Hospitali cum auctor Parisii agaret annis 1691 et 1692* (Mathematical lectures on the method of integrals, and on other topics written for the use of the Marquis de l'Hospital as the author gave them in Paris in the years 1691 and 1692).[159] Here Bernoulli developed what would eventually be known as the method of substitution; and taught it correctly, which is unusual in today's calculus textbooks.

L'Hospital's book became very popular and went through multiple editions. It was not until the mid eighteenth century that the competition caught up. In Italy, Maria Gaetana Agnesi (1718–1799) published the *Instituzioni analitiche ad uso della gioventù italiana* in 1748, two years before she was

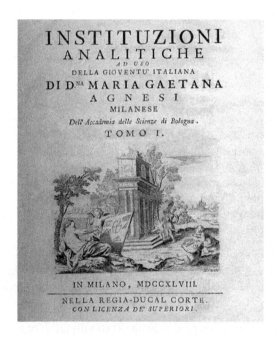

INSTITUZIONI
ANALITICHE
AD USO
DELLA GIOVENTU' ITALIANA
DI D.ᴺᴬ MARIA GAETANA
AGNESI
MILANESE
Dell' Accademia delle Scienze di Bologna.

TOMO I.

IN MILANO, MDCCXLVIII.

NELLA REGIA-DUCAL CORTE.
CON LICENZA DE' SUPERIORI.

---

[158] *Der Briefwechsel von Johann Bernoulli*, vol. I, 1955 pp. 235–236.
[159] They appeared first in *Johannis Bernoulli, Opera omnia*. The integral calculus is in volume III, pp. 385–558. Some selections, containing the method of substitution, are translated into English in Struik, *A source book in mathematics, 1200–1800*, pp. 324–328.

appointed a professor at the University of Bologna. However, Agnesi, without refusing the appointment, never took the position. Book 2 is devoted to the *Calcolo differenziale* (including maxima and minima and the study of curves), Book 3 to the *Calcolo integrale*, and Book 4 to *Equazioni differenziali*. Here is an example of differentiation from page 465:

Let $\sqrt{ax + xx + \sqrt[4]{a^4 - x^4}}$, that is $\overline{ax + xx + \sqrt[4]{a^4 - x^4}}^{\frac{1}{2}}$, the difference [differential] will be

$$\frac{a\,dx + 2x\,dx - \dfrac{x^3\,dx}{\overline{a^4 - x^4}^{\frac{3}{4}}}}{2\sqrt{ax + xx + \sqrt[4]{a^4 - x^4}}},$$

that is

$$\frac{\overline{a\,dx + 2x\,dx} \times \overline{a^4 - x^4}^{\frac{3}{4}} - x^3\,dx}{2 \times \overline{a^4 - x^4}^{\frac{3}{4}} \sqrt{ax + xx + \sqrt[4]{a^4 - x^4}}}.$$

This textbook contains no original mathematics, but it is a lucid exposition of the subject, which is mainly taught by example. Agnesi became well known for this book, which was translated into French and English, but chose to remain at home for the rest of her life. Born into a wealthy family, which made her education possible, she died in poverty after spending her fortune on works of charity.

And then Euler entered the picture, with the publication of *Institutiones calculi differentialis* in 1755. Euler shows us in this text that he could differentiate anything in sight, and we can too if only we follow his lead. Let the differentiation of the logarithm, on page 143, serve as an example of his style. Together with his four-volume *Institvtionvm calcvli integralis*, published from 1768 to 1770, these books formed a comprehensive treatise of everything that was known up to then about the calculus, including many new results.

However, none of the authors mentioned so far understood the foundations of the calculus. Nobody really knew what those differentials, those infinitesimals, those nascent and ultimate ratios, really were. As an example of late-in-the-game misconceptions we shall pick on Euler. Such was the caliber of his genius, second to none in mathematics, that he could have accepted a little picking in good humor.

In the preface to the 1755 book, Euler started by defining [pp. VIII; vii]:[160]

---

[160] For this and the next quotation, this page reference is to the original first and, after a semicolon, to Blanton's translation.

The Differential Calculus, *which is* a method for determining the ratio of the vanishing increments, that any functions take on when the variable quantity, of which they are functions, is given a vanishing increment . . .

Note the use of the word function in our present sense. It is the first time that it was used that way by any of the authors whose work has been discussed so far. The given definition is correct but leaves much to be explained, in particular the vanishing increments. This is what Euler had to say:

Accordingly they are called differentials, which seeing that they are without quantity, they are also said to be infinitely small; therefore by their nature they are interpreted, as nothing at all or they are to be thought of as equal to nothing.

Euler was very insistent throughout his book that differentials are equal to zero, which doesn't sit well with taking their quotients. But he explained this, already in the text proper, in Chapter III: *On the infinite and the infinitely small* [Art. 85], by stating that

when zero is multiplied by any number it gives zero, so that $n \cdot 0 = 0$, and thus $n : 1 = 0 : 0$. Hence it is clear that any two zeros can be in a geometric ratio [quotient], although, from an arithmetical point of view, that ratio is always of equals.

Then he added [Art. 86]:

If therefore, we accept the notation used in the Analysis of the infinite, dx denotes an infinitely small quantity, so that $dx = 0$, as well as $a\,dx = 0$, where a denotes any finite quantity. In spite of this, the geometric ratio $a\,dx : dx$ is finite, namely as $a : 1; \ldots$ In a similar manner, if we must deal with separate infinitely small quantities dx & dy, although these are both $= 0$, still their ratio is not apparent.

However, the quotient $dy : dx$ is not considered in the rest of Chapter III. It appears again in Chapter IV, but the next quotation is so out of context that it is difficult to understand the appearance of the function P that is used below [Art. 120]:

Whatever function y is of x, its differential dy is expressed by a certain function of x, for which we write P, multiplied by the differential of x, that is by dx. Thus although the differentials of x & y are in fact infinitely small, and therefore equal to zero; still there is a finite ratio between them: namely $dy : dx = P : 1$.

This is enough to realize that Euler could not bring home the bacon with all these gyrations. From our point of view, he was trying to explain what he did not fully understand. Without realizing it, he was trying to avoid something that is a must: the concept of limit.

6

CONVERGENCE

6.1 TO THE LIMIT

In our days the definition of limit is well known and used appropriately. In the past even a genius of the first water like Newton could only intuit the concept. And that in itself was a giant leap, even though such a leap is insufficient to bridge the infinite gap toward the ultimate ratio that he sought. The need for a definition was soon felt, and Jean le Rond d'Alembert, of Paris, tried to provide it but without success. As coeditor of the famous *Encyclopédie* he contributed the *Discours préliminaire* and numerous articles, including the following assessment of Newton's view of the calculus in the article DIFFÉRENTIEL of volume 4 (CON – DIZ) [p. 985]:

> He never regarded the *differential* calculus as the calculus of infinitely small quantities, but as the method of first & last ratios, that is the method of finding the limits of the ratios.

Hence, it is imperative to define limit, and d'Alembert did so in volume 9 of the *Encyclopédie* (JU – MAM) [p. 542]:

> LIMITE, s. f. (*Mathémat.*) It is said that a magnitude is the *limit* of another magnitude, when the second may approach the first closer than a given magnitude, as small as it can be imagined, and yet without the approaching magnitude, ever being able to surpass the magnitude that it approaches; ...

Then to make sure that his position on the essence of the calculus was properly explained and understood, he took on the metaphysics of the differential calculus in DIFFÉRENTIEL and expressed himself as follows [p. 987]:

The *differential* calculus is not about infinitely small quantities, as it is still ordinarily said; it is concerned only with the limits of finite quantities. ... We shall not say, along with many mathematicians, that a quantity is infinitely small, not before it vanishes, not after it vanishes, but at the very same instant at which it vanishes;[1] ... We shall say that there are no infinitely small quantities in the *differential* calculus.

We get the message, but from our present point of view, d'Alembert's definition imposes the excessive requirement that the approaching magnitude should not surpass the target magnitude. Actually, it is not a proper mathematical definition but just a play with words. What does it mean to "approach"? Is time involved? How does one measure this approach? Mathematicians were not used to absolute precision at that time, nor did they appreciate its fruitfulness outside the realms of synthetic geometry or number theory.

But then the need for a deeper understanding of the idea of convergence was made evident by the attempts to figure out the solutions of new applied problems involving infinite series, and it was in this context that it was defined first.

6.2 THE VIBRATING STRING MAKES WAVES

When a taut piece of string, tied down at both ends, is either plucked or hit, it emits audible vibrations. Their frequency, that is, the number of vibrations per second, was first determined by Brook Taylor in his 1713 paper "De motu nervi tensi," but he believed that the shape of the string at any given instant of its motion was a sine wave. There is, however, more than one audible frequency of vibration when such a string is in motion. The one found by Taylor is called *fundamental* because it is the lowest frequency, and the other vibrations that make up the motion of the string are called *harmonics*. Intuitively, then, we expect the shape of the string to be a sum of sine waves; perhaps an infinite series of sine waves.

D'Alembert was the first to determine the shape of a vibrating string at a given time, and his success was based on the fact that in 1747 he derived a partial differential equation that accurately describes the phenomenon in question and then solved it.[2] Neither d'Alembert's notation nor his derivations

---

[1] This is almost exactly what Newton had stated in the *Principia*, as quoted on page 311.

[2] "Recherches sur la courbe que forme une corde tenduë mise en vibration," and "Suite des recherches," both of 1747.

can be considered appealing to modern tastes, so we shall limit ourselves to stating his equation and the solution. He considered a string, initially stretched between two fixed points along the axis of abscissas. Then, if the vertical displacement of a point of the string at time t at a point of abscissa s is denoted by $y = \varphi(t, s)$ [p. 215], and if α and β denote what we now call the second partial derivatives of y with respect to t and s, respectively, the differential equation is [p. 216]

$$\alpha = \beta \frac{2aml}{\theta^2}.$$

The fraction on the right contains several physical constants that are of no interest to us. Today we would write this equation, which is now called the *wave equation*, as

$$\frac{\partial^2 y}{\partial t^2} = b \frac{\partial^2 y}{\partial s^2}$$

and simply say that b is a positive constant. The unit of time can be chosen so that $b = 1$, and doing so, d'Alembert then found the general solution of this equation to be [p. 217]

$$y = \Psi(t + s) + \Gamma(t - s),$$

where Ψ and Γ are "as yet unknown functions" that must be determined from the facts that the string is tied down at its endpoints, that it is given no initial velocity, and that its initial displacement is known.[3]

D'Alembert found these functions [pp. 228–230], but the problem with his solution is that it shows none of the vibrations in time that we expected, the fundamental and the harmonics. It was Euler who, two years later, in his paper "De vibratione chordarum exercitatio," brought up the fact that the motion of the string is periodic in time and made up of many individual vibrations. He started by rederiving the differential equation and its general solution by his own method, and then, if $y = f(x)$ represents the initial displacement of the string, he obtained the following equation for its displacement at a point of abscissa x (his choice of variable instead of d'Alembert's s) and at time $t > 0$ [Art. XXII]:

$$y = \tfrac{1}{2}f\left(x + t\sqrt{b}\right) + \tfrac{1}{2}f\left(x - t\sqrt{b}\right).[4]$$

As an example, if the initial displacement of a string of length a is given by [Art. XXXI]

$$y = \alpha \sin \frac{\pi x}{a} + \beta \sin \frac{2\pi x}{a} + \gamma \sin \frac{3\pi x}{a} + \&c,[5]$$

and if we use this right-hand side as $f(x)$ in the preceding general solution, then, simplifying the result and writing v for $t\sqrt{b}$, the displacement of the string at a point x and at time $t > 0$ is given by

$$y = \alpha \sin \frac{\pi x}{a} \cos \frac{\pi v}{a} + \beta \sin \frac{2\pi x}{a} \cos \frac{2\pi v}{a} + \gamma \sin \frac{3\pi x}{a} \cos \frac{3\pi v}{a} + \&c.$$

---

[3] An English translation of the relevant parts of d'Alembert's paper containing these derivations can be seen in Struik, *A source book in mathematics, 1200–1800*, 1969, pp. 352–357.

[4] Euler wrote $f:(x + t\sqrt{b})$ [p. 521] instead of $f(x + t\sqrt{b})$. Note, in any event, that $x - t\sqrt{b} < 0$ for t large, so that it is necessary to extend the function f to the negative x-axis. Furthermore, if we assume that the left endpoint of the string is at $x = 0$, the fact that this point is tied down implies that $f(t\sqrt{b}) + f(-t\sqrt{b}) = 0$ for all $t > 0$, which implies that the extension of f must be odd. Then, if the length of the string is a, this extension is an odd periodic function of period $2a$.

[5] As before, I have suppressed the period in Euler's abbreviation sin. for the sine.

Since v is a constant times t, we now have vibrations, but in order to produce them Euler had to choose what looks like a highly artificial initial displacement as an infinite sum of sines. The obvious question is: how representative is this of an arbitrary generating curve (*courbe génératrice*), as d'Alembert called the initial displacement [p. 220]?

With the preceding statements, Euler had opened a big can of worms, one that we cannot ignore, for it is at the very center of mathematical analysis. Here is how the situation looks before careful examination: if a tight string is given an arbitrary initial displacement and then let go, there is little doubt that there will be vibrations. Is this reason enough to conjecture that an arbitrary function can be represented by the sum of an infinite series of sines?

The first to pronounce himself on this subject was Daniel Bernoulli (1700 – 1782), son of Johann. Upon reading d'Alembert's and Euler's papers, he

DANIEL BERNOULLI CIRCA 1750
Portrait by Johann Niklaus Grooth.
From *Die Werke von Daniel Bernoulli, Band 2: Analysis Wahrscheinlichkeitsrechnung.*
in *Die Gesammelten Werke der Mathematiker und Physiker der Familie Bernoulli*,
ed. by David Speiser, Birkhäuser-Verlag, Basel, 1982.

published his own ideas on the subject in 1753.[6] Elaborating on a previously published belief that the shape of the string at a given instant is the superposition of individual vibrations, now he stated [Art. XII] that its displacement at any given time[7]

shall have in the general case the following equation for the same abscissa x:

$$y = \alpha \sin \frac{\pi x}{a} + \beta \sin \frac{2\pi x}{a} + \gamma \sin \frac{3\pi x}{a} + \delta \sin \frac{5\pi x}{a} + \text{etc.},$$

in which the quantities α, β, γ, δ, etc. are positive [*affirmatives*] or negative quantities.

Bernoulli, who did not write the dependence on time, based his solution on physical considerations alone and provided no mathematical reasons whatsoever to back it up. Euler pounced on it immediately, the very same year, refusing to accept that it could represent the general case [Art. VII].[8] D'Alembert was of the same opinion as Euler on this point: an arbitrary curve cannot be represented by the sum of an infinite series of sines, an opinion that he published in the article FONDAMENTAL in volume 7 of the *Encyclopédie* (FOA–GYT). But Bernoulli did not surrender his position, for, he said[9]

the last resulting curve [the one represented by the preceding infinite series] will include an infinitude of arbitrary quantities [the coefficients], which can be used to make the final curve pass through as many points given in position as one wishes, & to identify through them this curve with the proposed curve, with any degree of precision that one wishes.

6.3 FOURIER PUTS ON THE HEAT

Not too long after these times, great events took place in France that, indirectly, would come to settle the argument after the death of its principals. On July 12, 1789, the people of Paris—then a walled, dirty, unhealthy town of 550,000 with dusty, narrow streets—rioted after the dismissal of the prime minister Necker. What started as a riot, fueled by discontent about food shortages

---

[6] "Réflexions et éclaircissements sur les nouvelles vibrations des cordes exposées dans les mémoires de l'Académie de 1747 et 1748," 1753.

[7] From the translation in Struik, *A source book in mathematics, 1200–1800*, p. 364.

[8] "Remarques sur les mémoires précédens de M. Bernoulli," 1753.

[9] On page 165 of "Lettre de Monsieur Daniel Bernoulli, ... à M. Clairaut ... sur la vibrations des cordes tendues," 1758.

and low incomes, soon turned into a full-fledged revolution. A revolutionary council was formed that night, and then the Bastille was attacked and destroyed on July 14.

The monarchy was abolished in 1792, and the French republic proclaimed on September 22. A new revolutionary calendar, beginning with 1 Vendémiaire, Year I, had been established to mark time from that day; and on 7 Germinal Year VI—March 27, 1798—a young professor at the newly founded *École Polytechnique*, Jean Joseph Fourier (1768–1830), was made a member

JOSEPH FOURIER
Portrait by Julien Léopold Boilly.
Engraving by Amédée Félix Barthélémy Geille.
From *Œuvres de Fourier*, Tome Second, 1890.

of the Commission of Arts and Sciences of General Bonaparte's expedition to Egypt.

Upon his return to France in November 1801, after a British victory in Egypt, Fourier resumed his post at the *École Polytechnique* but only briefly. In February 1802, Bonaparte appointed him *Préfet* of the Department of Isère in the French Alps, and it was here, in the city of Grenoble, that Fourier returned to his physical and mathematical research. But Fourier's stay in Egypt had left a permanent mark on his health, which was to influence, perhaps, the direction of his research. He claimed to have contracted chronic rheumatic pains during the siege of Alexandria and that the change of climate from that of Egypt to that of the Alps was too distressful for him. He lived in overheated rooms, wore an excessive amount of clothing even in the heat of summer, and it was on the subject of heat that he concentrated his research efforts.

Fourier had a complete manuscript in 1807, in which he found the general equation for the diffusion of heat in a three-dimensional solid.[10] If v is the temperature, this equation is [Art. 28, p. 126]

$$\frac{dv}{dt} = \frac{K}{CD}\left[\frac{d^2v}{dx^2} + \frac{d^2v}{dy^2} + \frac{d^2v}{dz^2}\right],[11]$$

where "v is a function of the three coordinates x, y, z and of time t," and K, C, and D are physical constants with which we need not be concerned. It is now called the *heat equation*. Then he posed the following problem [Art. 32, p. 134]:

> We assume that a lamina or a rectangular surface, of infinite length, is heated at its end 1, and is kept at a constant temperature at all points of this edge, and that each of the other edges 0 and 0 [12] are also kept at a constant temperature 0 at all their points. It is a question of determining what the stationary temperatures at each point of the lamina must be.

To be precise, a lamina of negligible thickness in the z direction is located in the plane strip $0 \leq x < \infty$, $-1 \leq y \leq 1$. The edge at $x = 0$ is kept at a constant temperature 1 while the edges at $y = \pm 1$ are kept at a constant temperature 0 for all $t \geq 0$. It is desired to find the temperature at each point of the lamina after a sufficient amount of time has passed.

It is assumed that after some time the temperatures have reached a steady state and are no longer dependent on time. Thus, the left-hand side of the heat

---

[10] *Mémoire sur la propagation de la chaleur* = Grattan-Guinness and Ravetz, *Joseph Fourier 1768–1830*, 1972, pp. 30–440. Page references are to this edition.

[11] All these are partial derivatives, but there was no special notation for them in Fourier's time. The copyist who wrote this passage (the page shown in the photograph is mostly in the hand of a professional copyist but finished in Fourier's own hand), appears to have used partial derivatives, but such is not the case. It is simply that all his d's look like ∂'s.

[12] Fourier had a rather confusing way to refer to the edges of the lamina. Later in his paper it became clear what those edges are, and it is explained below this quotation.

A portion of page 60 of Fourier's 1807 manuscript.
Reproduced from Grattan-Guinness and Ravetz, *Joseph Fourier 1768–1830*.

equation vanishes. Also, the negligible thickness of the lamina means that there is no appreciable variation in temperature in the z direction, and then the last derivative also vanishes. Since the letter z can then be reused, Fourier switched his notation for the temperature from v to z, and all this reduces the heat equation to [p. 135]

$$\frac{d^2 z}{dx^2} + \frac{d^2 z}{dy^2} = 0.$$

Now this equation must be solved, and Fourier resorted to a method that was rare in his time, a method previously used by d'Alembert in 1750 for the vibrating string,[13] but after a few lines [p. 356] d'Alembert made nothing out of it. It was Fourier who put this method on the map, and as used by him it consists in assuming that there is a solution of the form $z = \phi(x)\psi(y)$, where ϕ and ψ are unknown functions to be determined [Art. 33, p. 137]. Taking this proposed solution to the equation for z, he obtained

$$\phi''(x)\psi(y) + \phi(x)\psi''(y) = 0,$$

and then, dividing by the product $\phi''\psi''$ (implicitly assuming that this can be done),[14]

$$\frac{\phi(x)}{\phi''(x)} + \frac{\psi(y)}{\psi''(y)} = 0.$$

Essentially, this means that a function of x is equal to a function of y, and the only way this can be is for each of them to be constant. That is,

$$\frac{\phi(x)}{\phi''(x)} = A \quad \text{and} \quad \frac{\psi(y)}{\psi''(y)} = -A,$$

where A is a constant. This simplifies the problem by transforming a partial differential equation in two variables into two ordinary differential equations, each in one variable. Then Fourier made the following statement:

One sees by this that we can take for $\phi(x)$ any quantity of this form e^{mx} and for $\psi(y)$ the quantity $\cos(ny)$. Therefore we shall assume that $z = ae^{mx} \cdot \cos(ny)$, and substituting in the proposed [simplified heat equation] we shall have the condition equation $m^2 = n^2$.

This means that we have found solutions of our particular heat equation of the form $z = ae^{nx}\cos(ny)$ and $z = ae^{-nx}\cos(ny)$. Fourier's carelessness in dividing by $\psi''(y)$, which is shown now to have zeros, becomes inconsequential because it is easily verified by differentiation that, for each n, any of the two functions stated here is indeed a solution of the given equation.

---

[13] "Addition au mémoire sur la courbe que forme une corde tenduë, mise en vibration," 1750, pub. 1752.

[14] This manuscript is in the hand of a professional copyist, who did not use parentheses around the variable of a function. For whatever reason, the rest of page 60 was finished in Fourier's own hand, and he used parentheses. I have used them throughout for consistency. However, I write cos and sin instead of Fourier's cos. and sin.

If we take $n > 0$, which does not make any difference in the cosine factor, the first of the two solutions above must be rejected because it grows without bound as x grows. This is impossible with the edges of the lamina kept at temperatures 1 and 0. Fourier continued as follows [p. 138]:

> Therefore the preceding solution is reduced to $z = ae^{-nx}\cos ny$, n being any positive number and a an undetermined constant. Now the general solution will be formed by writing
>
> $$z = a_1 e^{-n_1 x}\cos n_1 y + a_2 e^{-n_2 x}\cos n_2 y + a_3 e^{-n_3 x}\cos n_3 y + \cdots \text{\&c.}$$

The last assertion is a leap to say the least. It is easy to see by direct substitution that the sum of a finite numbers of solutions of the heat equation is also a solution. But is it clear that this holds for an infinite number of terms? If we humor Fourier and accept his claim, the constants $n_1, n_2, n_3, \ldots$ can then be determined from the fact that $z = 0$ for $y = \pm 1$. Then these constants must be odd integral multiples of $\pi/2$, so that the cosine factors vanish. Then, the fact that the temperature at $x = 0$ is 1 requires that [Art. 34, p. 139]

$$1 = a_1 \cos \tfrac{1}{2}\pi y + a_2 \cos \tfrac{3}{2}\pi y + a_3 \cos \tfrac{5}{2}\pi y + a_4 \cos \tfrac{7}{2}\pi y + \cdots \text{\&c.}$$

Fourier concluded with the statement [p. 140]: "It remains to find the values that we must give the coefficients $a_1 \ldots a_2 \ldots a_3 \ldots$ to satisfy the equation." He did find them, but their values are not relevant to us at this moment. Later [Art. 50, p. 194] he stated that this was just a particular case

> of a more general problem that consists in finding out whether an entirely arbitrary function can always be expanded in a series of terms that contain sines and cosines of multiple arcs.

Fourier began [Art. 51, p. 194] with the particular case in which the expansion has only sines,

$$\phi(x) = a \sin x + b \sin 2x + c \sin 3s + d \sin 4x + \text{\&c.}$$

He found the coefficients, first by some tedious and laborious method [Art. 61, pp. 211–213], and then by a much simpler one [Art. 63, p. 216] that is included in today's texts on the subject.

Fourier's belief that an arbitrary function can be represented by an infinite trigonometric series was rejected by Lagrange much as Bernoulli's had been rejected by Euler. The main obstacle in accepting that an infinite series could add up to an arbitrary function was the fact that mathematicians were accustomed to thinking of functions as being given by analytical expressions such as polynomials, roots, logarithms, and so on. It is clear that such aperiodic functions cannot be represented by an infinite series of periodic functions over the whole real line, but somehow they failed to realize that they could be equal over a finite interval.

Fourier gave numerous examples of the expansions of functions in series of sines or of cosines [Arts. 64–74, pp. 218–236], and, in particular, he gave three different expansions for $\frac{1}{2}x$. Here is one [Art. 65, p. 221]:

$$\tfrac{1}{2}x = \sin x - \tfrac{1}{2}\sin 2x + \tfrac{1}{3}\sin 3x - \tfrac{1}{4}\sin 4x + \tfrac{1}{5}\sin 5x - \cdots \ \&c.$$

Fourier explained as follows:

> The true sense in which one must take this result consists in the fact that the sine-like line [curve] that has the equation:
>
> $$y = \sin x - \tfrac{1}{2}\sin 2x + \tfrac{1}{3}\sin 3x + \cdots,$$
>
> and the straight line whose equation is $y = \frac{1}{2}x$, coincide in that part of their course that is placed above the axis, from 0 to π. Beyond this point, the first line departs from the second and moves down to cut the axis perpendicularly.

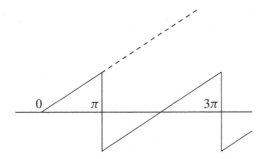

Thus the equation

$$y = \sin x - \tfrac{1}{2}\sin 2x + \tfrac{1}{3}\sin 3x - \tfrac{1}{4}\sin 4x + \cdots$$

belongs effectively to the line $01\pi\,1013\pi\,1\ldots$, which is made up of inclined lines and vertical lines.

Eventually, in the study of heat diffusion in an annulus, Fourier needed a more general expansion in terms of both sines and cosines [Art. 79, pp. 259–260] and was able to find the corresponding coefficients.

The propagation of heat was the subject chosen by the *Institut de France* for a prize essay in 1811, and on January 6, 1812, the prize was awarded to Fourier's *Théorie du mouvement de la chaleur dans les corps solides*, an expanded version of his 1807 memoir, which was finally printed in 1824 and 1826. This work was incorporated into Fourier's monumental work *Théorie analytique de la chaleur* of 1822.[15]

6.4 THE CONVERGENCE OF SERIES

The work on applied topics described in the preceding two sections presented mathematicians with the task of seriously studying the convergence of series.

This was first accomplished by José Anastácio da Cunha (1744–1787), of the University of Coimbra and the College of Saint Luke in Portugal. He was well acquainted with the works of d'Alembert and Euler, and his *Principios mathematicos* of 1790 was the first book containing the principles of analysis written in the modern fashion, using the now classic sequence Definition–Theorem–Proof–Corollary. Conciseness and rigor were da Cunha's trademarks.

---

[15] It is in this book that we find frequent use of two notations that gained favor rapidly. The first is the use of the symbol Σ for infinite series. It had been originally introduced by Euler, stating "summan indicabimus signum Σ" in his *Institutiones calculi differentialis*, 1755, Chapter 1, Art. 26. Fourier initiated the systematic use of Σ in the *Théorie analytique*, in Art. 235, with the statement: "Donc, en désignant par Σ cos. $\overline{ix - \alpha}$ la somme de la série précédente" (Thus, denoting by Σ cos. $\overline{ix - \alpha}$ the sum of the preceding series). Note the use of an overline rather than parentheses. Soon thereafter, he started writing the sum limits above and below the letter sigma. By the time he wrote the table of contents he was using the almost modern notation, as in

$$\frac{\pi}{2}\psi x = \sum_{i=\infty}^{i=-\infty} \cos. ix \int_0^\pi d\alpha \cos. i\alpha\, \psi.$$

Notice the definite integral on the right-hand side. This way of denoting the limits of integration was also introduced by Fourier. In Art. 222 we encounter "L'intégrale $\int_0^\pi x \sin. ix\, dx$" and in Art. 231 he stated: "Nous désignons en général par le signe $\int_a^b$ l'intégrale qui commence lorsque la variable équivaut à a, et qui est complète lorsque la variable équivaut à b" (In general we denote by $\int_a^b$ the integral that starts when the variable equals a and ends when the variable equals b); however, Fourier had already used the definite integral sign in his 1811 prize essay, but it was first published in the *Théorie analytique*.

The work of Fourier presented in this section is essentially contained in the *Théorie analytique* in Arts. 166–169 for separation of variables and the cosine series, Art. 190 for the special solution for the lamina, Arts. 219–223 for the sine series, and Arts. 240–241 for the full series. See the bibliography for English translations of these articles.

PRINCIPIOS MATHEMATICOS
PARA INSTRUCÇAÕ
DOS ALUMNOS DO COLLEGIO
DE
SAÕ LUCAS,
DA REAL CASA PIA DO CASTELLO
DE
SAÕ JORGE:
OFFERECIDOS
AO SERENISSIMO SENHOR
D. JOAÕ,
PRINCIPE DO BRAZIL:
COMPOSTOS PELO DOUTOR
JOSE' ANASTACIO DA CUNHA,
DE ORDEM DO DESEMBARGADOR DO PAÇO
DIOGO IGNACIO DE PINA MANIQUE,
Intendente Geral da Policia da Corte,
e Reino, &c, &c, &c.

LISBOA
NA OFFIC. DE ANTONIO RODRIGUES GALHARDO,
Impreſſor do Eminentiſſimo Senhor Cardeal Patriarca.
ANNO M.DCC.XC.
Com licença da Real Meza da Commiſſaõ Geral
ſobre o Exame, e Cenſura dos Livros.

Book VIIII starts with the definition of convergent series [p. 106; 117]:[16]

DEFINITION I.

Mathematicians call a convergent series that whose terms are similarly deter-
mined, each one by the number of the preceding terms, in a way that the series
can always be continued, and finally it comes to be indifferent to continue it or
not, because however many terms one wants to join to those already written or
indicated their sum can be neglected without noticeable error: and the latter
are indicated by writing &c. after the first two, or three, or however many one
wants: it is therefore necessary that the written terms show how the series
could be continued, or that this be known in some other manner.

Note the modesty shown by this statement. Mathematicians would eventually call such a series convergent, but da Cunha was the first to give this definition, more than 30 years before anyone else rediscovered it. However, the
clarity of the statement (from which we have omitted an impossible comma)

---

[16] Page references are to the Portuguese edition first and then to the French translation.

is impaired by the fact that it begins and ends by emphasizing at length that the law of continuation of the series must be clearly stated or possible to deduce from the observation of a few terms. For us that goes without saying. The heart of the definition is in the following words:

> finally it comes to be indifferent to continue it or not, because however many terms one wants to join to those already written or indicated their sum can be neglected without noticeable error.

What this means, using current notation, is that a series

$$u_1 + u_2 + u_3 + \cdots + u_n + \cdots$$

converges if for some n we can neglect without noticeable error the sum

$$u_{n+1} + u_{n+2} + u_{n+3} + \cdots + u_{n+m}$$

for every positive integer m. This is, except for some precision added in its present-day statement, the modern definition in mathematical analysis, although not the one included in calculus textbooks.

Da Cunha then illustrated the use of his definition in the proof of his first proposition [pp. 106–107; 117–119]:

PROPOSITION I.

A series of continued proportions is convergent, if the first is larger than the second.

He explained that this means a series of the form

$$A + B + B\frac{B}{A} + B\frac{B}{A}\times\frac{B}{A} + B\frac{B}{A}\times\frac{B}{A}\times\frac{B}{A} + \&c.,$$

in which $A > B$. He also assumed, but did not state, that $B > 0$.[17] Then he began the proof by writing $c = B/A$, so that his series becomes

$$A(1 + c + cc + ccc + cccc + \&c.),$$

---

[17] In spite of his modernity in other respects, Da Cunha seems to be working with positive numbers only. He made similar assumptions in the corollary and the definitions quoted in this section. Proposition I is valid if $B < 0$ provided that $A > |B|$.

and chose a number $O < A$ as small as desired, "so that it can be neglected without noticeable error."

Note that the assumptions on A and B imply that $0 < c < 1$. Then the terms of the series

$$1 + \frac{1}{c} + \frac{1}{cc} + \frac{1}{ccc} + \frac{1}{cccc} + \&c.,$$

which Da Cunha wrote next, are well defined and grow without bound. Therefore, there is a term, which he denoted by d, such that

$$d > \frac{A}{O - cO}$$

(he used an argument based on an axiom from Book III [p. 21; 24], but to us this is clear). The term that corresponds to d in the series $1 + c + cc + ccc + cccc + \&c.$ is $1/d$, so that the continuation of this series from this term on is

$$\frac{1}{d} + \frac{c}{d} + \frac{cc}{d} + \frac{ccc}{d} + \&c.,$$

and this sum can never equal

$$\frac{1}{d - dc}$$

no matter how far it is continued. Indeed, if it is continued to the term c^m/d, where m is a positive integer, then we have

$$\left(\frac{1}{d} + \frac{c}{d} + \frac{cc}{d} + \cdots + \frac{c^m}{d} \right)(d - dc) = 1 - c^{m+1} < 1.$$

But the choice of d gives $dO - cdO > A$, and then

$$\frac{O}{A} > \frac{1}{d - dc}.$$

Thus,

$$\frac{O}{A} > \frac{1}{d} + \frac{c}{d} + \frac{cc}{d} + \&c.$$

(the sum on the right is finite in spite of the symbol $\&c.$), and

$$O > \frac{A}{d} + \frac{cA}{d} + \frac{ccA}{d} + \&c.$$

But the right-hand side is the continuation of the original series. Since O is arbitrarily small, it follows from Definition I that the given series is convergent.

Then da Cunha turned to the defining series for e^a and stated the following corollary [pp. 107–108; 119]:

COROL. I.

Let a represent any [positive] *number. I say that the series*

$$1 + a + \frac{aa}{2} + \frac{aaa}{2\times3} + \frac{aaaa}{2\times3\times4} + \frac{aaaaa}{2\times3\times4\times5} + \&c.$$

is convergent.

For the proof he let b be an integer larger than a, denoted the bth term in this series by c, and wrote the continuation from c:

$$c + \frac{ac}{b+1} + \frac{aac}{(b+1)(b+2)} + \frac{aaac}{(b+1)(b+2)(b+3)} + \&c.$$

Then he observed that each term in this continuation does not exceed the corresponding term of the series

$$c + \frac{ac}{b+1} + \frac{aac}{(b+1)(b+1)} + \frac{aaac}{(b+1)(b+1)(b+1)} + \&c.,$$

which converges by Proposition I. Therefore, the preceding two series also converge by comparison.

Then, on this basis, da Cunha was able to define the general exponential function [pp. 108–109; 120].

DEFINITION II.

Let a [> 0] *and b represent any two numbers, and let c be the number that makes*

$$1 + c + \frac{cc}{2} + \frac{ccc}{2\times3} + \frac{cccc}{2\times3\times4} + \&c. = a:$$

the expression a^b will mean a number

$$= 1 + bc + \frac{bc}{2} + \frac{bbcc}{2\times3} + \frac{bbbccc}{2\times3\times4} + \&c.;$$

and it will be called the number a^b the power of a indicated by the exponent b:

Using Euler's number e, which da Cunha did not use in this chapter, the sums of these two series are e^c and $e^{bc} = (e^c)^b$, and it is clear that $a^b = e^{bc}$.

The remarkable thing is that while Euler was originally content with such vagueness as "$a^{\sqrt{7}}$ will be a certain value comprised between the limits a^2 and a^3," da Cunha was able to give a correct, rigorous, and general definition. From it he developed, in the rest of Book VIIII, a correct and rigorous theory of powers and logarithms. As an example we quote the following property of the exponential function [p. 113; 123]:

V. *Let a, b, c represent any three numbers: it will be* $(a^b)^c = a^{bc}$.

With arbitrary powers on a solid foundation, da Cunha went on to define logarithms [p. 119; 128]:

DEFINITIONS.

III. *Considering all numbers as powers of a given number, this is called base; and the exponents are called logarithms of the numbers to which they belong.*

IIII. *The logarithms are called hyperbolic, and also natural, when the base is*

$$1 + 1 + \frac{1}{2} + \frac{1}{2 \times 3} + \frac{1}{2 \times 3 \times 4} + \&c.$$

Then, using Euler's l to denote any logarithm, he concluded this chapter as follows [p. 120; 128]:

VIIII. $la^n = nla$.

Let b represent the base, it will be $a = b^{la}$ [9. def. 3], and then $a^n = b^{nla}$ [9. 5], and then $la^n = nla$ [9. def. 3].

When Fourier presented his prize essay to the *Institut* in 1811, he was not aware of da Cunha's work on series. He could not have been because the French translation of the *Principios* that appeared in Bordeaux in that year, although largely an excellent translation, erroneously misstated the definition by pronouncing a series convergent

> provided that on arriving at a given number of first terms, it is possible to neglect the others without considerable error. In such a case, and after having written enough terms to indicate the law of continuation, we denote the sum of those that are neglected, by an etc. after the sequence of first terms.

What this essentially says is that a series is convergent if it is still convergent after removing a finite number of first terms, and this is reinforced by the explanation of the meaning of the abbreviation etc. Thus, this definition is useless. It could not have been used by Fourier, and da Cunha's work was largely ignored by the mathematical establishment.[18]

One of the first definitions of convergence in France was given by Fourier himself in his prize paper of 1811, later incorporated into his book. Near the end of Article 228, he made the following statement about how to establish the convergence of a series:

> It is necessary that the values at which we arrive, on increasing continually the number of terms, should approach more and more a fixed limit, and should differ from it only by a quantity which becomes less than any given magnitude: this limit is the value of the series.

This definition is along the lines of the one usually included in calculus texts, but it has two main disadvantages over da Cuhna's. It is necessary to guess the limit before attempting to show that it is indeed the limit, and it must be possible to sum an arbitrary number of terms of the series in order to compare this sum with the guessed limit.

Working in relative isolation in Bohemia, Bernardus Placidus Johann Nepomuk Bolzano (1781–1848), a priest and professor of theology at Karl Ferdinand University in Prague (suspended in 1819 due to his liberal views on social issues and pacifist beliefs, put under house arrest, and forbidden to publish) discovered the better definition of convergence independently of da Cunha. It appeared at the start of Article 7 in a pamphlet proving what we now call the intermediate value theorem.[19] In Article 6 he established the notation by letting $\overset{n}{F}x, \overset{n+1}{F}x, \overset{n+2}{F}x, \ldots, \overset{n+r}{F}x$ be "the values of the sums of the first $n, n+1$, $n+2, \ldots, n+r$ terms of a series" of given functions of x. However, for the

---

[18] Although the sharp eye of Gauss took notice, and in a letter to Bessel of 21 November 1811, he stated: "Likewise, all the Paradoxes that some Mathematicians *have found* about Logarithms just disappear, when one does not start with the usual Definition base$^{\text{logar.}}$ = Number, which is really satisfied only, when the Exponent is a whole Number, and does not make any Sense, when the Exponent is imaginary—but each Quantity that substituted for x in the Series $1 + x + \frac{1}{2}xx + \frac{1}{6}x^3 +$ etc. $\cdots$ gives the Value A, is called the Logarithm of A; as I see with Pleasure, the Portuguese Acunha had actually selected this Definition" (Definitions II to IIII in Book VIIII of the *Principios* [p. 119; 128]), in *Briefwechsel zwischen Gauss und Bessel*, 1880, p. 153 = *Werke*, **101**, p. 364.

[19] "Rein analytischer Beweis des Lehrsatzes dass zwischen je zwey Werthen, die ein entgegensetztes Resultat gewähren, wenigstens eine reele Wurzel der Gleichung liege," 1817. The theorem quoted below is on pp. 266–267 of Russ' English translation of 2004.

BERNARD BOLZANO IN 1839
Portrait by Áron Pulzer.

purposes of this article and the next, the value of x is supposed to be fixed, so that we are actually dealing with a series of numbers rather than functions. Then Bolzano stated his result as follows:

Theorem. If a series [sequence] of quantities

$$\overset{1}{F}x, \overset{2}{F}x, \overset{3}{F}x, \ldots, \overset{n}{F}x, \ldots, \overset{n+r}{F}x, \ldots,$$

has the property that the difference between its nth term $\overset{n}{F}x$ and every later one $\overset{n+r}{F}x$, however far this latter term may be from the former, remains smaller than any given quantity if n has been taken large enough, then there is always a certain *constant quantity*, and indeed only *one*, which the terms of the series approach and to which they can come as near as we please, if the series is continued far enough.

If the nth term of Bolzano's sequence is the nth partial sum

$$\overset{n}{F}x = u_1 + u_2 + u_3 + \cdots + u_n$$

of an infinite series, then

$$\overset{n+r}{F}x - \overset{n}{F}x = u_{n+1} + u_{n+2} + u_{n+3} + \cdots + u_{n+r}$$

for every positive integer r, which shows that Bolzano's result is equivalent to da Cunha's definition.

The rigorous approach adopted by Bolzano in his paper might have helped solidify the foundations of the calculus if his work had reached the mathematical centers of the time. As it is, it went largely ignored. Thus it was left to Augustin-Louis Cauchy (1789–1857), who had just been appointed professor

AUGUSTIN-LOUIS CAUCHY IN 1821
Portrait by Julien Léopold Boilly.
From Smith, *Portraits of Eminent Mathematicians*, II, 1938.

of analysis and mechanics at the *École Polytechnique* in Paris, to rediscover the condition for the convergence of a series and to place the calculus on a solid foundation. His work on infinite series is contained in Chapter VI of

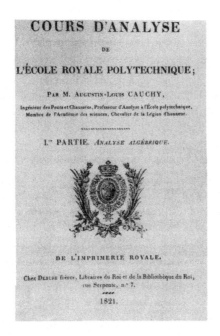

his first analysis textbook, published in 1821,[20] in which he announced his intention to endow analysis with "all the rigor that is demanded in geometry" [p. ij].

Cauchy began by giving the following definition of limit in the *Préliminaires* [p. 4; 19]:

> When the values successively attributed to the same variable indefinitely approach a fixed value, so that they finally differ from it by as little as one wishes, this latter [value] is called the *limit* of all the others.

This definition may be disappointing to those who regard Cauchy as the father of modern analysis and the initiator of the ϵ-δ proof. But this was early in Cauchy's work; his use of ϵ was yet to come, and we shall report on it in due course. For the time being, he relied on this notion of limit to state the following definition of the convergence of a series [p. 123; 114]:

---

[20] *Cours d'analyse de l'École Royale Polytechnique. 1re partie, Analyse algébrique*, 1821. Page references are to this edition and, after a semicolon, to the *Œuvres*. Quotations are from the *Cours d'analyse*. For English translations see the bibliography.

Let

$$s_n = u_0 + u_1 + u_2 + \cdots + u_{n-1}$$

be the sum of the first n terms, n denoting any whole number. If, for ever increasing values of n, the sum s_n approaches indefinitely a certain limit s, the series will be said to be *convergent*, and the limit in question will be called the *sum* of the series. If, on the contrary, while n increases indefinitely, the sum s_n does not approach any fixed limit, the series will be said to be *divergent*, and will have no sum.

These two definitions put together accomplish as much as Fourier's one definition, in which s_n is to differ from the limit of the series by "less than any given magnitude" on increasing n continually. Again, it is still necessary to guess the limit of a series before this definition can be used. But a couple of pages later Cauchy stated that it is necessary that [pp. 125–126; 115–116]

for increasing values of n, the different sums

$$u_n + u_{n+1},$$
$$u_n + u_{n+1} + u_{n+2},$$
$$\&c. \ldots$$

that is, the sums of the quantities

$$u_n, \quad u_{n+1}, \quad u_{n+2}, \quad \&c. \ldots$$

taken, from the first, in any number that one wishes, end up by constantly obtaining numerical [absolute] values smaller than any assignable limit. Conversely, when these several conditions are satisfied, the convergence of the series is ensured.

This is what da Cunha and Bolzano had already stated. Cauchy, who gave a definition of convergence requiring knowledge of the sum, made this into a necessary and sufficient condition that does not require it (but he did not prove the sufficiency). Then he used this result, which has become known as the *Cauchy criterion*, to examine the convergence of several series [pp. 126–129; 116–118]: the geometric series, the harmonic series, and the series for the exponential (which he dealt with in the same manner as da Cunha).

But what is truly remarkable about Cauchy's contribution is that he did not limit himself to stating the definition of convergence, but immediately started proving theorems giving tests for convergence: the root test [p. 132; 121],

the ratio test [pp. 134–135; 123],[21] the condensation test [p. 135; 123], the
logarithmic test [pp. 137–138; 125], and the alternating series test [p. 144;
130].[22] Later, in a separate publication and after he had already defined the

---

[21] Edward Waring (1734–1798), Lucasian professor of mathematics at Cambridge, had
already stated a ratio test in his *Meditationes analyticae* of 1776. In PROB. III. he made the
following obscure statement [pp. 298–299; PROB. II., pp. 349–350 of the 1785 edition, in
which there are minor variations and one error]:

> Let z be the distance from an arbitrary term to the first term of the series, and also T
> & T' two successive terms of the given series, & let there be given an equation of the
> relation between T & T', viz. $T = T' \times (az^r + \dfrac{b}{z^s} + \dfrac{c}{z^t} + \&c.)$ where $az^r + \dfrac{b}{z^s} + \dfrac{c}{z^t} + \&c.$
> is a series following the descending dimensions [powers] of the quantity z, then if
> $1 : -z + az^{r+1} + \dfrac{b}{z^{s-1}} + \dfrac{c}{z^{t-1}} + \&c.$ where z is assumed to be an infinite quantity,
> does not have a ratio [that is] positive & smaller than the ratio of equality by a finite
> [quantity], then the sum of the preceding series will be an infinite quantity, but if the
> opposite is true finite.

Waring was, of course, thinking of a series of positive numbers and by "the ratio of
equality" he must have meant 1. Then his inequality is $1 : -z + z(T/T') \not< r$, where $r > 0$
is smaller than 1 by a finite quantity, in which case, he claimed, the series has an infinite
sum. If, however, $1 : -z + z(T/T') < r < 1$ then the series has a finite sum. If we think
that z means our n, if we write u_n instead of T and u_{n+1} instead of T', and if we define
$\epsilon = (1/r) - 1$, then Waring's statement may be interpreted as saying that if

$$-n + n\frac{u_n}{u_{n+1}} > 1 + \epsilon > 1$$

for all sufficiently large n, then the series $\sum_{n=1}^{\infty} u_n$ converges, and otherwise it diverges.

Waring paid a price for his lack of clarity, since it was apparently not widely known that
this test was his when it was essentially rediscovered by Joseph Ludwig Raabe (1801–1859),
a German mathematician working in Zurich, and it is now known as Raabe's test.

In the 1785 edition [p. 350] Waring showed how to use this test to prove the convergence
of what we now call a p-series for $p > 1$ (using the binomial theorem to expand $(n + 1)^p$),
thus showing that this test is more powerful than Cauchy's root or ratio tests. Then he
gave a simpler ratio test for convergence [p. 351]: "If at an infinite distance $T : T'$ have a
ratio larger than equality by a finite [quantity], then the series is finite ..." and then gave
three examples. Even this test is more general than Cauchy's in that it does not require the
quotient u_n/u_{n+1} to have a limit as $n \to \infty$, which makes it applicable to series such as

$$1 + \frac{1}{2} + \frac{1}{2 \cdot 3} + \frac{1}{2^2 \cdot 3} + \frac{1}{2^2 \cdot 3^2} + \frac{1}{2^3 \cdot 3^2} + \cdots.$$

[22] First fully given by Leibniz as Proposition XLIX of *De quadratura arithmetica* [f. 33^v;
p. 115; 323; 125]. Later, in a letter to Johann Bernoulli of 25 October 1713, he stated that a
series converges whose terms "decrease endlessly and are alternately positive and negative."
In Gerhardt, *Leibnizens Mathematische Schriften*, Sec. 1, III, vol. 2, 1856, p. 923.

integral, he gave the integral test.[23]

Next we reproduce the alternating series test and Cauchy's "proof by example," which uses the so-called Cauchy criterion [pp. 144–145; 139–131].

3.rd THEOREM. *If in the series* (1) $[u_0 + u_1 + u_2 + \cdots + u_n + \&c. \ldots]$ *the absolute value of the general term u_n decreases continually and indefinitely, for increasing values of n, if in addition the different terms are alternately positive and negative, the series will be convergent.*

Consider, for example, the series

$$1, \ -\frac{1}{2}, \ +\frac{1}{3}, \ -\frac{1}{4}, \ +\&c.. \ \pm\frac{1}{n}, \ \mp\frac{1}{n+1}, \ +\&c.$$

The sum of terms whose order exceeds n, if taken in number equal to m, will be

$$\pm\left(\frac{1}{n+1} - \frac{1}{n+2} + \frac{1}{n+3} - \frac{1}{n+4} + \&c. \ldots \pm \frac{1}{n+m}\right).$$

Now the absolute value of this sum, namely,

$$\frac{1}{n+1} - \frac{1}{n+2} + \frac{1}{n+3} - \frac{1}{n+4} + \&c. \ldots \pm \frac{1}{n+m}$$

$$= \frac{1}{n+1} - \left(\frac{1}{n+2} - \frac{1}{n+3}\right) - \left(\frac{1}{n+4} - \frac{1}{n+5}\right) - \&c. \ldots$$

$$= \frac{1}{n+1} - \frac{1}{n+2} + \left(\frac{1}{n+3} - \frac{1}{n+4}\right) + \left(\frac{1}{n+5} - \frac{1}{n+6}\right) + \&c. \ldots,$$

being evidently between

$$\frac{1}{n+1} \quad \text{and} \quad \frac{1}{n+1} - \frac{1}{n+2},$$

will decrease indefinitely for increasing values of n, whatever m may be, which suffices to establish the convergence of the proposed series. The same reasoning can evidently be applied to all the series of this kind.

A proof of the convergence of Fourier's trigonometric series was attempted by several mathematicians, including Fourier himself throughout his life. He was never able to give a rigorous proof, but one of his sketches [24] would be

---

[23] Theorem 2 in "Sur la convergence des séries," 1827, p. 226 = *Œuvres*, Ser. II, vol. VII, pp. 272–273.

[24] *Théorie analytique de la chaleur*, Art. 423.

useful to the man who finally did. In 1822 a young Prussian, Johann Peter Gustav Lejeune Dirichlet (1805–1859), came as a student to Paris, then the mathematical center of the world. He attended a scientific salon hosted by

J. P. G. L. DIRICHLET IN 1853
Portrait by Julius Schrader.
From Dirichlet's *Mathematische Werke* of 1889.

Fourier, and became acquainted with trigonometric series and Fourier's sketch of the convergence proof. It was not until 1829, however, that Dirichlet could complete a valid proof.[25] After replacing a certain trigonometric identity in Fourier's sketch of proof with one of his own, he was able to give sufficient conditions for the convergence of a Fourier series: that the function to be expanded have a finite number of jump discontinuities and a finite number of maxima and minima.

---

[25] "Sur la convergence des séries trigonométriques qui servent a représenter une fonction arbitraire entre des limites données," 1829.

6.5 THE DIFFERENCE QUOTIENT

With the matter of the convergence of series already settled and the concept of limit on the table, the time was ripe for a clarification of Newton's ultimate ratios. This could be done in two ways: by putting the concept of infinitesimal on a firm ground and then using it correctly, or by properly defining the limit of a ratio of vanishing quantities.

Da Cunha adopted the first approach in Book XV of the *Principios*, which starts as follows [p. 193; 196]:

DEFINITIONS.

I. *If an expression admits more than one value, when another expression admits only one, this one is called constant, and that one, variable.*

II. ... *and a variable that can always admit a value smaller than any proposed quantity, is called infinitesimal.*

This is all he needed to address the problem of ultimate ratios, although he did it in a notation and language that may well appear archaic and confusing to us. To soften the blow he explained that he would denote a function of a root [26] x by a Greek capital letter, such as Γ. Then, Γx denotes the value of Γ at x, with no parentheses enclosing the variable. This takes care of the notation, but we keep his confusing language in the translation of the next two definitions, and then explain below [p. 194; 197].

DEFINITIONS.

IIII. *Having chosen any quantity, homogeneous to a root x, to be called the fluxion of this root, and denoted thus dx; we call the fluxion of Γx, and will be denoted thus $d\Gamma x$, the quantity that would make*

$$\frac{d\Gamma x}{dx}$$

constant, and

$$\frac{\Gamma(x+dx) - \Gamma x}{dx} - \frac{d\Gamma x}{dx}$$

---

[26] As we saw in the preceding section, da Cunha referred to the independent variable as the "root" of a function.

infinitesimal or cipher [zero], if dx were infinitesimal, and constant all that
does not depend on dx.

V. *Every quantity is called fluent of its fluxion, and it is denoted by writing*
this sign ∫ immediately before the fluxion.

What IIII means is that dx is just a new variable, expressible in the same
units as x; that $d\Gamma x$ is the product of a function of x times dx (and thus the first
quotient in the definition is "constant" in the sense that it "does not depend
on dx"), and that the last expression is, for each fixed x, infinitesimal as a
function of dx.

To show that he knew what he was talking about, he immediately stated
and proved the following propositions [pp. 194–195; 197–198]:

PROPOSITIONS.

I. *x infinitesimal will make* $Ax + Bx^2 + Cx^3 + Dx^4 + \&c.$ *infinitesimal,*
if the coefficients $A, B, C, D, \&c.$ *were constants.*

Let n be the number of coefficients A, B, C, D, &c., and P any quantity
larger than each of them:[27] let Q be any proposed quantity: take $x < Q/nP$,
and < 1: it will be

$$\frac{1}{n}Q > Px, \quad \frac{1}{n}Q > Px^2, \quad \frac{1}{n}Q > Px^3,$$

and so on; and then

$$n \times \frac{1}{n}Q,$$

that is

$$Q > Ax + Bx^2 + Cx^3 + Dx^4 + \&c.$$

This may be the first ϵ-δ argument in history. That is, what we now call an
ϵ-δ proof. Just write ϵ instead of Q and choose $\delta = \epsilon/nP$. Da Cunha showed
that if $x < \min\{\delta, 1\}$ then $Ax + Bx^2 + Cx^3 + Dx^4 + \&c. < \epsilon$. This is modern
analysis and shows that his definition of infinitesimal works. He used this
proposition and the binomial theorem to prove the next, finding the fluxion of
$\Gamma x = x^n$.

II. $d(x^n) = nx^{n-1} dx.$

---

[27] Da Cunha continued to assume that all his numbers are positive, but the modern reader
can easily generalize his argument.

For dx infinitesimal, and what does not depend on dx, constant, make

$$\frac{nx^{n-1}\,dx}{dx}\quad[=nx^{n-1}]$$

constant and

$$\frac{(x+dx)^n - x^n}{dx} - \frac{nx^{n-1}\,dx}{dx}$$

$$\left[= n\frac{n-1}{2}x^{n-2}\,dx + n\frac{n-1}{2}\times\frac{n-2}{3}x^{n-3}\,dx^2 + \&c.\right]$$

infinitesimal.

In a similar manner, da Cunha then proved the following propositions [pp. 195–196; 198–200]:

III. $d(x+\Gamma x) = dx + d\Gamma x.$

IIII. $d(a+bx) = b\,dx.$

V. $d((a+x)^n)\;\cdots\;= n(a+x)^{n-1}\,dx.$ [28]

VI. dx infinitesimal, and what does not depend on dx constant, make $\Gamma(x+dx) - \Gamma x$ infinitesimal.

VII. $d(x\Gamma x) = dx\,\Gamma x + x\,d\Gamma x.$

VIII. Let x represent any [positive] number, and let l indicate hyperbolic logarithms: it will be $dx = x\,dlx.$

Using the fact that $x = e^{lx}$ and the series for the exponential, da Cunha proved this proposition as follows:

For it is

$$dx = d\left(1 + lx + \frac{1}{2}(lx)^2 + \frac{1}{6}(lx)^3 + \frac{1}{24}(lx)^4 + \frac{1}{120}(lx)^5 + \&c.\right)$$

$$= dlx + \frac{2}{2}(lx)dlx + \frac{3}{6}(lx)^2dlx + \frac{4}{24}(lx)^3dlx + \frac{5}{120}(lx)^4dlx + \&c.$$

$$= \left(1 + lx + \frac{1}{2}(lx)^2 + \frac{1}{6}(lx)^3 + \frac{1}{24}(lx)^4 + \&c.\right)dlx$$

$$= x\,dlx.$$

---

[28] In the original, the last exponent is $n-2$, but this is an obvious typo that was corrected in the French translation.

After stating two corollaries, he turned his attention to higher-order differentials, stated his version of Taylor's theorem, and proved some equalities for what we now call mixed partial derivatives [pp. 197–200; 200–203]. With one exception, the rest of Book XV is about arcs, areas, and volumes. The one exception is on indefinite integration [p. 202; 205]:

XVI. Between two fluents of equal fluxions there can be no difference that depends on the root [independent variable].

Let $d\Gamma x = d\Delta x$, and if it is possible, let $\Gamma x - \Delta x = \Theta x$: it will be $d\Theta x = d\Gamma x - d\Delta x = 0$: which shows that Θx cannot be a function of the root.

To sum up, da Cunha gave us the so-called difference quotient

$$\frac{\Gamma(x + dx) - \Gamma x}{dx}$$

for the first time in print, and built on it what can be viewed as a quasi-rigorous but incomplete differential theory. However, it is still quite different from today's presentations in calculus textbooks.

The difference quotient was independently introduced as the fundamental notion of the differential calculus by Simon Antoine Jean L'Huilier (1750–1840), of Geneva. In 1786, his essay *Exposition élémentaire des principes des calculs supérieurs* won the prize offered by the Berlin Academy for "a clear and precise theory of what is called *Infinity* in mathematics."[29] In §XIV [p. 31] of this text, L'Huilier expressed himself as follows:

To abbreviate & to facilitate the calculus with a more convenient notation, it has been agreed to otherwise denote lim. $\dfrac{\Delta P}{\Delta x}$, the limit of the quotient of the simultaneous changes of P and x, namely by $\dfrac{dP}{dx}$; so that lim. $\dfrac{\Delta P}{\Delta x}$ and $\dfrac{dP}{dx}$ denote the same thing.

Adding a few lines below [p. 32]:

... one must not think of this symbol as composed of two terms dP, & dx; but as a unique & unbreakable expression giving the limit quotient of ΔP over Δx.

---

[29] Page references are to the Berlin edition of 1787.

While these statements are valuable and the notation "lim" is used to this day, the fact is that l'Huilier's work contains very little of real value beyond some elementary examples. It is also based on an antiquated definition of limit, with which he began §I of Chapter 1 [p. 7]:

> 1st *Definition*. Let a variable quantity, always smaller or always larger than a proposed constant quantity; but that can differ from the latter less than any given quantity no matter how small: this constant *quantity* is called the *limit* in largeness or in smallness of the variable quantity.

This is a regression to the d'Alembertian idea that a variable quantity cannot "surpass the magnitude that it approaches." [30]

Eventually, l'Huilier would reach some wrong conclusions, such as the following [§LXVIII, p. 167]:

> *If a variable quantity, susceptible of a limit, enjoys constantly a certain property, its limit enjoys the same property.*

Not so. We know, for example, about an irrational number as the limit of a sequence of rational numbers.

Another contestant for the 1786 Berlin prize was the Frenchman Lazare Nicolas Marguérite Carnot (1753–1823). Since his 1785 essay *Dissertation sur la théorie de l'infini mathématique* did not win the prize, it was not published until much later. A revised and expanded version was published in Carnot's own time, but it did not really advance the calculus as we know it.

As president of the Berlin Academy, Joseph-Louis Lagrange (1736–1813), born in Turin as Giuseppe Lodovico LaGrangia, knew that the entries in the 1786 competition were weak, and decided to revise an earlier attempt of his own to make the calculus rigorous in an algebraic manner.[31] This revision was published in 1797 after he moved to France, already a member of the Academy of Sciences in Paris, with a title that clearly expressed his intention: *Theory of analytic functions, containing the principles of the differential calculus, free of any consideration of infinitesimals or vanishing [quantities], of limits or fluxions, and reduced to the algebraic analysis of finite quantities.*[32]

---

[30] l'Huilier would relax this requirement in the 1795 edition [pp. 17–18], but only for alternating series. His general definition [p. 1] was one-sided.

[31] "Sur une nouvelle espèce de calcul relatif a la différentiation et a l'intégration des quantités variables,"1772.

[32] *Théorie des fonctions analytiques*, 1797, 1813. Quotations are from the second edition of 1813, except for enclosing the variable x in parentheses. Page references are to this edition and to its reproduction in *Œuvres de Lagrange*, vol. 9, 1881.

For obvious reasons, Lagrange did not like infinitesimals, and he thought that fluxions, defined in terms of motion in time, were not admissible in mathematics. He also refused to accept Euler's zero over zero quotients, but found something very promising in the use that Euler had made of infinite series in the *Introductio*.

Lagrange based his study on his belief that a function can always be expanded in a power series in the following manner [pp. 2; 15–16]:

> When the variable of a function is given any increment whatever, by adding to this variable an undetermined quantity, we can by the ordinary rules of algebra, if the function is algebraic, expand it according to the powers of this undetermined [quantity]. The first term in the expansion will be the proposed function, which will be called the *primitive function*; the following terms will be formed from functions of the same variable, multiplied by the successive powers of the undetermined [quantity]. These new functions will depend only on the primitive function from which they are derived, and can be called *derived functions*. In general, whatever the primitive function may be, algebraic or not, it can always be expanded or supposed to be expanded in the same manner, and thus originate the derived functions.

This statement is reproduced in full because it contains terminology that is in use to this day, although Lagrange's derived functions are now called derivatives.[33] However, it is best to give the expansion in the following notation. If f represents the primitive function and i is the increment of the variable x, Lagrange expanded $f(x + i)$ [pp. 8; 21–22]

> in a series of this form

$$f(x) + pi + qi^2 + ri^3 + \text{etc.},$$

> in which the quantities p, q, r, etc., coefficients of the powers of i, will be new functions of x, derived from the primitive function of x, and independent of the undetermined i [for *indéterminée*].

In this formulation, it is possible to consider the difference quotient and show that there is no problem in putting $i = 0$ [p. 24]:

---

[33] As for these terms, John Collins' use of the words "derived" and "Derivative Æquation" in the present technical sense, in a letter to James Gregory of 8 November 1672, p. 244, is perhaps the earliest in the English language.

Thus we shall have

$$f(x + i) = f(x) + iP;$$

and therefore $f(x + i) - f(x) = iP$, and consequently divisible by i; once the division is performed, we shall have

$$P = \frac{f(x + i) - f(x)}{i}.$$

Here $P = p + qi + ri^2 + \cdots$, so that $P = p$ for $i = 0$. Thus, there is no need for limits, and no differentials or fluxions have been used. Eventually, Lagrange introduced a new notation for the derived functions [p. 32]:

$$p = f'(x), \quad q = \frac{f''(x)}{2}, \quad r = \frac{f'''(x)}{2 \cdot 3}, \quad \ldots, \text{ }^{34}$$

so that his general expansion is a Taylor series, and carried on with his program of making calculus rigorous.

Unfortunately, there is a hole in the bucket, and it is precisely his basic belief that every function can be expanded in the stated manner. We know that this is not the case, as was plainly demonstrated by Cauchy in a paper of 1822.[35] Here he showed that if

$$f(x) = e^{-1/x^2}$$

(completed with the value $f(0) = 0$), then $f'(0) = 0$, $f''(0) = 0, \ldots$, so that this nontrivial function cannot equal its Maclaurin series [pp. 49–50; 277–278]. The same is true of

$$f(x) = e^{-1/\sin^2 x} \quad \text{and} \quad f(x) = e^{-1/x^2(a+bx+cx^2+\cdots)},$$

where a is a positive constant and $a + bx + cx^2 + \cdots$ an entire function of x. As a consequence, the functions

$$e^{-x^2} \quad \text{and} \quad e^{-x^2} + e^{-1/x^2}$$

have the same Maclaurin series, but its sum can equal only one of them [p. 278].

Thus Lagrange's approach is also unsatisfactory, and the answer to the problem of providing rigor to the calculus must lie elsewhere.

---

[34] In the original 1797 edition Lagrange did not use parentheses about the variable x [p. 14].

[35] "Sur le développement des fonctions en séries et sur l'intégration des équations différentielles ou aux differences partielles," 1822 = Œuvres, Ser. II, II, Page references are to the original paper first and then to the Œuvres.

6.6 THE DERIVATIVE

Cauchy took a different path and established his calculus on the concept of limit, adopting l'Huilier's notation for it. He presented the differential part in the first twenty lessons of his second book: *Résumé des leçons donnés à l'École Royale Polytechnique sur le calcul infinitésimal* of 1823.

As da Cunha had done before him, Cauchy began by defining variable and constant, and then limit [p. 1; 13],[36] using word for word, in each case, the definitions previously given in the *Cours d'Analyse*. Then he defined infinitesimal as a variable whose numeric values [p. 4; 16]

> decrease indefinitely in such a manner as to become smaller than any given number.

This is a less precise version of what da Cunha had said, but Cauchy added: "A variable of this kind has zero for a limit." As examples, Cauchy evaluated the limits of

$$\frac{\sin \alpha}{\alpha} \quad \text{and} \quad (1 + \alpha)^{\frac{1}{\alpha}}$$

as $\alpha \to 0$, which he would need later.[37]

In the second lesson he introduced the terms independent variable and function [p. 5; 17], and later defined continuous function much as Bolzano had already done, all of these with the meanings in use today. And then, in the third lesson [p. 9; 22], we find a breakthrough. Referring to a continuous function $y = f(x)$ and denoting an increment in x values by $\Delta x = i$, he considered the *difference quotient* (his italics)

$$\frac{\Delta y}{\Delta x} = \frac{f(x + i) - f(x)}{i}$$

in the case in which i is infinitesimal, so that both numerator and denominator are infinitesimal.[38] Here is the complete quotation containing Cauchy's discovery of the derivative, in which he adopted some of Lagrange's notation and terminology [pp. 9; 22–23]:

---

[36] Page references are to the *Résumé* first and, after a semicolon, to *Œuvres*, Ser. II, IV. Quotations are from the *Résumé*.

[37] He had already found the latter in the *Cours d'Analyse* [pp. 166–167; 147–148].

[38] Cauchy had already considered this difference quotient in his *Cours d'analyse* and evaluated its limit in some examples [pp. 62–64; 65–67].

But, while these two terms indefinitely and simultaneously approach the limit zero, the quotient itself could converge to another limit, either positive, or negative. This limit, when it exists, has a specific value, for each particular value of x; but it varies with x. Thus, for example, if we take $f(x) = x^m$, m denoting a whole number, the quotient of the infinitely small differences will be

$$\frac{(x+i)^m - x^m}{i} = mx^{m-1} + \frac{m(m-1)}{1 \cdot 2}x^{m-2}i + \cdots + i^{m-1}$$

and it will have the quantity mx^{m-1} as a limit, that is, a new function of the variable x. It will be the same in general; only, the form of the new function that will serve as the limit of the quotient $\dfrac{f(x+i) - f(x)}{i}$ will depend on the form of the proposed function $y = f(x)$. To indicate this dependence, the new function is given the name *derived function*, and it is denoted, with the aid of an accent, by the notation

$$y' \quad \text{or} \quad f'(x).$$

In typical fashion, Cauchy then proceeded to compute the derivatives (derived functions) of "the simple functions that produce the operations of algebra and trigonometry" [p. 10; 23], that is, $a + x$, $a - x$, ax, a/x, x^a, A^x, Lx (the logarithm to base A in this context), $\sin x$, $\cos x$, $\arcsin x$, and $\arccos x$. To deal with some of these he used the limits found in lesson one, but for the last two he needed a new result on composite functions, and he produced it as follows [p. 11; 25]:

Now let z be a second function of x, related to the first $y = f(x)$ by the formula
$$z = F(y).$$
z or $F(fx)$ will be what one calls a *function of a function* of the variable x; and, if we denote by Δx, Δy, Δz, the infinitely small and simultaneous increments of the three variables x, y, z, we shall find
$$\frac{\Delta z}{\Delta x} = \frac{F(x+\Delta y) - F(x)}{\Delta x} = \frac{F(x+\Delta y) - F(x)}{\Delta y} \cdot \frac{\Delta y}{\Delta x},$$
and then, passing to the limit,
$$z' = y'.F'(y) = f'(x).F'(fx).$$

In this way the chain rule appeared for the first time in print, and then Cauchy used it to compute the remaining derivatives of simple functions. For example [pp. 12; 25–26], for $y = \arcsin x$,

$$\sin y = x, \quad y' \cos y = 1, \quad y' = \frac{1}{\cos y} = \frac{1}{\sqrt{1 - x^2}};$$

for $y = \arccos x$,

$$\cos y = x, \quad y' \times -\sin y = 1, \quad y' = \frac{-1}{\sin y} = \frac{1}{\sqrt{1 - x^2}}.$$

After this, Cauchy took flight, and we can only feel like the cameraman at Kitty Hawk watching Orville Wright getting farther and farther away. In the fourth lesson he defined differential and differentiation, in the fifth he differentiated sums, differences, products, and quotients and introduced logarithmic differentiation, in the sixth he dealt with maxima and minima, the seventh contains a treatment of undetermined limits (∞/∞ and the like) and the mean value theorem,[39] the eighth deals with partial derivatives, in the ninth he gave us the chain rule for functions of several variables and differentiated implicit functions, in the tenth he explored maxima and minima of functions of several variables, in the twelfth he told us about higher-order derivatives for functions of one variable and in the thirteenth for functions of several variables, the fourteenth dealt with total differentials, and in the fifteenth and the sixteenth he returned to maxima and minima. He finished at the twentieth with partial fraction decomposition. Thus, almost the entire differential calculus sequence was covered in the first 121 pages of this book.

Cauchy used the ϵ-δ notation and this type of proof, but not quite as often or as precisely as we would like to think. The following example from Lesson 7 is probably his clearest use of this method [pp. 27–28; 44–45]:

Theorem. *If, the function $f(x)$ being continuous between the limits $x = x_0$, $x = X$, we denote by A the smallest, and by B the largest of the values that the derived function $f'(x)$ attains in this interval, the ratio of the finite differences*

(4)
$$\frac{f(X) - f(x_0)}{X - x_0}$$

will be necessarily between A and B.

---

[39] First obtained by Lagrange in the form "If a prime function of z, such that $f'z$ is always positive for all the values of z, from $z = a$ to $z = b$, b being $> a$, the difference of primitive functions corresponding to these two values of z, namely, $fb - fa$, will necessarily be a positive quantity," *Théorie des fonctions analytiques*, 1797, Art. 48, p. 45 = 1813, Art. 38, p. 63 = *Œuvres de Lagrange*, vol. 9, Art. 38, p. 78.

Demonstration. Let δ, ϵ, denote two very small numbers, the first chosen in such manner that, for numeric [absolute] values of i smaller than δ, and for any value of x between the bounds x_0, X, the ratio

$$\frac{f(x+i) - f(x)}{i}$$

will always remain larger than $f'(x) - \epsilon$,[40] and smaller than $f'(x) + \epsilon$. If we insert $n - 1$ new values of the variable x between the bounds x_0, X, namely,

$$x_1, \quad x_2, \quad \ldots, \quad x_{n-1},$$

so as to divide the difference $X - x_0$ into elements

$$x_1 - x_0, \quad x_2 - x_1, \quad \ldots, \quad X - x_{n-1},$$

which, all of them having the same sign, have numeric [absolute] values smaller than δ; then the fractions

$$(5) \qquad \frac{f(x_1) - f(x_0)}{x_1 - x_0}, \quad \frac{f(x_2) - f(x_1)}{x_2 - x_1}, \quad \ldots \quad \frac{f(X) - f(x_{n-1})}{X - x_{n-1}},$$

are bound, the first between the limits $f'(x_0) - \epsilon$, $f'(x_0) + \epsilon$, the second between the limits $f'(x_1) - \epsilon$, $f'(x_1) + \epsilon$, &c. ... will all be larger than the quantity $A - \epsilon$, and smaller than the quantity $B + \epsilon$. Moreover, since the fractions (5) have denominators of the same sign, if we divide the sum of their numerators by the sum of their denominators, we shall obtain a *mean* fraction, that is, bound between the smallest and the largest of those considered [*see* l'Analyse algébrique, note II, 12.e theorem]. The expression (4), with which this mean coincides, will thus itself be bound between the limits $A - \epsilon$, $B + \epsilon$; and, since this conclusion holds, no matter how small the number ϵ, it can be asserted that the expression (4) will be between A and B.

In 1829, with the *Résumé* out of print, Cauchy published a modified edition of the differential calculus separately.[41]

---

[40] There is an obvious typo in the *Œuvres*, where the last expression appears as $f(x) - \epsilon$.
[41] *Leçons sur le calcul différentiel*, 1829.

6.7 CAUCHY'S INTEGRAL CALCULUS

The second half of the *Résumé des leçons donnés à l'École Royale Polytechnique sur le calcul infinitésimal* is devoted to the integral calculus, which Cauchy developed as extensively as the differential calculus.

Mathematicians of the early nineteenth century had abandoned the Leibnizian concept of integral as a sum, perhaps having some difficulty accepting it as the sum of infinitely many infinitely small quantities, in favor of the integral as a primitive or prederivative. Cauchy reversed that approach in the twenty-first lesson of the *Résumé*. He considered a continuous function $y = f(x)$ and $n - 1$ increasing (or decreasing) values of the variable x between the bounds x_0, X, namely [p. 81; 122]

$$x_1, \quad x_2, \quad \ldots, \quad x_{n-1}.$$

Then he formed the sum

$$S = (x_1 - x_0)f(x_0) + (x_2 - x_1)f(x_1) + \cdots + (X - x_{n-1})f(x_{n-1}).$$

Assuming that f was what we now call uniformly continuous, Cauchy proved that [p. 83; 125]

if we make the numeric values of these elements [the lengths $x_i - x_{i-1}$] decrease indefinitely, while increasing their number, the value of S will become practically [*sensiblement*] constant, or, in other words, will ultimately attain a certain limit that depends uniquely on the form of the function $f(x)$, and on the extreme values x_0, X of the variable x. This limit is what we call a *definite integral*.

For this definite integral he proposed three notations [p. 84; 126], the first of which,

$$\int_{x_0}^{X} f(x)\, dx,$$

for which he credited Fourier, he considered the simplest.

In the next two lessons Cauchy proved some of the elementary theorems for definite integrals that are familiar to us from calculus textbooks and evaluated some examples. In the twenty-fourth and twenty-fifth lessons he discussed improper integrals and defined the principal value. Then, in the twenty-sixth lesson, he defined a new function [p. 101; 151]

$$\mathscr{F}(x) = \int_{x_0}^{x} f(x)\, dx$$

of the variable upper limit of integration, but without thinking of using what we call a dummy variable in the integrand. From the mean value theorem for definite integrals, which he had established in the twenty-third lesson, Cauchy obtained

$$\mathscr{F}(x) = (x - x_0)f[x_0 + \theta(x - x_0)], \qquad \mathscr{F}(x_0) = 0,$$

"θ being a number smaller than unity" (and not smaller than zero). From one of the elementary theorems previously proved,

$$\int_{x_0}^{x+\alpha} f(x)\,dx - \int_{x_0}^{x} f(x)\,dx = \int_{x}^{x+\alpha} f(x)\,dx = \alpha f(x + \theta\alpha),$$

or

$$\mathscr{F}(x + \alpha) - \mathscr{F}(x) = \alpha f(x + \theta\alpha).$$

Thus, if f is continuous, dividing by α [p. 101; 152],

we shall conclude, passing to the limit, that

$$\mathscr{F}'(x) = f(x).$$

Therefore the [stated] integral, considered as a function of x, has as its derivative the function $f(x)$ under the sign $\int$ in this integral.

This is Cauchy's version of the fundamental theorem of calculus. Later in the same lesson he gave the formula [p. 104; 155]

$$\int_{x_0}^{x} f(x)\,dx = F(x) - F(x_0),$$

where F is a function such that $F'(x) = f(x)$.

In the twenty-seventh lesson Cauchy turned to indefinite integration, prominently displaying what he termed the *arbitrary constant* [p. 105; 157] of integration, and gave a method that he called *integration by substitution* as follows [pp. 107; 159–160]: if $y = f(x)$ and if instead of x we substitute another variable z related to the first by $x = \chi(z)$, then

$$\int f(x)\,dx = \int \mathfrak{f}(z)\,dz,$$

where $\mathfrak{f}(z) = f[\chi(z)]\chi'(z)$.

Then *integration by parts* was given in the form that has become standard [p. 108; 162]:

$$\int u\, dv = uv - \int v\, du.$$

It is not necessary to fully describe the contents of the rest of this work. In general terms, it can be said that, with the notable exception of vector analysis (gradient, divergence, curl), the entire calculus sequence as taught today is contained in the *Résumé*. The converse is not necessarily true, as shown by formulas such as [p. 119; 179]

$$\int \left[l\left(x + \sqrt{x^2+1}\right) \right]^n dx$$

$$= \left[l\left(x + \sqrt{x^2+1}\right) \right]^n \left\{ x - \frac{n\sqrt{x^2+1}}{l\left(x + \sqrt{x^2+1}\right)} + \frac{n(n-1)x}{\left[l\left(x + \sqrt{x^2+1}\right) \right]^2} \right.$$

$$\left. - \frac{n(n-1)(n-2)\sqrt{x^2+1}}{\left[l\left(x + \sqrt{x^2+1}\right) \right]^3} + \cdots \right\} + \mathcal{C}.$$

6.8 UNIFORM CONVERGENCE

It may have come as an unwelcome surprise to Cauchy, once the convergence of series was established on a firm ground, that the Norwegian mathematician Niels Henrik Abel (1802–1829) noticed an incorrect theorem in the *Cours d'analyse*: the sum of a convergent series of continuous functions is continuous.[42] Abel wrote:[43]

> But it seems to me that this theorem has exceptions. For example the [sum of the] series
> $$\sin x - \tfrac{1}{2}\sin 2x + \tfrac{1}{3}\sin 3x - \cdots$$
> is discontinuous for every value $(2m+1)\pi$ of x, where m is an integer.

Indeed, this is Fourier's series expansion of $y = x/2$ valid for $-\pi < x < \pi$ (see the example on page 379), and its convergence is guaranteed by Dirichlet's

---

[42] *Cours d'analyse de l'École Royale Polytechnique*, Chapter VI, 1.st THEOREM, [pp. 131–132; 120].

[43] "Untersuchungen über die Reihe $1 + (m/1).x + (m \cdot (m-1)/1 \cdot 2).x^2 + (m \cdot (m-1) \cdot (m-2)/1 \cdot 2 \cdot 3).x^3 + \cdots$ u.s.w.," 1826. French translation in *Œuvres complètes*. The stated quotation is in the footnote on page 225.

NIELS HENRIK ABEL IN 1826
Portrait by Johan Görbitz.

1829 theorem. But Cauchy did not take Abel's objection to heart at that time, for he restated the incorrect theorem in his *Résumés analytiques*, published in Turin in 1833.[44]

We shall then seek additional conditions for the sum of a convergent series of continuous (alternatively, differentiable or integrable) functions

$$u_1 + u_2 + \cdots + u_n + \cdots$$

to be continuous (or, respectively, differentiable or integrable). Now let s denote the sum of the series and let s_n denote the sum of the first n terms. Using Cauchy's letter ϵ, to be fashionable, the definition of convergence of the stated series can be rephrased as follows. Imitating Cauchy's style in proving the theorem of Section 6.6, the convergence of the series

$$u_1(x) + u_2(x) + \cdots + u_n(x) + \cdots$$

---

[44] 7.th Theorem, p. 46 = *Œuvres*, Ser. II, X, 1895, p. 56.

to a sum $s(x)$ for a given x can be expressed as follows. Given any small positive number ϵ, there is a positive integer N such that for $n > N$ the partial sum

$$s_n(x) = u_1(x) + u_2(x) + \cdots + u_n(x)$$

will always remain larger than $s(x) - \epsilon$ and smaller than $s(x) + \epsilon$. Since this definition does not guarantee the validity of Cauchy's theorem on the continuity of s, a natural question is whether it can be modified so as to render Cauchy's theorem valid.

The fact that a convergent series of continuous functions can have a discontinuous sum was examined, about 1848, by Phillip Ludwig von Seidel (1821–1896) [45] and, independently, by Sir George Gabriel Stokes (1819–1903).[46] They noticed that, for a given ϵ, no N can be chosen so that the inequalities in the preceding definition of convergence are satisfied for all x near a point of discontinuity of s. That is, while for each such x there is an N such that the stated inequalities hold, *the same N does not work for all x*. We could then conjecture that if we replace the definition of convergence with one that requires N to be valid for all x, then this more demanding kind of convergence should make it impossible for the sum of the series to be discontinuous.

It turns out that in 1838, Christof Gudermann (1798–1852) had already introduced a kind of convergence at the same rate (*einem im Ganzen gleichen Grad der Convergenz*), which is the precursor of the one that we seek,[47] but the importance of the concept was realized by Karl Theodor Wilhelm Weierstrass (1815–1897), who would later become one of the main forces in making analysis rigorous. Weierstrass was a student at the University of Bonn, who in 1839 transferred to Münster to attend Gudermann's lectures (it was then the custom in Germany for students to switch universities at some time). It is quite likely that Gudermann and Weierstrass discussed the new concept of convergence. Weierstrass never earned his doctorate, and became a *Gymnasium* (high school) teacher in 1841. Such was the amount of firstrate research that he produced from this moment on that it eventually earned him a position at the University of Berlin in 1856. This research remained unpublished for a long time, but the fact that he referred to a power series that converges uniformly—*gleichmässig convergirt*—in an 1841 manuscript, *Zur Theorie der Potenzreihen*, supports the idea that he may have learned about it from Gudermann. This manuscript contains a definition of uniform

---

[45] "Note über eine Eigenschaft der Reihen, welche discontinuirliche Functionen darstellen," 1847–49.

[46] "On the critical values of the sums of periodic series," 1848.

[47] "Theorie der Modular-Functionen und der Modular-Integrale," 1838, pp. 251–252.

KARL WEIERSTRASS
From Weierstrass' *Mathematische Werke* of 1894.

convergence for a series of functions of several variables, which Weierstrass
restated in a more polished form in "Zur Functionenlehre," a paper of 1880.[48]

An infinite series

$$\sum_{\nu=n}^{\infty} f_\nu,$$

whose members are functions of arbitrarily many variables, converges uniformly in a given part (B) of its region of convergence, if after the assumption
of an arbitrarily small positive quantity δ a whole number m can always be
determined, such that the absolute value of the sum

$$\sum_{\nu=n}^{\infty} f_\nu,$$

---

[48] With very minor modifications, this translation is from Grattan-Guinness, "The emer-
gence of mathematical analysis and its foundational progress, 1780–1880," in Grattan-
Guinness, *From the calculus to set theory, 1630–1910. An introductory history*, 1980, 2000,
p. 134.

is smaller than δ for each value of n, which is $\geq m$, and for each collection of values of the variable belonging to the region B.

Cauchy was unaware of these facts and, quite independently, when he finally saw the light on this point, introduced the concept of uniform convergence (but not this name) in 1853.[49] Then he proved the corrected version of his earlier theorem that motivated the preceding discussion. In current terminology, the sum of a uniformly convergent series of continuous functions is continuous.[50]

As in Cauchy's case regarding ordinary convergence, the importance of Weierstrass' contribution stems from the fact that he realized the usefulness of uniform convergence and incorporated it in theorems on the integrability and differentiability of series of functions term by term.

---

[49] "Note sur les séries convergentes dont les divers terms sont des fonctions continues d'une variable réelle ou imaginaire entre des limites données," 1853 = *Œuvres*, Ser. I, XII, 1900.

[50] Cauchy's own statement of this result (Theorem II) is not very clear. He required the sum $u_n + u_{n+1} + \cdots + u_{n'-1}$ to become "always infinitely small for infinitely large values of the integers n and $n' > n$" [pp. 456–457; 34–35], but did not mention any values of x for which this should be true. Did he rely on the word "always" to mean "for all x in the domain of the stated functions"?

BIBLIOGRAPHY

Abel, Niels Henrik

"Untersuchungen über die Reihe:

$$1 + \frac{m}{1} \cdot x + \frac{m \cdot (m-1)}{1 \cdot 2} \cdot x^2 + \frac{m \cdot (m-1) \cdot (m-2)}{1 \cdot 2 \cdot 3} \cdot x^3 + \cdots \text{u.s.w.,}"$$

Journal für die reine und angewandte Mathematik, **1** (1826) 311–339. Pages 311–315 are reproduced with English translation in Stedall, *Mathematics emerging*, pp. 516–524. Translation of an extract in Birkhoff, *A source book in classical analysis*, pp. 68–70; reprinted in Calinger, *Classics of mathematics*, pp. 594–596. French version: "Recherches sur la série

$$1 + \frac{m}{1}x + \frac{m(m-1)}{1 \cdot 2}x^2 + \frac{m(m-1)(m-2)}{1 \cdot 2 \cdot 3}x^3 + \cdots"$$

in *Œuvres complètes de Niels Henrik Abel*, 2nd. ed., Vol. 1, *Contenant les mémoirs publiés par Abel*, pp. 219–250.

Œuvres complètes de Niels Henrik Abel. Nouvelle édition publiée aux frais de l'état norvégien par MM. L. Sylow et S. Lie, 2 vols., Grøndahl & Søn, Christiania, Norway, 1881, 1965; facsimile reproduction by Éditions Jacques Gabay, Sceaux, 1992.

Agnesi, Maria Gaetana

Instituzioni analitiche ad uso della gioventù italiana, 4 vols., Nella Regia-Ducal Corte, Milan, 1748. English translation as *Analytical institutions, in four books*: *originally written in Italian, by Donna Maria Gaetana Agnesi, Professor of the Mathematicks and Philosophy in the University of Bologna. Translated into English by the late Rev. John Colson, M.A. F.R.S. and Lucasian Professor of the Mathematicks in the University of Cambridge. Now first printed, from the translators manuscript, under the inspection of the Rev. John Hellins, B.D. F.R.S. and vicar of Potter's-Pury, in Northhamptonshire*, Taylor and Wilks, Chancery-Lane, London, 1801 (Colson changed Agnesi's differential notation to fluxonial notation).

Al-Masudi, Ali

Muruj al-Zahab wa al-Maadin al-Jawahir, originally written in 947, revised 956. Edited by Charles Pellat, 7 vols., Publications de l' Université Libanaise, Section des Études Historiques, Beirut, 1965–1979. English translation of volume 1 by Aloys Sprenger as *El-Masudi's historical encyclopaedia entitled "Meadows of gold and mines of gems,"* Oriental Translation Fund of Great Britain and Ireland, London, 1841.

Al-Tusi, Mohammed

Kitab shakl al-qatta; facsimile reproduction of the Arabic version and French translation as *Traité du quadrilatère attribué a Nassiruddin-el Toussy* by Alexandros Carathéodory, Typographie et Litographie Osmanié, Constantinople, 1891.

d'Alembert, Jean le Rond

"Addition au mémoire sur la courbe que forme une corde tenduë, mise en vibration," *Histoire de l'Académie Royale des Sciences et des Belles-Lettres de Berlin*, **6** (1750, pub. 1752) 355–360.

Encyclopédie, ou Dictionnaire raisonné des sciences, des arts et des métiers, par une société de gens de lettres. Mis en ordre et publié par M. Diderot, . . . & quant à la partie mathématique, par M. d'Alembert, . . ., 17 volumes, Briasson, David l'aîné, Le Breton, Durand, Paris, 1751–1765, and 18 additional volumes in later years.

"Recherches sur la courbe que forme une corde tenduë mise en vibration," *Histoire de l'Académie Royale des Sciences et des Belles-Lettres de Berlin*, **3** (1747) 214–219 (pages 214–217 are reproduced with translation in Stedall, *Mathematics emerging*, pp. 265–271), and "Suite des recherches," **3** (1747) 220–249.

Archimedes of Syracuse

The quadrature of the parabola, in Heath *The works of Archimedes with The method of Archimedes*, pp. 233–252.

The sand-reckoner, in Heath, *The works of Archimedes with The method of Archimedes*, pp. 221–232; reprinted in Hawking, *God created the integers. The mathematical breakthroughs that changed history*, pp. 200–208; selections also reprinted in Calinger, *Classics of mathematics*, pp. 147–151.

Argand, Robert

Essai sur une manière de représenter les quantités imaginaires, dans les constructions géométriques, privately printed, Paris, 1806. Second edition with a preface by Guillaume-Jules Hoüel, Gauthier-Villars, Paris, 1874; reproduced by Albert Blanchard, Paris, 1971. Articles 3 to 7 reproduced with translation in Stedall, *Mathematics emerging*, pp. 462–467. Translated into English by Arthur Sherburne Hardy as *Imaginary quantities: Their geometrical interpretation*, D. Van Nostrand, New York, 1881.

"Philosophie mathématique. Essay sur une manière de représenter les quantités imaginaires dans les constructions géométriques," *Annales de mathématiques pures et apliquées*, **4** (1813–1814) 133–147.

Aristarchos of Samos

Ἀριστάρχου περὶ μεγεθῶν καὶ ἀποστημάτων ἡλίου καὶ σελήνης. The complete text in Greek with an English translation entitled *On the sizes and distances of the sun and moon* can be seen in Heath, *Aristarchus of Samos, the ancient Copernicus*, pp. 352–411.

Aryabhata

The *Aryabhatiya*, translated into English with notes by William Eugene Clark as *The Aryabhatiya of Aryabhata. An ancient Indian work on mathematics and astronomy*, The University of Chicago Press, Chicago, Illinois, 1930. Critical edition in Sanskrit with English translation and notes by Kripa Shankar Shukla and K.

Venkateswara Sarma, *Aryabhatiya of Aryabhata*, Indian National Science Academy, New Delhi, 1976.

Bardi, Jason Socrates

The calculus wars. Newton, Leibniz, and the greatest mathematical clash of all time, Thunder's Mouth Press, New York, 2006.

Baron, Margaret E.

The origins of the infinitesimal calculus, Pergamon Press, Oxford, 1969; reprinted by Dover Publications, New York, 1987, 2004.

Barrow, Isaac

Lectiones geometricæ; in quibus (præsertim) generalia curvarum linearum symptomata declarantur, William Godbid, London 1670; 2nd. ed. 1674; first translated into English by Edmund Stone as *Geometrical lectures: explaining the generation, nature and properties of curve lines. Read in the University of Cambridge by Isaac Barrow; translated from the Latin edition, revised, corrected and amended by the late Sir Isaac Newton*, Printed for S. Austen, London, 1735; second edition of 1674 reprinted by Georg Olms Verlag, Hildesheim · New York, 1976. Abridged translation with notes, deductions, proofs omitted by Barrow, and further examples of his method by Child as *The geometrical lectures of Isaac Barrow*.

Bateman, Harry

"The correspondence of Brook Taylor," *Bibliotheca Mathematica*, (3) **7** (1906–1907) 367–371.

Bhaskara

Mahabhaskariya, edited and translated by Kripa Shankar Shukla, Lucknow University, Lucknow, 1960.

Beaugrand, Jean de

"De la manière de trouver les tangentes des lignes courbes par l'algèbre et des imperfections de celle du S. des C[artes];" reproduced in *Œuvres de Fermat*, V, *Supplément*, 1922, pp. 102–113.

Bell, Eric Temple

Men of mathematics, Simon and Schuster, New York, 1937; reprinted by Touchstone Books, New York, 1986.

Berggren, J. Lennart

Episodes in the mathematics of medieval Islam, Springer-Verlag, 1986, 2003.

"Mathematics in medieval Islam," in Katz, ed., *The mathematics of Egypt, Mesopotamia, China, India, and Islam*.

Berkeley, George

The analyst; or, a discourse addressed to an infidel mathematician. Wherein it is examined whether the object principles, and inferences of the modern analysis are more distinctly conceived, or more evidently deduced, than religious mysteries and points of faith, Jacob Tonson, London, 1734; reproduced in pp. 63–102 of Vol. 4

of *The works of George Berkeley Bishop of Cloyne*, ed. by Arthur Aston Luce and Thomas Edmund Jessop, 9 volumes, Thomas Nelson and Sons, Ltd., London, 1951, 1964. Reprinted by Kessinger Publishing Company, Kila, Montana, 2004; reprinted by Nabu Press, Berlin, 2010.

Bernoulli, Daniel

"Lettre de Monsieur Daniel Bernoulli, de l'Académie Royale des Sciences, à M. Clairaut de la même Académie, au sujet des nouvelles découvertes faites sur la vibrations des cordes tendues," *Journal des Sçavans*, (March 1758) 157–166.

"Réflexions et éclaircissements sur les nouvelles vibrations des cordes exposées dans les mémoires de l'Académie de 1747 et 1748," *Histoire de l'Académie Royale des Sciences et des Belles-Lettres de Berlin*, **9** (1753) 147–172 and 173–195.

Bernoulli, Jakob

"Analysis problematis antehac propositi," *Acta Eruditorum*, **9** (1690) 217–219 = *Jacobi Bernoulli, Basileensis, Opera*, Tomus primus, N°. XXXIX, pp. 421–426.

Jacobi Bernoulli, Basileensis, Opera, Gabriel and Philibert Cramer, Geneva, 1744; reprinted by Birkhäuser Verlag, 1968.

Tractatus de seriebus infinitis earumque summa finita, et usu in quadraturis spatiorum & rectificationibus curvarum, in pp. 241–306 of *Ars conjectandi, opus posthumum. Accedit Tractatus de seriebus infinitis, et epistola Gallicè scripta de ludo pilæ reticularis*, Impensis Thurnisiorum, Fratrum, Basel, 1713; reprinted by Culture et Civilisation, Brussels, 1968. The *Tractatus* can also be seen in *Jacobi Bernoulli, Basileensis, Opera*, 1744. Parts I and II are in *Tomus primus* with the titles *Positiones arithmeticæ [Positionum arithmeticarum] de seriebus infinitis, earumque summa finita*, N°. XXXV, pp. 375–402, and N°. LIV, 517–542. Parts III to V are in *Tomus secundus* with the title *Positionum de seriebus infinitis, earumque usu in quadraturis spatiorum & rectificationibus curvarum* (there is a slight variation in the title of Part III), N°. LXXIV, pp. 745–767, N°. XC, pp. 849–867, and N°. CI, pp. 955–975.

Bernoulli, Johann

"Additamentum effectionis omnium quadraturarum & rectificationum curvarum per seriem quandam generalissimam," *Acta Eruditorum*, **13** (November 1694) 437–441 = *Opera omnia*, Vol. I, N°. XXI, pp. 125–128.

Der Briefwechsel von Johann Bernoulli, Otto Spiess, ed., Birkhäuser Verlag, Basel, Vol. I, 1955.

Lectiones de calculo differentialium, Paul Schafheitlin, Basel 1922. German translation as *Die Differentialrechnung aus dem Jahre 1691/92* in Ostwald's *Klassiker der exakten Wissenschaften*, Engelmann, Leipzig, No. 211, 1924.

Letter to Leibniz, May 25, 1712. In Gerhardt, *Leibnizens Mathematische Schriften*, Sec. 1, III **2**, pp. 885–888.

Letter to the Marquis de l'Hospital, July 22, 1694. In *Der Briefwechsel von Johann Bernoulli*, Vol. I, 1955, pp. 231–236.

Opera omnia, tam antea sparsim edita, quam hactenus inedita, 4 vols., Marci-Michaelis Bousquet, Lausannæ & Genevæ, 1742; reprinted by Georg Olms Verlagsbuchhandlung, with an introduction by Joseph Ehrenfried Hofmann, Hildesheim, 1968. "Perfectio regulæ suæ edita in libro Gall. *Analyse des infiniment petits, Art. 163.* pro determinando valore fractionis, cujus Numerator & Denominator certo casu evanescunt," *Acta Eruditorum*, **23** (1704) 375–380 = *Opera Omnia*, Vol. I, N°. LXXI, pp. 401–405.

Birkhoff, Garrett, ed.
A source book in classical analysis, Harvard University Press, Cambridge, Massachusetts, 1973.

Boas Hall, Marie
See Hall, Alfred Rupert.

Bodemann, Eduard
Die Leibniz-Handschriften der Königlichen Öffentlichen Bibliothek zu Hannover, Hahn'schen Buchhandlung, Hanover and Leipzig, 1895; reprinted by Georg Olms Verlag, Hildelsheim, 1966. This is a catalog of Leibniz' works, and references to it are by part (Mathematics is Part XXXV), volume, item number, and page.

Bolzano, Bernardus Placidus Johann Nepomuk
"Rein analytischer Beweis des Lehrsatzes dass zwischen je zwey Werthen, die ein entgegengesetztes Resultat gewähren, wenigstens eine reele Wurzel der Gleichung liege," Gottlieb Haase, Prague, 1817 = *Abhandlungen der Königlichen Böhmischen Gesellschaft der Wissenschaften*, **5** (1814–1817, pub. 1818) 1–60 = *Ostwald's Klassiker der exakten Wissenschaften* # 153, ed. by Philip Edward Bertrand Jourdain, Wilhelm Engelmann, Leipzig, 1905, pp. 3–43; facsimile reproduction in "Early mathematical works," with introduction by Luboš Nový and Jaroslav Folta, *Acta Historiæ Rerum Naturalim necnon Technicarum*, Special Issue 12, Prague, 1981. French translation in Jan Sebestik, "Démonstration purement analytique du théorème: entre deux valeurs quelconques qui donnent deux résultats de signes opposés se trouve au moins une racine réele de l'équation," *Révue d'Histoire des Sciences*, **17** (1964) 136–164. First English translation in Stephen Bruce Russ, "A translation of Bolzano's paper on the intermediate value theorem," *Historia Mathematica*, **7** (1980) 156–185. Latest English translation in Stephen Bruce Russ, *The mathematical works of Bernard Bolzano*, Oxford University Press, Oxford, 2004, pp. 250–277. Articles 6 and 7 are reproduced with translation in Stedall, *Mathematics emerging*, pp. 497–499.

Bombelli, Rafael
L'Algebra. Opera di Rafael Bombelli da Bologna, diuisa in tre libri, con la quale ciascuno da se potrà venire in perfetta cognitione della teorica dell'aritmetica. Con vna tauola copiosa delle materie, che in essa si contengono. Posta hora in luce à beneficio delli studiosi di detta professione, Giouanni Rossi, Bologna, 1572 and

1579. The complete manuscript is contained in Codice B. 1569 of the *Biblioteca comunale dell'Archiginnasio* in Bologna. It was first edited and printed by Bortolotti as *Rafael Bombelli da Bologna. L'Algebra. Prima edizione integrale*, 1966.

Bortolotti, Ettore, ed.
"La scoperta e le successive generalizzazioni di un teorema fondamentale di calcolo integrale," *Archeion*, **5** (1924) 204–227.

Rafael Bombelli da Bologna. L'Algebra. Prima edizione integrale, Feltrinelli Editore, Milano, 1966. It has an introduction by Umberto Forti and a preface and analysis of the work by Bortolotti (his justification of the 1550 dating of the manuscript is on pages XXXI and XXXIX). The notation has been modernized in this edition, which uses + and − signs, and the pagination is different from that of the 1572 edition.

Boyer, Carl Benjamin
The concepts of the calculus. A critical and historical discussion of the derivative and the integral, 2nd. ed., Hafner Publishing Company, 1949; reprinted by Dover Publications as *The history of the calculus and its conceptual development*, New York, 1959.

A history of mathematics, John Wiley & Sons, New York, 1968; reprinted by Princeton University Press, Princeton, New Jersey, 1985. Second edition, revised by Uta C. Merzbach, John Wiley & Sons, New York, 1989, 1991.

Brahe, Tycho
Tychonis Brahe, Dani De nova et nvllivs ævi memoria privs visa stella, iam pridem anno à nato Christo 1572. mense Nouembrj primùm conspecta, contemplatio mathematica, Lavrentivs Benedictj, Copenhagen, 1573; reproduced in Dreyer, ed., *Tychonis Brahe Dani scripta astronomica*, 1913, following the Prolegomena Editoris.

Tychonis Brahe, mathim, eminent, Dani Opera omnia, sive Astronomiæ instauratæ progymnasmata, in duas partes distributa, quorum prima de restitutione motuum solis & lunæ, stellarumque inerrantium tractat. Secunda autem de mundi ætherei recentioribus phænomensis agit. Editio ultum nunc cum indicibus & figuris prodit, Ioannis Godofredi Schönwetteri, Franchfurt, 1648; reprinted by Georg Olms Verlag, Hildesheim, 2001; reprinted by Kessinger Publishing Company, Kila, Montana, 2009.

Briggs, Henry
Arithmetica logarithmica sive logarithmorvm chiliades triginta, pro numeris naturali serie crescentibus ab vnitate ad 20,000: et a 90,000 ad 100,000, William Jones, London, 1624. There is a Latin transcription with complete (except for the Preface) annotated translation into English by Ian Bruce as *Briggs' Arithmetica Logarithmica*, available online at URL www-history.mcs.st-andrews.ac.uk/Miscellaneous/Briggs/index.html. August 2006. Second edition: *Arithmetica logarithmica, sive logarithmorvm chiliades centvm, pro numeris naturali serie crescentibus ab vnitate ad*

100000, Petrus Rammasenius for Adriaan Vlacq, Gouda, Holland, 1628; reprinted by Georg Olms Verlag, Hildesheim and New York, 1976.

Letter to James Usher, March 10, 1615. Letter XVI in Parr, *The life of the most reverend father in God, James Usher*, pp. 35–36 of the collection of three hundred letters appended at the end of the volume.

Trigonometria britannica: sive de doctrina triangulorum libri duo. Quorum prior continet constructionem canonis sinuum tangentium & secantium, una cum logarithmis sinuum & tangentium ad gradus & graduum centesimas & ad minuta & secunda centesimis respondentia: a Henrico Briggio compositus; Posterior verò usum sive applicationem canonis in resolutione triangulorum tam planorum quam sphaericorum e geometricis fundamentis petita, calculo facillimo, eximiisque compendiis exhibet: ab Henrico Gellibrand constructus, Petrus Rammasenius for Adriaan Vlacq, Gouda, Holland, 1633. There is a Latin transcription with complete annotated translation into English by Ian Bruce as *Briggs' Trigonometria britannica*, available online at URL www-history.mcs.st-andrews.ac.uk/history/Miscellaneous/Briggs2/index.html. August 2006.

Brouncker, William, Viscount

"The squaring of the hyperbola, by an infinite series of rational numbers, together with its demonstration, by that eminent mathematician, the Right Honourable the Lord Viscount Brouncker," *Philosophical Transactions*, **3** (1668) 645–649. Pages 647–649 are reproduced in Stedall, *Mathematics emerging*, pp. 86–88.

Burgess, Ebenezer

English translation of the *Surya-Siddhanta* with introduction, notes (enriched by William Dwight Whitney) and an appendix as "Surya-Siddhanta. A text-book of Hindu astronomy," *Journal of the American Oriental Society*, **6** (1860); reprinted by Kessinger Publishing Company, Kila, Montana, 1998, 2007.

Bürgi, Joost

Aritmetische vnd geometrische Progress Tabulen, sambt gründlichem vnterricht, wie solche nützlich in allerley Rechnungen zugebrauchen vnd verstanden werden sol, Paul Sessen, Löblichen Universitet Buchdruckern, Prague, 1620.

Burn, Robert P.

"Alphonse Antonio de Sarasa and logarithms," *Historia Mathematica*, **28** (2001) 1–17.

Calinger, Ronald

Classics of mathematics, Prentice-Hall, Upper Saddle River, New Jersey, 1995.

Cajori, Florian

A history of mathematical notations, vols. I and II, Open Court, Chicago, 1928, 1929; reprinted by Dover Publications, New York, 1993, in two volumes bound as one; reprinted in two volumes by Cosimo Classics, New York, 2007.

A history of mathematics, Macmillan and Company, New York, 1893; revised edition 1919; third, fourth, and fifth (revised) editions, Chelsea Publishing Company, New York, 1980, 1985, 1991; reprinted by the American Mathematical Society, Providence, Rhode Island, 1999, 2000; reprinted by BiblioLife, Charleston, South Carolina, 2010.

Cardano, Girolamo

Artis magnæ, sive de regvlis algebraicis, lib. unus. Qui & totius operis de arithmetica, quod opvs perfectvm inscripsit, est in ordine decimus, Iohannis Petreius, Nuremberg, 1545 = *Opera omnia,* Vol. 4: *Qvo continentvr arithmetica, geometrica, musica,* pp. 221–302; translated by T. Richard Witmer as *The great art or the rules of algebra,* including additions from the 1570 and 1663 editions, The MIT Press, Cambridge, MA, 1968; reprinted as *Ars magna or the rules of algebra,* by Dover Publications, New York, 1993, 2007.

Opera omnia: tam hactenvs excvsa; hîc tamen aucta & emendata; quàm nunquam aliàs visa, ac primùm ex auctoris ipsius autographis eruta: Curâ Caroli Sponii, doctoris medici collegio medd. Lugdunaorum aggregati, 10 vols., Ioannis Antonii Hvgvetan & Marci Antonii Ravavd, Lyon, 1663; facsimile reproduction by Friedrich Frommann Verlag, Stuttgart-Bad Cannstatt, 1966; facsimile reproduction by Johnson Reprint Corporation, New York and London, 1967.

Carnot, Lazare Nicolas Marguérite

Dissertation sur la théorie de l'infini mathématique, a 1785 manuscript, in Gillispie, *Lazare Carnot, savant,* 1971, pp. 171–262. Revised and expanded version as *Réflexions sur la métaphysique du calcul infinitésimal,* Duprat, Paris, 1797; English translation by William Dickson in *Philosophical Magazine,* **8** (1800) 222–240 and 335–352; **9** (1801) 39–56. Enlarged second edition published by Mme. Vve. Courcier, Paris, 1813; English translation by William Robert Browell as *Reflexions on the infinitesimal principles of the mathematical analysis,* Parker, Oxford, 1832. There are Portuguese, German, and Italian translations.

Carra de Vaux, Baron Bernard

"L'Almageste d'Abû'lwéfa Albûzdjâni," *Journal Asiatique,* Ser. 8, **19** (May–June 1892) 408–471 = Sezgin, *Islamic mathematics and astronomy,* 61 II, pp. 12–75.

Cauchy, Augustin-Louis

Cours d'analyse de l'École Royale Polytechnique; Ire partie. Analyse algébrique, Debure Frères, Paris, 1821; facsimile reproduction by Sociedad Andaluza de Educación Matemática «Thales», Sevilla, 1998 = *Œuvres complètes d'Augustin Cauchy,* Ser. 2, Vol. 3. The spelling and the notation were slightly modernized in the *Œuvres.* Page 4 of the *Cours d'analyse,* showing the definition of limit, and portions of Chapter 6 are reproduced with translation in Stedall, *Mathematics emerging,* pp. 299–300, 501–515 and 524–526. An extract from the beginning of Chapter VI, translated into English, can be seen in Birkhoff, *A source book in classical analysis,* p. 3; reprinted in Calinger, *Classics of mathematics,* pp. 599–600.

Leçons sur le calcul différentiel, Chez de Bure Frères, Paris, 1829 = *Œuvres complètes*, Ser. II, Vol. IV, pp. 263–609.

"Mémoire sur une nouvelle théorie des imaginaires, et sur les racines symboliques des équations et des équivalences," *Comptes Rendus Hebdomaires des Séances de l'Academie*, **24** (1847) 1120–1130 = *Œuvres*, Ser. I, Vol. X, 1897, pp. 312–323.

"Note sur les séries convergentes dont les divers terms sont des fonctions continues d'une variable réelle ou imaginaire entre des limites données," *Comptes Rendus Hebdomaires des Séances de l'Academie*, **36** (1853) 454–459 = *Œuvres complètes*, Ser. I, Vol. XII, 1900, pp. 30–36.

Œuvres complètes d'Augustin Cauchy, Gauthier-Villars et fils, Paris, 1882–1974.

Résumé des leçons donnés à l'École Royale Polytechnique sur le calcul infinitésimal, Imprimerie Royale, Paris, 1823; facsimile reproduction by Ellipses, Paris, 1994 = *Œuvres complètes*, Ser. II, Vol. IV, pp. 5–261. English translation of Lessons 3, 4 and 21–24 in Hawking, *God created the integers. The mathematical breakthroughs that changed history*, pp. 643–667. Pages 9–11 of Lesson 3, 27–28 of Lesson 7, 81–84 of Lesson 21, and 101 of Lesson 26 reproduced from the 1823 edition with translation in Stedall, *Mathematics emerging*, pp. 416–420, 422–424, 436–444, and 445–446. Translation of extracts from Lessons 7 and 21 in Grabiner *The origins of Cauchy's rigorous calculus*, 1981, pp. 168–175.

Résumés analytiques, L'Imprimerie Royale, Turin, 1833 = *Œuvres complètes*, Ser. II, Vol. X, 1895, pp. 7–184.

"Sur la convergence des séries," *Exercices de mathématiques*, **2** (1827) 221–232 = *Œuvres complètes*, Ser. II, Vol. VII, pp. 267–279.

"Sur le développement des fonctions en séries et sur l'intégration des équations différentielles ou aux differences partielles," *Bulletin des Sciences, par la Société Philomatique de Paris*, **9** (1822) 49–54 = *Œuvres complètes d'Augustin Cauchy*, Ser. II, Vol. II, pp. 276–282.

Child, James Mark

The early mathematical manuscripts of Leibniz; translated from the Latin texts published by Carl Immanuel Gerhardt with critical and historical notes, Open Court, Chicago and London, 1920; reprinted by Dover Publications, New York, 2005; reprinted by Kessinger Publishing Company, Kila, Montana, 2007. There are additional recent editions.

The geometrical lectures of Isaac Barrow. Translated, with notes and proofs, and a discussion of the advance made therein on the work of his predecessors in the infinitesimal calculus, Open Court, Chicago and London, 1916; reprinted by Kessinger Publishing Company as *Geometrical lectures of Isaac Barrow (1916)*, Kila, Montana, 1998, 2007. There are additional recent editions.

Clagett, Marshall

Nicole Oresme and the medieval geometry of qualities and motions: A treatise on the uniformity and difformity of intensities known as Tractatus de configurationibus qualitatum at motuum, The University of Wisconsin Press, Madison, 1968.

The science of mechanics in the Middle Ages, The University of Wisconsin Press, Madison, 1959; reprinted 1961.

Collins, John

Letter to James Gregory, December 24, 1670. In, *James Gregory tercentenary memorial volume*, pp. 153–159 and Turnbull, *The correspondence of Isaac Newton*, I, pp. 52–58.

Letter to James Gregory, November 8, 1672. In, Turnbull, *The correspondence of Isaac Newton*, I, pp. 244–245.

Letter to Henry Oldenburg, September 30, 1676. In, Hall and Hall, eds., *The correspondence of Henry Oldenburg*, Vol. 13, p. 89.

Copernicus, Nicolaus

De lateribvs et angvlis triangulorum, tum planorum rectilineoreum, tum sphæricorum, libellus eruditissimus & utilissimus, cum ad plerasque Ptolemæi demonstrationes intelligendas, tum uero ad alia multa, scriptus à clarissimo & doctissimo uiro D. Nicolao Copernico Toronensi, Johannem Lufft, Wittemberg, 1542.

Nicolai Copernici Torinensis de revolvtionibvs orbium cœelestium, Libri vi., Iohannis Petreium, Nuremberg, 1543; reprinted by Culture et Civilisation, Brussels, 1966. English translation by Charles Glenn Wallis as *On the revolutions of heavenly spheres*, Prometheus Books, Amherst, New York, 1995. Facsimile reproduction of the 1543 original with an introduction by Owen Gingerich and English translation by Edward Rosen available in CD ROM, Octavo Editions, Oakland, California, 1999.

Cotes, Roger

Harmonia mensurarum, sive analysis & synthesis per rationum & angulorum mensuras promotæ: accedunt alia opuscula mathematica, Robert Smith, ed., Cambridge, 1722.

"Logometria," *Philosophical Transactions*, **29** (1714) 5–45 = *Harmonia mensurarum*, pp. 4–41. English translation in Gowing, *Roger Cotes – Natural Philosopher*, pp. 143–185.

Da Cunha, José Anastácio

Principios mathematicos para instrucçaõ dos alumnos do Collegio de Saõ Lucas, da Real Casa Pia do Castello de Saõ Jorge: offerecidos ao Serenissimo Senhor D. Joaõ, Principe do Brazil, Antonio Rodrigues Galhardo, Lisbon, 1790; French translation by J. M. d'Abreu as *Principes mathématiques de feu Joseph-Anastase da Cunha, traduits·littéralement du Portugais*, Imprimerie d'André Racle, Bordeaux, 1811. Facsimile editions of both titles by Departamento de Matemática, Faculdade de Ciências e Tecnologia, Universidade de Coimbra, 1987.

De Moivre, Abraham

"De reductione radicalium ad simpliciores terminos, seu de extrahenda radice quaqunque data ex binomio $a + \sqrt{+b}$, vel $a + \sqrt{-b}$, Epistola," *Philosophical*

Transactions, **40** (1739) 463–478. Portions of this paper are translated into English by Raymond Clare Archibald in Smith, *A source book in mathematics*, pp. 447–450.

"De sectione anguli," *Philosophical Transactions*, **32** (1722) 228–230. Translated into English by Raymond Clare Archibald in Smith, *A source book in mathematics*, pp. 444–446.

Delambre, Jean-Baptiste Joseph

Histoire de l'astronomie ancienne, Vol. 1, Mme. Vve. Courcier, Paris, 1815; reprinted by Johnson Reprint Corporation, New York, 1965; reprinted by Éditions Jacques Gabay, Sceaux, 2005; reprinted by Nabu Press, Berlin, 2010.

Histoire de l'astronomie du moyen age, Mme. Vve. Courcier, Paris, 1819; reprinted by Nabu Press, Berlin, 2010. Pages 156–170 of this work are reprinted in Sezgin, *Islamic mathematics and astronomy*, 61I, pp. 1–17.

Descartes, René

La Géométrie, written as an appendix to *Discours de la méthode pour bien conduire sa raison, & chercher la verité dans les sciences, plvs La dioptriqve. Les météores. et La géométrie. Qui sont des essais de cete méthode* (Discourse on the method for guiding one's reason well and searching for truth in the sciences), Ian Maire, Leiden, 1637. Reprinted as *La géométrie* by Victor Cousin, ed., in *Œuvres de Descartes*, **5**, Levrault, Paris, pp. 309–428; also reprinted as *La géométrie*, Éditions Jacques Gabay, Sceaux, 1991; reprinted by Kessinger Publishing Company, Kila, Montana, 2009. There are other recent editions. Translated into English from the French and Latin as *The geometry of René Descartes* by David Eugene Smith and Marcia L. Latham, Dover Publications, New York, 1954.

Diophantos of Alexandria

Arithmetica, translated from the Greek by Wilhelm Holtzman (Guilielmus Xylander), Basel, 1575. Greek text with Latin translation by Clavdio Gaspare Bacheto Meziriaco Sebvsiano, Y. C., as *Diophanti alexandrini arithmeticorvm libri sex, et de nvmeris mvltangvlis liber vnvs. Nunc primùm græcè & latinè editi, atque absolutissimis commentariis illustrati*, Sebastiani Cramoisy for Ambrose Drouart, Lvtetiaæ Parisiorvm, 1621.

Dirichlet, Johann Peter Gustav Lejeune

"Sur la convergence des séries trigonométriques qui servent a représenter une fonction arbitraire entre des limites données," *Journal für die reine und angewandte Mathematik*, **4**, 1829, 157–169. Also in volume 1 of *G. Lejeune Dirichlet's Werke*, 2 vols., ed. by Lazarus Immanuel Fuchs and Leopold Kronecker, Berlin, 1889–1897, pp. 117–132; reprinted in one volume by Chelsea Publishing Company, New York, 1969; volume 1 reprinted by the Scholarly Publishing Office, University of Michigan Library, Ann Arbor, Michigan, 2006. The proof of the convergence of Fourier series, originally contained in pages 128–131, can be seen, translated into English, in Birkhoff, *A source book in classical analysis*, pp. 145–146.

Dreyer, Johan Ludvig Emil, ed.
Tychonis Brahe Dani scripta astronomica, Libraria Gyldendaliana, Nielsen & Lydiche, Hauniæ (Copenhagen), 1913.

Edwards, Charles Henry
The historical development of the calculus, Springer-Verlag, New York, 1979, 1994.

Euclid
The thirteen books of The Elements. Translated with an introduction and commentary by Sir Thomas L. Heath, 2nd. ed., Cambridge University Press, Cambridge, 1925; reprinted in three volumes by Dover Publications, New York, 1956.

Euler, Leonhard
"De formulis differentialibus angularibus maxime irrationalibus, quas tamen per logarithmos et arcus circulares integrare licet," presented to the Academy of St. Petersburg on May 5, 1777, but first published in his book *Institvtiovm calcvli integralis*, 2nd. ed., Vol. IV, St. Petersburg, 1794, pp. 183–194.

"De la controverse entre Mrs. Leibnitz et Bernoulli sur les logarithmes négatifs et imaginaires," *Mémoires de l'Académie Royale des Sciences et des Belles-Lettres de Berlin*, **5** (1749, pub. 1751) 139–179 = *Opera omnia*, Ser. 1, **17**, ed. by August Gutzmer, B. G. Teubner, Leipzig, 1915, pp. 195–232.

"De seriebus quibusdam considerationes," *Commentarii Academiæ Scientiarum Petropolitanæ*, **12** (1740, pub. 1750) 53–96 = *Opera omnia*, Ser. 1, **14**, ed. by Carl Boehm and Georg Faber, B. G. Teubner, Leipzig and Berlin, 1925, pp. 407–462.

"De summatione innumerabilum progressionum," *Commentarii Academiæ Scientiarum Petropolitanæ*, **5** (1730/31, pub. 1738) 91–105 = *Opera omnia*, Ser. 1, **14**, ed. by Carl Boehm and Georg Faber, B. G. Teubner, Leipzig and Berlin, 1925, pp. 25–41.

"De svmmis seriervm reciprocarvm," *Commentarii Academiæ Scientiarum Petropolitanæ*, **7** (1734/35, pub. 1740) 123–124 = *Opera omnia*, Ser. 1, **14**, ed. by Carl Boehm and Georg Faber, B. G. Teubner, Leipzig and Berlin, 1925, pp. 73–86.

"De vibratione chordarum exercitatio," *Nova Acta Eruditorum* (1749) 512–527 = *Opera omnia*, Ser. 2, **10**, ed. by Fritz Stüsi and Henri Favre, Orell Füssli, Zürich, 1947, pp. 50–62; translated into French as "Sur les vibrations des cordes," *Histoire de l'Académie Royale des Sciences et des Belles-Lettres de Berlin*, **4** (1748/1750) 69–85 = *Opera omnia*, Ser. 2, **10**, ed. by Fritz Stüsi and Henri Favre, Orell Füssli, Zurich, 1947, pp. 63–77.

Leonhardi Euleri opera omnia. The publication of Euler's complete works began in 1911. Published so far: Series 1, 29 vols.; Series 2, 31 vols.; Series 3, 12 vols.; Series 4, in progress. Ed. by Societatis Scientiarum Naturalium Helveticæ, Leipzig/Berlin/Zurich/Lausanne/Basel. There are more recent reprints of the original volumes by Birkhäuser Verlag, Basel.

Institutiones calculi differentialis, cum eius vsu in analysi finitorum ac doctrina serierum, Petropoli (St. Petersburg), Academiæ Imperialis Scientiarum Petropolitanæ, 1755 = *Opera omnia*, Ser. 1, **10**, ed. by Gerhard Kowalewski, B. G. Teubner, Leipzig, 1913. This is a work in two parts: theory and applications. The first part has been translated into English by John David Blanton as *Foundations of differential calculus*, Springer-Verlag New York, 2000.

Institvtionvm calcvli integralis: in qvo methodvs integrandi a primis principiis vsqve ad integrationem æqvationvm differentialivm primi gradvs pertractatvr, Petropoli (St. Petersburg), Academiæ Imperialis Scientiarum Petropolitanæ, 1768–1770 = *Opera omnia*, Ser. 1, **11**, **12** and **13**, ed. by Friedrich Engel and Ludwig Schlesinger, B. G. Teubner, Leipzig, 1913–1914; reprinted by Nabu Press, Berlin, 2010.

Introductio in analysin infinitorum, Marc-Michel Bousquet, Lausanne, 1748 = *Opera omnia*, Ser. 1, **8**, ed. by Adolf Krazer and Ferdinand Rudio and **9**, ed. by Andreas Speiser, B. G. Teubner, Leipzig and Berlin, 1922. Facsimile reproduction of the first volume by Francisco Javier Pérez Fernández in a two-volume boxed edition by Sociedad Andaluza de Educación Matemática «Thales», Sevilla, 2000. Translated into English by John David Blanton as *Introduction to analysis of the infinite*, Springer-Verlag New York, 1988, 2 Books. The first volume translated into French by Jean-Baptiste Labey with notes and clarifications as *Introduction à l'analyse infinitésimale. Tome premier*, Barrois, Paris, 1796; reprinted by ACL-Éditions, Paris, 1987. Translated into Spanish by José Luis Arantegui Tamayo as *Introducción al análisis de los infinitos. Libro primero*, annotated by Antonio José Durán Guardeño, in a two volume boxed edition by Sociedad Andaluza de Educación Matemática «Thales», Sevilla, 2000. Articles 114–116, 122, 132–134 and 138, referred to in the text, reproduced with translation in Stedall, *Mathematics emerging*, pp. 245–254.

Mechanica sive motvs scientia analytice exposita, 2 vols., Academiæ Scientiarvm, Petropoli (St. Petersburg), 1736 = *Opera omnia*, Ser. 2, **1** and **2**, ed. by Paul Stäckel, B. G. Teubner, Leipzig, 1912.

"Meditatio in Experimenta explosione tormentorum nuper instituta," in *Opera posthuma mathematica et physica*, ed. by Paul Heinrich Fuss and Nicolaus Fuss, Petropoli (St. Petersburg), 1862, II, pp. 800–804 = *Opera Omnia*, Ser. 2, **14**, ed. by Friedrich Robert Scherrer, B. G. Teubner, Leipzig and Berlin, 1922, pp. 468–477.

"Recherches sur les racines imaginaires des équations," *Mémoires de l'Académie Royale des Sciences et des Belles-Lettres de Berlin*, **5** (1749, pub. 1751) 222–288 = *Opera omnia*, Ser. 1, **6**, ed. by Ferdinand Rudio, Adolf Krazer, and Paul Stäckel, B. G. Teubner, Leipzig and Berlin, 1921, pp. 78–150. There is an English translation of §79–85 by Raymond Clare Archibald in Smith, *A source book in mathematics*, pp. 452–454.

"Remarques sur les mémoires précédens de M. Bernoulli," *Histoire de l'Académie Royale des Sciences et des Belles-Lettres de Berlin*, **9** (1753, pub. 1755) 196–222 = *Opera omnia*. Ser. 2, **10**, ed. by Fritz Stüsi and Henri Favre, Orell Füssli, Zurich, 1947, pp. 233–254. Some selections translated in Struik, *A source book in mathe-*

matics, 1200–1800, pp. 364–368.

Vollständige Anleitung zur Algebra, Gedruckt bey der Kays, Akademie der Wissenschaften, St. Petersburg, 1770 (first published in Russian translation as *Universal'naya arifmetika g. Leongarda Eilera, perevedennaya s nemetskogo podlinnika studentami Petrom Inokhodtsovym i Ivanom Yudinym. Tom pervyi, soderzhashchii v sebe vse obrazy algebraicheskogo vychislenia*. St. Petersburg, 1768) = Opera Omnia, Ser. 1, **1**, ed. by Heinrich Weber, B. G. Teubner, Leipzig, 1911. Translated into English by John David Blanton as *Elements of algebra*, Springer-Verlag New York, 1984.

Fermat, Pierre de

Ad eamdem methodum, c. 1638, in *Varia opera mathematica*, pp. 66–69 = *Œuvres de Fermat*, I, pp. 140–147. French translation in *Œuvres de Fermat*, III, pp. 126–130.

De æquationum localium transmutatione, & emendatione, ad multimodam curvilineorum inter se, vel cum rectilineis comparationem. Cvi annectitvr proportionis geometricæ in quadrandis infinitis parabolis & hyperbolis usus = Varia opera, pp. 44–57 = *Œuvres de Fermat*, I, pp. 255–285. Fermat's quadrature of hyperbolas has an English translation in Struik, *A source book in mathematics, 1200–1800*, pp. 219–222; reprinted in Calinger, *Classics of mathematics*, pp. 374–376.

Méthode de maximis et minimis expliquée et envoyée par M. Fermat a M. Descartes, in *Œuvres de Fermat*, II, pp. 154–162.

Methodus ad disquirendam maximam et minimam and *De tangentibus linearum curvarum*. The original manuscript is now lost and difficult to date with precision. In a 1636 letter to Roberval, Fermat stated that the copy of his manuscript that Etienne d'Espagnet gave to Roberval had been sent by Fermat to d'Espagnet about seven years before, so the manuscript may have been written about 1629. It was first printed, after Fermat's death, in his *Varia opera mathematica*, 1679, pp. 63–64 = *Œuvres de Fermat*, I, pp. 133–136. English translation in Struik, *A source book in mathematics, 1200–1800*, pp. 223–227; reprinted in Calinger, *Classics of mathematics*, pp. 377–380. Fermat's tangent method, reproduced from *Varia opera* with English translation, can be seen in Stedall, *Mathematics emerging*, pp. 73–74.

Methodus de maxima et minima, in *Œuvres de Fermat*, I, pp. 147–153. There is a French translation in *Œuvres de Fermat*, III, pp. 131–136.

Œuvres de Fermat, edited by Paul Tannery and Charles Henry, Gauthier-Villars, Paris, 4 vols., 1891, 1894, 1896, and 1912. *Supplément* by Cornelius de Waard, 1922.

Varia opera mathematica D· Petri de Fermat senatoris tolosani. Acceserunt selectæ quædam ejusdem epistolæ, vel ad ipsum à plerisque doctisimis viris Gallicè, Latinè, vel Italicè, de rebus ad mathematicas disciplinas aut physicam pertinentibus scriptæ, Toulouse, Joannem Pech, 1679; reprinted by Culture et Civilisation, Brussels, 1969; reprinted by Kessinger Publishing Company, Kila, Montana, 2009.

Feingold, Mordechai

"Newton, Leibniz, and Barrow too. An attempt at a reinterpretation," *Isis*, **84** (1993) 310–338.

Fincke, Thomas

Thomæ Finkii flenspurgensis geometriæ rotundi libri XIIII, Basileæ: Per Sebastianum Henricpetri, 1583.

Fontana, Nicolo

Qvesiti, et inventioni diverse de Nicolo Tartalea brisciano, stampata per V. Ruffinelli ad instantia et requisitione, & á proprio spese de Nicolo Tartalea brisciano autore, Venice, 1546.

Fourier, Joseph

Mémoire sur la propagation de la chaleur. It was not published until 1972 and can be seen, with commentary and notes (but no translation), in Grattan-Guinness and Ravetz, *Joseph Fourier 1768–1830*, 1972, pp. 30–440.

Théorie analytique de la chaleur, Firmin Didot, Père et Fils, Paris, 1822; facsimile reproduction by Éditions Jacques Gabay, Paris, 1988, 2003. Translated by Alexander Freeman as *The analytical theory of heat*, Cambridge University Press, Cambridge, 1878, 2009; reprinted by Dover Publications, New York, 1955; reprinted by Kessinger Publishing Company, Kila, Montana, 2008. There are additional recent editions. All the articles referred to here can be found in English translation in Birkhoff, *A source book in classical analysis*, pp. 132–144. Arts. 163–237 can also be found in English translation in Hawking, *God created the integers. The mathematical breakthroughs that changed history*, pp. 500–562. Art. 219 is reproduced with translation in Stedall, *Mathematics emerging*, pp. 224–227.

"Théorie du mouvement de la chaleur dans les corps solides," *Mémoires de l'Académie Royale des Sciences de l'Institut de France*, **4** (1819–1820, pub. 1824) 185–555 and "Suite du mémoire intitulé: Théorie du mouvement de la chaleur dans les corps solides," *Mémoires de l'Académie Royale des Sciences de l'Institut de France*, **5** (1821–1822, pub. 1826) 156–246.

Français, Jacques Frédéric

"Philosophie mathématique. Nouveaux principes de géométrie de position, et interprétation géométrique des symboles imaginaires," *Annales de mathématiques pures et apliquées*, **4** (1813–1814) 61–71.

Fuss, Paul Heinrich

Correspondence mathématique et physique de quelques célèbres géomètres du XVIIIème siècle, tome 1, St. Petersburg, 1843, pp. 379–383; reprinted by Johnson Reprint Corporation, New York, 1968.

Gardiner, William

Tables of logarithms for all numbers from 1 to 102100, and for the sines and tangents to every ten seconds of each degree in the quadrant; as also, for the sines of the first 72 minutes for every single second: with other useful and necessary tables, printed for the Author in Green Arbour Court, by G. Smith, London, 1742.

Gauss, Johann Carl Friedrich

Briefwechsel zwischen Gauss und Bessel, Verlag von Wilhelm Engelmann, Leipzig, 1880; reprinted as *Briefwechsel C. F. Gauss – F. W. Bessel*, by Georg Olms Verlag, Hildesheim · New York, 1975.

Carl Friedrich Gauss Werke, Königlichen Gesellschaft der Wissenschaften zu Göttingen, Gedruckt in der Dieterichschen Universitäts-Buchdruckerei (W. Fr. Kaestner), 14 vols., 1863–1933. Vol. 1: *Disquisitiones arithmeticae*, 1863; Vol. 2: *Theorematis arithmetici demonstratio nova*, 1863; Vol. 8: *Arithmetika und Algebra*, Supplements to Parts I–III, 1910; Vol. 101: *Kleinere Veröffentlichungen*, 1917, B. G. Teubner, Leipzig.

Carl Friedrich Gauss' Untersuchungen über höhere Arithmetik. (*Disquisitiones Arithmeticae. Theorematis arithmetici demonstratio nova. Summatio quarumdam serierum singularium. Theorematis fundamentalis in doctrina de residuis quadraticis demonstrationes et ampliationes novae. Theoria residuorum biquadraticorum, commentatio prima et secunda. Etc.*), Verlag von Julius Springer, Berlin, 1889; facsimile reproduction by the American Mathematical Society, Providence, Rhode Island, 1965, 2006; reprinted by Nabu Press, Berlin, 2010. This is a collection of German translations of the works stated in the title by Hermann Maser.

Disquisitiones arithmeticae, Gerhardt Fleischer, Leipzig, 1801 = *Werke*, **1**. Translated into English by Arthur A. Clarke as *Disquisitiones arithmeticae* (English edition), Yale University Press, New Haven and London, 1966; reprinted by Springer-Verlag, New York/Berlin/Heidelberg/Tokyo, 1986 (translation revised by William Charles Waterhouse). German translation in *Carl Friedrich Gauss' Untersuchungen über höhere Arithmetik*, pp. 1–453.

Letters to Friedrich Wilhelm Bessel, November 21 and December 18, 1811. In *Briefwechsel zwischen Gauss und Bessel*, pp. 151–155 and 155–160 = *Werke*, **101**, pp. 362–365 and 365–371. A portion of the second letter, containing the quotation included here, can also be seen in *Werke*, **8**, pp. 90–92.

"Theoria residuorum biquadraticorum. Commentatio secunda," *Commentationes Societatis Regiae Scientiarum Gottingensis Recentiores*, **7** (1832) = *Werke*, **2**, pp. 93–148. German translation in *Carl Friedrich Gauss' Untersuchungen über höhere Arithmetik*, pp. 534–586.

Gerhardt, Carl Immanuel, ed.

Der Briefwechsel von Gottfried Wilhelm Leibniz mit Mathematikern, Mayer & Müller, Berlin, 1899; reprinted by Georg Olms Verlag, Hildesheim, 1962.

Die Entdeckung der differentialrechnung durch Leibniz, H. W. Schmidt, Halle, 1848.

Die Geschichte der höheren Analysis; erste Abtheilung, Die Entdeckung der höheren Analysis, Druck und Verlag von H. W. Schmidt, Halle, 1855.

Historia et origo calculi differentialis, a G. G. Leibnitio conscripta, Im Verlage der Hahn'schen Hofbuchhandlung, Hanover, 1846; reprinted in *Leibnizens Mathematische Schriften*, Sec. 2, I, *Analysis infinitorum*, No. XXXI, pp. 392–410; reprinted by Nabu Press, Berlin, 2010.

"Leibniz in London," *Sitzungsberichte der Königlich Preussischen Akademie der Wissenschaften zu Berlin*, (1891) 157–176. English translation with critical notes by Child in *The Monist*, **22** (1917) 524–559; reprinted in *The early mathematical manuscripts of Leibniz*, pp. 159–195.

Leibnizens Mathematische Schriften, Erste Abtheilung (First Section), Band I (Volume I): Briefwechsel zwischen Leibniz und Oldenburg, Collins, Newton, Galloys, Vitale Giordano, Verlag von A. Asher und Comp., Berlin, 1849; Band III: Briefwechsel zwischen Leibniz, Jacob Bernoulli, Johann Bernoulli und Nicolaus Bernoulli, Druck und Verlag von H. W. Schmidt, Halle, **1** 1855, **2** 1856; Zweite Abtheilung (Second Section), Band I: Die mathematischen Abhandlungen, Druck und Verlag von H. W. Schmidt, Halle, 1858; reprinted by Georg Olms Verlag as *G. W. Leibniz Mathematische Schriften*, **5**, Hildesheim · New York, 1971.

Gillispie, Charles Coulston

Lazare Carnot, savant: A monograph treating Carnot's scientific work, Princeton University Press, Princeton, New Jersey, January 1971, May 1989.

Girard, Albert

Invention nouvelle en l'algèbre, Guillaume Iansson Blaeuw, Amsterdam, 1629; reprinted by Bierens de Haan, Leiden, 1884; reprinted by Kessinger Publishing Company, Kila, Montana, 2009. There are additional recent editions.

Glaisher, James Whitbread Lee

"The earliest use of the radix method for calculating logarithms, with historical notices relating to the contributions of Oughtred and others to mathematical notation," *The Quarterly Journal of Pure and Applied Mathematics*, **46** (1915) 125–197.

Goldstine, Herman Heine

A history of numerical analysis from the 16th through the 19th century, Springer-Verlag, New York, 1977.

González-Velasco, Enrique Alberto

"James Gregory's calculus in the *Geometriæ pars universalis*," *American Mathematical Monthly*, **114** (2007) 565–576.

Gowing, Ronald

Roger Cotes – Natural Philosopher, Cambridge University Press, Cambridge, 1983, 2002.

Grabiner, Judith Victor

The origins of Cauchy's rigorous calculus, The MIT Press, Cambridge, Massachusetts, and London, 1981; reprinted by Dover Publications, New York, 2005.

Grattan-Guinness, Ivor

From the calculus to set theory, 1630–1910. An introductory history, Gerald Duckworth & Co, London, 1980; reprinted by Princeton University Press, Princeton and Oxford, 2000.

Grattan-Guinness, Ivor and Jerome Raymond Ravetz

Joseph Fourier 1768–1830, The MIT Press, Cambridge, Massachusetts, 1972, pp. 30–440.

Gregory, James

Exercitationes geometricæ, William Godbid for Mosis Pitt, London, 1668.

Geometriæ pars vniversalis, inserviens quantitatum curvarum transmutationi & mensuræ, Heirs of Pauli Frambotti, Padua, 1668.

Letters to John Collins of November 23, 1670; December 19, 1670; February 15, 1671. In Turnbull, *James Gregory tercentenary memorial volume*, pp. 118–122; 148–150; 168–172 and in *The Correspondence of Isaac Newton*, I, pp. 45–48; 49–51; 61–64. Enclosure by Gregory with the November 23 letter, with English translation, pp. 131–132; the translation is reproduced in Struik, *A source book in mathematics, 1200–1800*, pp. 290–291.

Vera circvli et hyperbolæ qvadratvra, in propria sua proportionis specie, inuenta, & demonstrata, Iacobi de Cadorinis, Padua, 1667.

Gudermann, Christof

"Theorie der Modular-Functionen und der Modular-Integrale," *Journal für die reine und angewandte Mathematik*, **18** (1838) 1–54, 142–175, 220–258, 303–364.

Gunter, Edmund

Canon triangulorum, sive Tabulæ sinuum et tangentium artificialium ad radium 10000.0000. & ad scrupula prima quadrantis, Gulielmus Iones, London, 1620.

Gutas, Dimitri

Greek thought, Arabic culture, Routledge, London and New York, 1998, 1999.

Hall, Alfred Rupert

Philosophers at war: The quarrel between Newton and Leibniz, Cambridge University Press, Cambridge, 1980, 2002.

Hall, Alfred Rupert and Marie Boas Hall, eds.

The correspondence of Henry Oldenburg, 13 vols, University of Wisconsin Press, 1965–1975; Mansell, London, 1975–1977; Taylor and Francis, London, 1986.

Unpublished scientific papers of Sir Isaac Newton. A selection from the Portsmouth Collection in the University Library, Cambridge, Cambridge University Press, Cambridge, 1962.

Halley, Edmund

"A most compendious and facile method for constructing the logarithms, exemplified and demonstrated from the nature of numbers, without any regard to the hyperbola, with a speedy method for finding the number from the logarithm given," *Philosophical Transactions*, **19** (1695, pub. 1697) 58–67.

Halma, Nicolas

Composition mathématique de Claude Ptolémée, 2 vols. Vol. 1, Chez Henri Grand, Paris, 1813; Vol. 2, Jean-Michel Eberhart, Paris, 1816.

Hamilton, Sir William Rowan

The mathematical papers of Sir William Rowan Hamilton, 4 vols., Cambridge University Press for the Royal Irish Academy, Cambridge, 1931–2000.

"Theory of conjugate functions or algebraic couples; with a preliminary and elementary essay on algebra as a science of pure time," *Transactions of the Royal Irish Academy*, **17** (1837) 293–422 = *The mathematical papers of Sir William Rowan Hamilton*, Vol. 3, ed. by Heini Halberstam and R. E. Ingram, 1967, pp. 3–96. Electronic edition of the original paper by David R. Wilkins, European Mathematical Society, 2000.

Haskins, Charles Homer

The renaissance of the twelfth century, Harvard University Press, Cambridge, Massachusetts, 1927. Multiple reprints.

Hawking, Stephen, ed.

God created the integers. The mathematical breakthroughs that changed history, Running Press, Philadelphia, 2005.

Heath, Sir Thomas Little

Aristarchus of Samos, the ancient Copernicus. A history of Greek astronomy to Aristarchus, together with Aristarchus's treatise on the sizes and distances of the Sun and Moon, a Greek text with translation and notes by Sir Thomas Heath, The Clarendon Press, Oxford, 1913; reprinted by Dover Publications as *Aristarchus of Samos, the ancient Copernicus*, New York, 1981, 2004; reprinted by Elibron Classics, Adamant Media Corporation, Chestnut Hill, Massachusetts, 2007. There are additional recent editions.

The works of Archimedes with The method of Archimedes, edited in modern notation with introductory chapters, Cambridge University Press, Cambridge, 1912; reprinted by Dover Publications, New York, 1953. There are several recent reprints of both of these works edited by Sir Thomas Heath.

Heytesbury, William

Regule solvendi sophismata, 1335, Stadsbibliotheek, Bruges, 500, ff. 56^v–57^r. First printed in Venice, 1494. Translation by Ernest Moody, sligthly altered by Clagett, in *The science of mechanics in the Middle Ages*, 1959, pp. 238–242.

Hille, Einar Carl

Analytic function theory, Vol. 1, Blaisdell Publishing Company, New York, 1959, 1965; reprinted by Chelsea Publishing Company New York, 1973, 1998.

Hodgson, Marshall Goodwin Simms
The secret order of assassins, Moulton & Company, 1955; reprinted by University of Pennsylvania Press, Philadelphia, 2005.

l'Hospital, Guillaume François Antoine de
Analyse des inifiniment petits pour l'intelligence des lignes courbes, Imprimérie Royale, Paris, 1696; the 1768 edition reprinted by Kessinger Publishing Company, Kila, Montana, 2010. English translation by Edmund Stone as *The method of fluxions both direct and inverse. The former being a translation from the celebrated Marquis de l'Hospital's Analyse des inifiniments petits: and the latter suppl'd by the translator*, William Innys, London, 1730; reprinted by Kessinger Publishing Company, Kila, Montana, 2008. Some selections from this translation, with the notation restored to that in the French original, can be seen in Struik, *A source book in mathematics, 1200–1800*, pp. 313–316. Spanish translation by Rodrigo Cambray Núñez, as *Análisis de los infinitamente pequeños para el estudio de las líneas curvas*, UNAM, México, 1998.

L'Huilier, Simon Antoine Jean
Exposition élémentaire des principes des calculs supérieurs, qui a remporté le prix proposé par l'Académie Royale des Sciences et Belles-Lettres pour l'année 1786, Georges Jacques Decker, Berlin, 1787. A revised version was reissued, in Latin, as *Principiorum calculi differentialis et integralis expositio elementaris ad normam dissertationis ab Academia Scient. Reg. Prussica · Anno 1786. Præmii honore decoratæ elaborata*, Johannem Georgium Cottam, Tübingen, 1795; reprinted by Kessinger Publishing Company, Kila, Montana, 2009.

Hudde, Johann
"Extrait d'une lettre du feu M. Hudde à M. van Schooten, Professeur en Mathématiques à Leyde. Du 21. Novembre 1659," *Journal Literaire de la Haye*, **1** (1713) 460–464; reproduced in Gerhard, *Der Briefwechsel von Gottfried Wilhelm Leibniz mit Mathematikern*, pp. 234–237.

Jones, Sir William
Analysis per quantitatum series, fluxiones, ac differentias: cum enumeratione linearum tertii ordinis, Ex Officina Pearsoniana, London, 1711. Facsimile reproduction by Antonio José Durán Guardeño and Francisco Javier Pérez Fernández in a two-volume boxed edition by Sociedad Andaluza de Educación Matemática «Thales», Sevilla, 2003. Translated into Spanish by José Luis Arantegui Tamayo as *Análisis de cantidades mediante series, fluxiones y diferencias, con una enumeración de las líneas de tercer orden*. Introductions by José Manuel Sánchez Ron, Javier Echeverría and Antonio José Durán Guardeño and annotated by Antonio José Durán Guardeño, Sevilla, 2003.

Synopsis palmariorum matheseos: or, a new introduction to the mathematics: Containing the principles of arithmetic & geometry demonstrated, in a short and easy method; with their application to the most useful parts thereof: As, resolving of equations, infinite series, making the logarithms; Interest, simple and compound; The chief properties of the conic sections; Mensuration of surfaces and solids; The fundamental precepts of perspective; Trigonometry: The laws of motion apply'd to mechanic powers, gunnery, &c. Design'd for the benefit, and adapted to the capacities of beginners, John Matthews for Jeffery Wale, At the Angel in St. Paul's Church-Yard, London, 1706.

Jyesthadeva

Yuktibhasa, c. 1530. Edited with notes as *Yuktibhasa of Jyesthadeva* by Rama Varma Maru Tampuran and A. K. Akhilesvara Iyer, Mangalodayam, Trichur, 1950. Edited and translated by K. Venkateswara Sarma as *Ganita Yuktibhasa*, Indian Institute of Advanced Study, Shimla, 2009. Edited and translated by K. Venkateswara Sarma as *Ganita-Yukti-bhasa (Rationales in mathematical astronomy) of Jyesthadeva*, with explanatory notes by K. Ramasubramanian, M. S. Sriram and M. D. Srinivas, Springer-Verlag, New York, 2009.

Kaufmann-Bühler, Walter

Gauss. A Biographical Study, Springer-Verlag, Berlin/Heidelberg/New York, 1981.

Katz, Victor J., ed.

The mathematics of Egypt, Mesopotamia, China, India, and Islam, Princeton University Press, Princeton and Oxford, 2007.

Kepler, Johannes

Nova stereometria doliorum vinariorum, inprimis Austriaci, figurae omnium aptissimae; et usus et in eo virgae cubicae compendiosissimus et plane sigularis. Accessit stereometriae Archimedeae supplementum, Joannes Plancus, Linz, 1615 = *Opera omnia*, ed. by Christian Frisch, Heyder & Zimmer, Frankfort-on-Main, Vol. IV, 1863, pp. 551–646. German translation as *Ausszug aus der Uralten Messe-Kunst Archimedis*, Linz, 1616 = *Opera omnia*, Vol. V, 1864, pp. 497–610.

Knobloch, Eberhard

"Leibniz et son manuscrit inédite sur la quadrature des sections coniques," in *The Leibniz Renaissance*, International Workshop (Florence, June 2–5, 1986), Centro Fiorentino di Storia e Filosofia della Scienza, Biblioteca di Storia della Scienza, Vol. 28, Leo Samuel Olschki Editore, Florence, 1989, pp. 127–151.

Knott, Cargill Gilston

Napier tercentenary memorial volume, Longmans, Green and Company, London, 1915; reprinted by Nabu Press, Berlin, 2010. There are additional recent editions.

Lagrange, Joseph-Louis

Œuvres de Lagrange, ed. by Joseph Alfred Serret, Gauthier-Villars, Paris, 1867–1892; reprinted by Georg Olms Verlag, Hildesheim and New York, 1973.

"Sur une nouvelle espèce de calcul relatif a la différentiation et a l'intégration des quantités variables," *Nouveaux Mémoires de l'Académie Royale des Sciences et Belles-Lettres de Berlin*, (1772) 185–221 = *Œuvres de Lagrange*, Vol. 3, 1869, pp. 439–476.

    Théorie des fonctions analytiques, contenant les principes du calcul différentiel, dégagés de toute considération d'infiniment petits ou d'évanouissans, de limites ou de fluxions, et réduits a l'analyse algébrique des quantités finies, Journal de l'École Polytechnique, 9ᵉ Cahier, Tome III, Imprimerie de la République, year V (1797). Second edition, Mme. Vve. Courcier, Paris, 1813 = *Œuvres de Lagrange*, Vol. 9, 1881; reprinted by Éditions Jacques Gabay, Sceaux, 2007.

Laubenbacher, Reinhard and David Pengelley
    Mathematical expeditions: chronicles by the explorers, Springer-Verlag New York, 1998.

Leibniz, Gottfried Wilhelm
    Analysis tetragonistica ex centrobarycis, 1675. Bodemann XXXV, Vol. VIII / 18, p. 300. Published with notes and commentary in Gerhardt, *Die Geschichte der höheren Analysis; erste Abtheilung, Die Entdeckung der höheren Analysis*, pp. 117–131; reprinted with modifications in Gerhardt, *Der Briefwechsel von Gottfried Wilhelm Leibniz mit Mathematikern*, pp. 147–160. In *Sämtliche Schriften und Briefe*, Ser. VII, **5**, pp. 263–269, 288–295, and 310–316. English translation in Child, *The early mathematical manuscripts of Leibniz*, pp. 65–90.

    Calculus tangentium differentialis, adjecta sub finem machina construendi æquationes per logarithmicam, 1676. Bodemann XXXV, Vol. V / 12, p. 293. In Gerhardt, *Die Entdeckung der differentialrechnung durch Leibniz*, pp. 56–59 = Gerhardt, *Die Geschichte der höheren Analysis*, pp. 140–142, = Gerhardt, *Der Briefwechsel von Gottfried Wilhelm Leibniz mit Mathematikern*, pp. 229–231 = *Sämtliche Schriften und Briefe*, Ser. VII, **5**, pp. 614–620; translated in Child, *The early mathematical manuscripts of Leibniz*, pp. 124–127.

    Compendium quadraturæ arithmeticæ, c. 1680. Bodemann XXXV, Vol. II / 1, p. 287. In Gerhardt, *Leibnizens Mathematische Schriften*, Sec. 2, I, *De quadratura arithmetica, circuli, ellipseos et hyperbolæ*, No. III, pp. 99–112.

    "De geometria recondita et analysi indivisibilium atque infinitorum," *Acta Eruditorum*, **5** (1686) 292–300 = *Opera Mathematica*, No. XIX, pp. 188–194 = Gerhardt, *Leibnizens Mathematische Schriften*, Sec. 2, I, *Analysis infinitorum*, No. II, pp. 226–233. Partial English translation in Smith, *A source book in mathematics*, pp. 624–626. Partial English translation in Struik, *A source book in mathematics, 1200–1800*, pp. 281–282. Complete Spanish translation in Javier de Lorenzo and Teresa Martín Santos as "Sobre una geometría altamente oculta y el análisis de los indivisibles e infinitos" in *Gottfried Wilhelm Leibniz. Análisis infinitesimal*, 2nd. ed., pp. 17–29, Tecnos, Madrid, 1994.

    De quadratura arithmetica circuli ellipseos et hyperbolæ cujus corollarium est trigonometria sine tabulis. Autore G. G. L., 1675–1676. Bodemann XXXV,

Vol. II / 1, p.287. Critical edition with commentary by Eberhard Knobloch, Abhandlungen der Akademie der Wissenschaften in Göttingen, Mathematisch-Physikalische Klasse, Dritte Folge (Third Series), Nr. 43, Vandenhoeck & Ruprecht, Göttingen, 1993. Reproduced, with an introduction and French translation in Marc Parmentier, *Quadrature arithmétique du cercle, de l'ellipse et de l'hyperbole*, Librairie Philosophique J. Vrin, Paris, 2004. Reproduced, with an introduction and German translation in Otto Hamborg, *Über die arithmetische Quadratur des Kreises, der Ellipse und der Hyperbel, von der ein Korollar die Trigonometrie ohne Tafeln ist*, 2007, available online at URL http://www.hamborg-berlin.de/persona/interessen/ Leibniz_komplett.pdf, pp. 1–142.

De triangulo curvarum characteristico, January 1675. Bodemann XXXV, Vol. VIII / 20, p. 300. In *Sämtliche Schriften und Briefe*, Ser. VII, **5**, pp. 184–191.

"De vera proportione circuli ad quadratum circumscriptum in numeris rationalibus, à G. G. Leibnitio expressa," *Acta Eruditorum*, **1** (February, 1682) 41–46 = *Opera Mathematica*, No. VIII, pp. 140–144 = Gerhardt, *Leibnizens Mathematische Schriften*, Sec. 2, I, *De quadratura arithmetica, circuli, ellipseos et hyperbolæ*, No. VI, pp. 118–122.

Elementa calculi novi pro differentiis et summis, tangentibus et quadraturis, maximis et minimis, dimensionibus linearum, superficierum, solidorum, aliisque communem calculum transcendentibus, c. 1680. Bodemann XXXV, Vol. V / 25, p. 294. In Gerhardt, *Die Geschichte der höheren Analysis; erste Abtheilung, Die Entdeckung der höheren Analysis*, pp. 149–155; translated in Child *The early mathematical manuscripts of Leibniz*, pp. 136–144.

Historia et origo calculi differentialis, 1714. Bodemann XXXV, Vol. VII / 15, p. 298. In Gerhardt, *Historia et origo calculi differentialis, a G. G. Leibnitio conscripta* = Gerhardt, *Leibnizens Mathematische Schriften*, Sec. 2, I, *Analysis Infinitorum*, No. XXXI, pp. 392–410. English translation by Child as "History and origin of the differential calculus" in *The early mathematical manuscripts of Leibniz*, pp. 22–57. Reproduced with a German translation in Otto Hamborg, *Geschichte und Ursprung der Differentialrechnung*, 2007, available online at URL http://www.hamborg-berlin.de/persona/interessen/Leibniz_komplett.pd f

Letter to Jean-Paul de La Roque, end of 1675. Revised version in Gerhardt, *Leibnizens Mathematische Schriften*, Sec. 2, I, *De quadratura arithmetica, circuli, ellipseos et hyperbolæ*, No. I, pp. 88–92 = *Sämtliche Schriften und Briefe*, Ser. III, **1**, pp. 344–353.

Letters to Johann Bernoulli, June 20, 1712 and October 25, 1713. In Gerhardt, *Leibnizens Mathematische Schriften*, Sec. 1, III **2**, pp. 888–889 and pp. 922–923.

Letter to Henry Oldenburg, June 21, 1677. In Gerhardt, *Leibnizens Mathematische Schriften*, Sec. 1, I, pp. 154–162 = Gerhardt, *Der Briefwechsel von Gottfried Wilhelm Leibniz mit Mathematikern*, pp. 240–248 = *Sämtliche Schriften und Briefe*, Ser. III, **2**, 166–182 = Turnbull, *The correspondence of Isaac Newton*, II, pp. 212–219; English translation pp. 219–225.

Méthode générale pour mener les touchantes des Lignes Courbes sans calcul, et sans réduction des quantités irrationelles et rompues, 1677. Bodemann XXXV, Vol. V / 16, p. 293. In Gerhardt, *Die Entdeckung der differentialrechnung durch Leibniz*, pp. 59–61 = Gerhardt, *Die Geschichte der höheren Analysis*; *erste Abtheilung, Die Entdeckung der höheren Analysis*, pp. 143–145; translated in Child, *The early mathematical manuscripts of Leibniz*, pp. 128–131.

Methode nouvelle des tangentes, ou de maximis et minimis. ita ut non sit opus tollere irrationalitates, 1677. Bodemann XXXV, Vol. V / 16, p. 293. In Gerhardt, *Die Entdeckung der differentialrechnung durch Leibniz*, pp. 62–65 = Gerhardt, *Die Geschichte der höheren Analysis*; *erste Abtheilung, Die Entdeckung der höheren Analysis*, pp. 145–148; translated in Child, *The early mathematical manuscripts of Leibniz*, pp. 131–134.

Methodi tangentium inversæ exempla, 1675. Bodemann XXXV, Vol. V / 9, p. 293. In Gerhardt, *Die Entdeckung der differentialrechnung durch Leibniz*, pp. 32–40 = Gerhardt, *Die Geschichte der höheren Analysis, erste Abtheilung, Die Entdeckung der höheren Analysis*, pp. 132–139 = Gerhardt, *Der Briefwechsel von Gottfried Wilhelm Leibniz mit Mathematikern*, pp. 161–167 = *Sämtliche Schriften und Briefe*, Ser. VII, **5**, pp. 321–331; translated in Child, *The early mathematical manuscripts of Leibniz*, pp. 93–103.

Methodus tangentium inversa, 1676. Bodemann XXXV, Vol. V / 10, p. 293. In Gerhardt, *Die Entdeckung der differentialrechnung durch Leibniz*, pp. 51–54 = *Sämtliche Schriften und Briefe*, Ser. VII, **5**, pp. 600–604; translated in Child, *The early mathematical manuscripts of Leibniz*, pp. 120–122.

"Nova methodvs pro maximis et minimis, itemque tengentibus, quæ nec fractas, nec irrationales quantitates moratur, & singulare pro illis calculi genus, per G. G. L." (Leibniz' initials in this Latin title stand for Gottfredus Guilielmus Leibnitius), *Acta Eruditorum*, **3** (October 1684) 467–473; facsimile reproduction and English translation (except for the last example) in Stedall, *Mathematics emerging*, pp. 121–127 = *Opera mathematica*, No. XIII, pp. 167–172 = Gerhardt, *Leibnizens Mathematische Schriften*, Sec. 2, I, *Analysis Infinitorum*, No. I, pp. 220–226. There is an English translation in Struik, *A source book in mathematics, 1200–1800*, pp. 272–280; translation reprinted in Calinger, *Classics of Mathematics*, pp. 387–392. There is an English translation of selections from the first half of this paper in Smith, *A source book in mathematics*, pp. 620–623. Complete Spanish translation in Javier de Lorenzo and Teresa Martín Santos as "Un nuevo método sobre los máximos y los mínimos, así como para las tangentes, que no se detiene ante las cantidades fraccionarias o irracionales, y es un singular género de cálculo para estos problemas" in *Gottfried Wilhelm Leibniz. Análisis infinitesimal*, 2nd. ed., pp. 3–15, Tecnos, Madrid, 1994.

"Observatio, quod rationes sive proportiones non habeant locum circa quantitates nihilo minores, & de vero sensu methodi infinitesimalis," *Acta Eruditorum*, **31** (1712) 167–169 = *Opera mathematica*, No. LXXX, pp. 439–441 = Gerhardt, *Leibnizens*

Mathematische Schriften, Sec. 2, I, *Analysis infinitorum*, No. XXIX, pp. 387–389.

Opera mathematica, Vol. 3 in *Gothofredi Guillelmi Leibnitii opera omnia*, ed. by Louis Dutens, Fratres de Tournes, Geneva, 1768; reprinted by Georg Olms Verlag, Hildesheim, 1989.

"Responsio ad nonnullas difficultates, a Dn. Bernardo Niewentiit circa methodum differentialem seu infinitesimalen motas," *Acta Eruditorum*, **14** (1695) 310–316 = *Opera Mathematica*, No. LII, pp. 327–334 = Gerhardt, *Leibnizens Mathematische Schriften*, Sec. 2, I, *Analysis infinitorum*, No. XVIII, pp. 320–328.

Sämtliche Schriften und Briefe, Series III: *Mathematischer Naturwissenschaftlicher und technischer Briefwechsel*, Vol. 1: 1672–1676; edited by Joseph Ehrenfried Hofmann, 1976; Vol. 2: 1676–1679; edited by Heinz-Jürgen Heß, 1987; Series VII: *Mathematische Schriften*, Vol. 3: Differenzen, Folgen, Reihe 1672–1676, edited by Siegmund Probst · Eberhard Knobloch · Nora Gädeke, 2003; Vol. 5: *Infinitesimalmathematik 1674–1676*, edited by Siegmund Probst · Uwe Mayer · Heike Sefrin-Weis, 2008. Berlin-Brandenburgischen Akademie der Wissenschaften and Akademie der Wissenschaften zu Göttingen, Akademie-Verlag, Berlin.

"Specimen novum analyseos pro scientia infiniti circa summas & quadraturas," *Acta Eruditorum*, **21** (1702) 210–219 = *Opera mathematica*, No. LXV, pp. 373–381 = Gerhardt, *Leibnizens Mathematische Schriften*, Sec. 2, I, *Analysis infinitorum*, No. XXIV, pp. 350–361.

"Supplementum geometriæ dimensoriæ, seu generalissima omnium tetragonismorum effectio per motum: similiterque multiplex constructio lineæ ex data tangentium conditione," *Acta Eruditorum*, **12** (1693) 385–392 = *Opera Mathematica*, No. XLIV, pp. 287–293 = Gerhardt, *Leibnizens Mathematische Schriften*, Sec. 2, I, *Analysis infinitorum*, No. XIII (2), pp. 294–301. German translation in Ostwald's *Klassiker der exakten Wissenschaften*, Engelmann, Leipzig, No. 82, 1908, pp. 24–34. Partial English translation in Struik, *A source book in mathematics, 1200–1800*, pp. 282–284 (but Struik has introduced a new symbol in both the figure and the text that is not in the original); translation reprinted in Calinger, *Classics of mathematics*, pp. 393–394 and in Laubenbacher and Pengelley, *Mathematical expeditions: chronicles by the explorers*, pp. 133–134.

Theorema arithmeticæ infinitorum, 1674. Bodemann XXXV, Vol. III, B / 10, p. 290. In *Sämtliche Schriften und Briefe*, Ser. VII, **3**, 2003, pp. 361–364. There is an extract in Gerhardt, *Der Briefwechsel von Gottfried Wilhelm Leibniz mit Mathematikern*, p. 79.

Lilly, William

William Lilly's history of his life and times, from the year 1602 to 1681, Elias Ashmole, London, 1715; reprinted for Charles Baldwin, 1822; facsimile reproduction of this edition by Kessinger Publishing Company, Kila, Montana, 2004, 2007. There are additional recent editions.

MacLaurin, Colin

A treatise of fluxions. In two books, Thomas and Walter Ruddiman, Edinburgh, 1742. Article 751 (from p. 610, Ch. II, Book II of Vol. II), containing the Maclaurin series, reproduced in Stedall, *Mathematics emerging*, p. 207. Second edition printed by Knight & Compton, Middle Street, Cloth Fair, for William Baynes and William Davis, London, 1801. There are recent reprints of this edition.

Mahoney, Michael Sean

The mathematical career of Pierre de Fermat, 1601–1665, 2nd. ed., Princeton University Press, Princeton, New Jersey, 1994.

Mercator, Nicolaus

Logarithmotechnia: sive methodus construendi logarithmos nova, accurata, & facilis; scripto antehàc communicata, anno sc. 1667. Nonis augusti: cui nunc accedit. Vera quadratura hyperbolæ & inventio summæ logarithmorum, William Godbid for Mosis Pitt, London, 1668. Propositions XIV and XV, including the series for $a/(a + a)$ are reproduced with translation in Stedall, *Mathematics emerging*, pp. 96–99

Millás Vallicrosa, José María

Estudios sobre Azarquiel, Instituto «Miguel Asín» del Consejo Superior de Investigaciones Científicas, Madrid–Granada, 1943–1950; reprinted by Excelentísima Diputación Provincial de Toledo, Toledo, 1993.

Moore, Sir Jonas

A mathematical compendium, or, useful practices in arithmetick, geometry, and astronomy, geography and navigation, embattelling, and quartering of armies, fortification and gunnery, gauging and dyalling. Explaining the logarithms, with indices; Nepair's rods or bones; making of movements and the application of pendulums; with the projection of the sphere for an universal dyal, &c, collected out of the notes and papers of Sir Jonas Moore by Nicholas Stephenson. Printed and sold by Nathaniel Brooke at the Angel in Cornhil, London, 1674.

More, Louis Trenchard

Isaac Newton. A biography, Constable and Company Limited, London, 1934; reprinted by Dover Publications, New York, 1962.

Mourey, C. V.

La vrai théorie des quantités négatives et des quantités prétendues imaginaires, Mallet-Bachelier, Paris, 1828; reprinted 1861.

Murdoch, John Emery

"Nicole Oresme's *Quæstiones super geometriam Euclidis*," *Scripta Mathematica*, **27** (1964) 67–91.

Napier, John

Mirifici logarithmorvm canonis constrvctio; et eorvm ad natvrales ipsorum numeros habitudines; vnà cvm appendice, de aliâ eâque præstantiore logarithmorum specie condenda, quibis accessere propositiones ad triangula sphærica faciliore calculo resolvenda: vnà cum annotationibus aliquot doctissimi D. Henrici Briggii in eas & memoratam appendicem, Andreas Hart, Edinburgh, 1619; reprinted by Bartholomæum Vincentium, Lyons, 1620. English version as *The construction of the wonderful canon of logarithms by John Napier Baron of Merchiston. Translated from Latin into English with notes and a catalogue of the various editions of Napier's works,* by William Rae Macdonald, F.F.A, William Blackwood & Sons, Edinburgh and London, 1889; facsimile reproduction by Dawsons of Pall Mall, London, 1966; also in The Classics of Science Library, New York, 1997. There are additional recent editions. Ample extracts from Macdonald's translation can be found in the Appendix to Lord Moulton's inaugural address to the International Congress on the tercentenary of the publication of the *Descriptio,* "The invention of logarithms. Its genesis and growth," in Knott, *Napier tercentenary memorial volume,* pp. 25–32. A different selection of extracts by William Deweese Cairns can be found in Smith, *A source book in mathematics,* pp. 149–155. A third selection of extracts was made by Struik in *A source book in mathematics, 1200–1800,* pp. 12–21; reprinted in Calinger, *Classics of Mathematics,* pp. 282–289.

Mirifici logarithmorum canonis descriptio, ejusque usus, in utraque trigonometria; ut etiam in omni logistica mathematica, amplissimi, facilimi, & expeditissimi explicatio, Andreas Hart, Edinburgh, 1614; reprinted by Bartholomæum Vincentium, Lyons, 1620; reprinted by Francis Maseres in *Scriptores logarithmici,* Vol. 6, London, 1807, pp. 475–624. Translated into English by Edward Wright as *A description of the admirable table of logarithmes: with a declaration of the most plentifvl, easy, and speedy vse thereof in both kindes of trigonometrie, and also in all mathematicall calculations,* Nicholas Okes, London, 1616; facsimile reproduction by, Theatrum Orbis Terrarum, Amsterdam and Da Capo Press, New York, 1969. Second edition of Wright's translation printed for Simon Waterson, London, 1618.

Newton, Sir Isaac

Arithmetica universalis; sive de compositione et resolutione arithmetica liber. Cui accessit Halleiana æquationum radices arithmetice inveniendi methodus. In usum juventutis academicæ, William Whiston, ed., Typis Academicis, Cantabrigiæ; Impensis Benjamin Tooke, Bibliopolæ (bookseller), Londini, 1707. Translated into English by Joseph Raphson as *Universal arithmetick: or, a treatise of arithmetical composition and resolution. To which is added, Dr. Halley's method of finding the roots of æquations arithmetically. Translated from the Latin by the late Mr. Raphson, and revised and corrected by Mr. Cunn.* Printed for J. Senex, W. Taylor, T. Warner, and J. Osborn, London, 1720. There is another translation in Whiteside, *The mathematical papers of Isaac Newton,* V, pp. 51–491.

"Counter observations on Leibniz' review [of W. Jones, *Analysis per quantitatum series, fluxiones, ac differentias: cum enumeratione linearum tertii ordinis,*" in Whiteside, *The mathematical papers of Isaac Newton,* II, pp. 263–273.

De analysi per æquationes numero terminorum infinitas, a 1669 manuscript first printed by William Jones in a selection of Newton's mathematical papers under the title *Analysis per quantitatum series, fluxiones, ac differentias: cum enumeratione linearum tertii ordinis*, 1711, pp. 1–21. Facsimile reproduction of the Latin edition by Antonio José Durán Guardeño and Francisco Javier Pérez Fernández and translated into Spanish by José Luis Arantegui Tamayo as *Análisis de cantidades mediante series, fluxiones y diferencias, con una enumeración de las líneas de tercer orden*. Introductions by José Manuel Sánchez Ron, Javier Echeverría and Antonio José Durán Guardeño and annotated by Antonio José Durán Guardeño. Two volume boxed edition by Sociedad Andaluza de Educación Matemática Thales, Sevilla, 2003. Translated into English by John Stewart in *Sir Isaac Newton's two treatises of the quadrature of curves and analysis by equations of an infinite number of terms, explained: containing the treatises themselves, translated into English, with a large commentary; in which the demonstrations are supplied where wanting, the doctrine illustrated, and the whole accommodated to the capacities of beginners, for whom it is chiefly designed*, James Bettenham, London, 1745; reproduced in Whiteside, *Mathematical works of Isaac Newton*, I, pp. 3–25; reprinted by Kessinger Publishing Company, Kila, Montana, 2009. There are additional recent editions. The original Latin paper with a new English translation can be seen in Whiteside, *The mathematical papers of Isaac Newton*, II, pp. 206–247.

De quadratura curvarum, a November 1691 manuscript, revised in December 1691, now published with an English translation in Whiteside, *The mathematical papers of Isaac Newton*, VII, pp. 24–48. The completed December revision is on pp. 48–129. Two 1692 extracts that Newton sent to Wallis can be seen in his *Opera mathematica*, **2**, 1693, pp. 390–396; reproduced, with notes but without English translation in Whiteside, *The mathematical papers of Isaac Newton*, VII, pp. 170–182. There is an English translation in Turnbull, *The correspondence of Isaac Newton*, III, pp. 220–221 and 222–228. Newton wrote a severely truncated version of *De quadratura* in 1693 (without the Taylor or Maclaurin expansions) which, retitled *The rational quadrature of curves*, became Book 2 of his projected, but never published, treatise on Geometry. This Book 2 is reproduced, with English translation, in Whiteside, *The mathematical papers of Isaac Newton*, VII, pp. 507–561. In 1703, Newton wrote an Introduction and a Scholium and appended them to the abridged *De quadratura* as the second of *Two treatises of the species and magnitude of curvilinear figures* that appeared, with the title *Tractatus de quadratura curvarum*, in the *Opticks*, 1704, pp. 165–211. A slightly modified version of this tract, incorporating Newton's corrections, under the title *De quadratura curvarum*, was included by William Jones in his *Analysis per quantitatum series, fluxiones, ac differentias: cum enumeratione linearum tertii ordinis*, 1711, pp. 41–68. Jones' version was further recast in 1712 as *Analysis per quantitatis fluentes et earum momenta*, and this version is reproduced with English translation in Whiteside, *The mathematical papers of Isaac Newton*, VIII pp. 258–297. This tract on quadrature was first translated into

English by John Stewart as *Sir Isaac Newton's two treatises on the quadrature of curves and analysis by equations of an infinite number of terms, explained* (see the previous reference for full title and details). This English translation was reprinted in Turnbull, *The mathematical works of Isaac Newton*, I, pp. 141–160. Extracts from this translation, containing the essential parts of Newton's presentation, can be found in Struik, *A source book in mathematics, 1200–1800*, pp. 303–311; partially reprinted in Calinger, *Classics of mathematics*, pp. 413–417.

Further development of the binomial expansion (Whiteside's title), a 1665 manuscript now published in Whiteside, *The mathematical papers of Isaac Newton*, I, pp. 122–134.

Letter to Henry Oldenburg, June 13, 1676. In Turnbull, *The correspondence of Isaac Newton*, II, pp. 20–47. Extracts of this letter, only in English translation, are reproduced in Smith, *A source book in mathematics*, pp. 224–225; in Struik, *A source book in mathematics, 1200–1800*, pp. 285–297; and in Calinger, *Classics of mathematics*, pp. 400.

Letter to Henry Oldenburg, October 24, 1676. In Turnbull, *The correspondence of Isaac Newton*, II, pp. 110–161 = Leibniz, *Sämtliche Schriften und Briefe*, Ser. III, **2**, 83–116. Extracts of this letter, only in English translation, are reproduced in Smith, *A source book in mathematics*, pp. 225–228; in Struik, *A source book in mathematics, 1200–1800*, pp. 287–290; and in Calinger, *Classics of Mathematics*, pp. 401–405.

Methodus differentialis, the last selection in Jones, *Analysis per quantitatum series, fluxiones, ac differentias: cum enumeratione linearum tertii ordinis*, pp. 93–101. The *Methodus* is reprinted with an English translation in Whiteside, *The mathematical papers of Isaac Newton*, VIII, pp. 244–257.

Opticks: or, a treatise of the reflexions, refractions, inflexions and colours of light. Also two treatises of the species and magnitude of curvilinear figures, Samuel Smith and Benjamin Walford, printers to the Royal Society at the Prince's Arms in St. Paul's Church-yard, London, 1704; facsimile reproduction by Culture et Civilisation, Brussels, 1966.

Philosophiæ naturalis principia mathematica, printed by Josephi Streater for the Royal Society, London, 1686. Facsimile editions by Culture et Civilisation, Brussels, 1965 and by I. Bernard Cohen and Alexandre Koyré, from the third edition of 1726, Harvard University Press, Cambridge, Massachusetts, 1972. Translated into English by Andrew Motte as *The mathematical principles of natural philosophy*, printed for Benjamin Motte at the Middle-Temple-Gate in Fleetstreet, London, 1729; facsimile reproduction with an introduction by I. Bernard Cohen, Dawsons of Pall Mall, London, 1968. There are several reprints of this translation; for instance, by Prometheus Books, Amherst, New York, 1995 (reproduced from the first American edition by N. W. Chittenden, Daniel Adee, New York, 1848). Lemmas I to XI and the Scholium of Book I, Section I, are translated into English in Hawking, *God created the integers. The mathematical breakthroughs that changed history*, pp. 374–382.

Regula differentiarum, a 1676 manuscript now published with an English translation in Whiteside, *The mathematical papers of Isaac Newton*, IV, pp. 36–69.

Tractatus de methodis serierum et fluxionum, a 1671 manuscript eventually published in English translation by John Colson as *The method of fluxions and infinite series; with its application to the geometry of curve-lines. By the inventor Sir Isaac Newton, K^{t.} Late President of the Royal Society. Translated from the author's Latin original not yet made publick. To which is subjoin'd a perpetual comment upon the whole work, consisting of annotations, illustrations, and supplements, in order to make this treatise a compleat institution for the use of Learners*, Henry Woodfall, London, 1736; reprinted by Kessinger Publishing Company, Kila, Montana, 2009. There are additional recent editions. The original work with a new English translation can be seen in Whiteside, *The mathematical papers of Isaac Newton*, III, pp. 32–353.

Newton, John

Trigonometria Britanica: or, The doctrine of triangles, in two books, the first of which sheweth the construction of the naturall, and artificiall sines, tangents and secants, and table of logarithms: with their use in the ordinary questions of arithmetick, extraction of roots, in finding the increase and rebate of money and annuities, at any rate or time propounded. The other, the use or application of the canon of artificiall sines, tangents and logarithms, in the most easie and compendious wayes of resolution of all triangles, whether plain or spherical. The one composed, the other translated, from the latine copie written by Henry Gellibrand, sometime Professor of Astronomy in Gresham-Colledge London. A table of logarithms to 100.000, thereto annexed. With the artificial sines and tangents, to the hundred part of every degree; and the three first degrees to a thousand parts. Robert & William Leybourn, London, 1658.

Nieuwentijt, Bernard

Analysis infinitorum, seu curvilineorum proprietates ex polygonorum natura deductæ, Joannem Wolters, Amstelædami, 1695.

Considerationes circa analyseos ad quantitates infinite parvas applicatæ principia, & calculi differentialis usum in resolvendis problematibus geometricis, Joannem Wolters, Amstelædami, 1694.

Oresme, Nicole

Tractatus de configurationibus qualitatum et motuum, also known as *De uniformitate et difformitate intensionum*, or *Tractatus de figuratione potentiarum et mensurarum*, or by various other titles that depend on the manuscript copy. Edited with an English translation by Clagett in *Nicole Oresme and the medieval geometry of qualities and motions*. Selections in Clagett, *The science of mechanics in the Middle Ages*, pp. 367–381; English translation of these selections and commentary, pp. 347–367.

Quæstiones super geometriam Euclidis, edited with an English translation by

Hubert Lambert Ludwig Busard and Evert Jan Brill, Leiden, 1961. Corrected version by Murdoch in "Nicole Oresme's *Quæstiones super geometriam Euclidis.*"

Oughtred, William

An addition vnto the vse of the instrument called the circles of proportion, for the working of nauticall questions. Together with certaine necessary considerations and advertisements touching navigation. All which, as also the former rules concerning this instrument are to bee wrought not onely instrumentally, but with the penne, by arithmeticke, and the canon of triangles. Hereunto is also annexed the excellent vse of two rulers for calculation. And is to follow after the 111 page of the first part, Avgvstine Mathewes, London, 1633.

The circles of proportion and the horizontall instrvment. Both invented, and the vses of both written in Latine by Mr. W. O. Translated into English, and set forth for the publique benefit by William Forster, Avgvstine Mathewes for Elias Allen, London, 1632, 1633, 1639.

Clavis mathematicæ, or, more specifically, *Arithmeticæ in numeris et speciebus instituti quæ tum logisticæ, tum analyticæ, atque adeo totius mathematicæ, quasi clavis est*, Thomam Harpervm, London, 1631. There are several editions in Latin under the title *Clavis mathematicæ denvo limata, sive potius fabricata*; reprint of the 1667 edition by Kessinger Publishing Company, Kila, Montana, 2009. English translation by Robert Wood as *The key of the mathematicks new forged and filed: together with a treatise of the resolution of all kinde of affected æquations in numbers. With the rule of compound usury; and demonstration of the rule of false position. And a most easie art of delineating all manner of plaine sun-dyalls. Geometrically taught by Will. Oughtred*, Printed by Tho. Harper for Rich. Whitaker, Pauls Church-yard, London, 1647. There is also a 1694 translation by Edmund Halley.

Pappos of Alexandria

Synagoge (Collection). Translated into Latin with commentaries by Federico Commandino as *Pappi Alexandrini Mathematicæ collectiones à Federico Commandino Vrbinate in latinvm conversæ, et commentariis illvstratæ*, Hieronymum Concordiam, Pesaro, 1588; critical Greek-Latin edition by Friedrich Hultsch as *Pappi Alexandrini collectionis quæ supersunt*, Berlin, 1877; French translation by Paul VerEeck as *La collection mathématique*, 2 vols., Albert Blanchard, Paris, 1933, 1982. Book VII of the Collection, which contains the relevant Proposition 61, has been edited and translated into English by Alexander Jones as *Pappus of Alexandria. Book 7 of the* Collection. *Part 1: Introduction, Text, and Translation; Part 2: Commentary, Index, and Figures*, Springer-Verlag New York, 1986.

Parmentier, Marc, ed. and trans.

La naissance du calcul différentiel, Librairie Philosophique J. Vrin, Paris, 1989, 2000. Contains French translations of 26 papers by Leibniz from the *Acta Eruditorum*.

Parr, Richard

The life of the most reverend father in God, James Usher, late Lord Arch-Bishop of Armagh, primate and metropolitan of all Ireland. With a collection of three hundred letters, between the said Lord Primate and most of the eminentest persons for piety and learning in his time, both in England and beyond the seas. Printed for Nathanael Ranew, at the Kings-Arms in St. Pauls Church-Yard, London, 1686.

Pengelley, David

See Laubenbacher, Reinhard.

Pitiscus, Bartholomäus

Trigonometria: sive de solvtione triangvlorvm tractatus breuis & perspicuus. It was published as the final part of Abraham Scultetus' *Sphæricorum libri tres methodice conscripti et utilibus scholiis expositi,* Abrahami Smesmanni for Matthaei Harnisch, Heidelberg, 1595. A second edition appeared with the title *Trigonometriæ, siue, de dimensione triangulo̲r. libri qvinqve. Item problematvm variorv̄. nempe geodæticorum, altimetricorum, geographicorum, gnomonicorum, et astronomicorum: libri decem. Trigonometriæ svbivncti ad vsvm eivs demonstrandvm,* Dominci. Custodis chalcographi, August[æ] Vindeli[corum] (Augsburg), 1600. English translation by Raph Handson as *Trigonometry, or, the doctrine of triangles,* Edward Allde for Iohn Tap, London, 1614. There were editions of 1630 and c. 1642.

Prag, Adolf

"On James Gregory's Geometriæ pars universalis," in Turnbull, *James Gregory tercentenary memorial volume,* pp. 487–509.

Ptolemy of Alexandria

Mathematike syntaxis or *The Almagest.* Gherardus Cremonensis, or Gherardo de Cremona, made a translation of *The Almagest* in 1175 that formed the basis of the first printed Latin translation, *Almagestum Cl. Ptolemei,* Venice, Petrus Liechtenstein, 1515. Greek version by Johan Ludwig Heiberg, Leipzig, 1898 (2nd. ed. 1903). English translation by Gerald James Toomer as *Ptolemy's Almagest,* Duckworth, London, 1984; reprinted by Princeton University Press, Princeton, New Jersey, 1998. An earlier Greek version and French translation can be found in Halma, *Composition mathématique de Claude Ptolémée.* Article 10, containing the construction of Ptolemy's table of chords, is reproduced in Greek (Heiberg edition) and with English translation in Thomas, *Selections illustrating the history of Greek mathematics,* II, pp. 412–443. It is partially reproduced in Calinger, *Classics of mathematics,* pp. 167–170.

Rajagopal, Cadambathur Tiruvenkatacharlu and M. S. Rangachari

"On an untapped source of medieval Keralese mathematics," Arch. History Exact Sci., **18** (1978) 89–102.

Ravetz, Jerome Raymond

See Grattan-Guinness, Ivor.

Recorde, Robert

The whetstone of witte, whiche is the seconde parte of arithmetike: containyng the extraction of rootes, the cossike practise, with the rule of equation: and the woorkes of surde nombers, Ihon Kyngstone, London, 1557. Facsimile edition by Theatrum Orbis Terrarum, Amsterdam, and Da Capo Press, New York, 1969; quasi-facsimile edition by TGR Renascent Books, Mickleover, Derby, England, 2010.

Regio Monte, Iohannis de

De triangvlis omnimodis libri qvinqve: Quibus explicantur res necessariaæ cognitu, uolentibus ad scientiarum astronomicarum perfectionem deuenire: quæ cum nusquã alibi hoc tempore expositæ habeantur, frustra sine harum instructione ad illam quisquam aspirarit. Accesserunt huc in calce pleraq[ue] D. Nicolai Cusani de quadratura circuli, Deq́[ue] recti ac curui commensuratione: itemq́[ue] Io. de monte Regio eadem de re ἐλενκτικὰ, hactenus à nemine publicata, Iohannis Petreus, Nuremberg, 1533. Facsimile reproduction and English translation, with introduction and notes in Barnabas Hughes, *Regiomontanus on triangles*, The University of Wisconsin Press, Madison, Wisconsin, 1967. Second edition of *De triangulis*, Daniel Santbech, Basel, 1561.

Epytoma Joãnis de mõte regio In almagestũ ptolemei, Iohannis Ha[m]man de Landoia, Venice, 1496.

Tabule directionũ profectionũq[ue] famosissimi viri Magistri Joannis Germani de Regiomonte in natiuitatibus multum vtiles, Erhard Ratdolt, Augsburg, 1490.

Rheticus, Georg Joachim

Opvs palatinvm de triangvlis a Georgio Ioachimo Rhetico coeptvm: L. *Valentinvs Otho principis palatini Friderici IV. Electoris mathematicvs consvmmavit*, Matthaeus Harnisius, Neustadt an der Haardt, 1596.

Thesaurus mathematicus: sive, Canon sinuum ad radium 1.00000.00000.00000. et ad dena quaeque scrupula secunda quadrantis: una cum sinibus primi et postremi gradus, ad eundem radium, et ad singula scrupula secunda quadrantis: adjunctis ubique differentiis primis et secundis: atq[ue] ubi res tulit, etiam tertijs, Nicolaus Hoffmannus, Frankfurt am Main, 1613.

Ricci, Michelangelo

Michaelis Angeli Riccii geometrica exercitatio, Nicolaum Angelum Tinassium, Rome, 1666. Reprinted with the expanded title *Michaelis Angeli Riccii geometrica exercitatio de maximis & minimis* as an appendix to Mercator's *Logarithmotechnia*.

Robins, Gay and Charles Shute,

The Rhind mathematical papyrus, an ancient Egyptian text, The British Museum Press, London, 1987, 1998.

Rome, Adolphe

Commentaires de Pappus et de Théon d'Alexandrie sur l'Almageste, Tome II *Théon d'Alexandrie, Commentaire sur les livres 1 et 2 de l'Almageste*, Studi e Testi 72, Città del Vaticano, 1936.

Sarasa, Alfonso Antonio de

Solvtio problematis a R P Marino Mersenno Minimo propositi: datis tribus quibscumq̄ magnitudinibus, rationalibus vel irrationalibus, datisque duarum ex illis logarithmis, tertiæ logarithmum geometricè inuenire: Duo a proponente de hac propositione pronuntiantur, unum quod forsitan longè difficiliorem quam ipsa quadratura solutionem requirat alterum quòd quadratura circuli à R. P. Gregorio a S^to Vincentio exhibita, abeat in illud necdum solutum problema: quibus videtur indicare solutionem problematis de quadratura circuli expeditam fore, si defectus suppleatur, quem in solutione problematis à se propositi iudicavit, Ioannem & Iacobvm Meursios, Antwerp, 1649.

Saint-Vincent, Grégoire de

Opvs geometricvm quadraturæ circvli et sectionvm coni. Decem libris comprehensum, Ioannem & Iacobvm Mevrsios, Antwerp, 1647.

Sastri, K. Sambasiva, ed.

Aryabhatiya with the bhashya of Nilakantha, Trivandrum Sanskrit Series Nos. 101 and 110, Trivandrum, 1930, 1931.

Schubring, Gert

"Argand and the early work on graphical representation: New sources and interpretations." In *Around Caspar Wessel and the geometric representation of complex numbers*, Proceedings of the Wessel symposium at The Royal Danish Academy of Sciences and Letters, Copenhagen, 1998, pp. 125–146. Edited by Jesper Lützen. Matematisk-fysiske Meddelelser 46:2, Det Kongelige Danske Videnskabernes Selskab, Copenhagen, 2001.

Seidel, Phillip Ludwig von

"Note über eine Eigenschaft der Reihen, welche discontinuirliche Functionen darstellen," *Abhandlungen der Königlich Bayerischen Akademie der Wissenschaften (München)*, **5** (1847–49) 379–393.

Sezgin, Fuat, ed.

Islamic mathematics and astronomy, Vol. 61 I and II, Institute for the History of Arabic-Islamic Science, Johann Wolfgang Goethe University, Frankfurt am Main, 1998.

Sharma, P. D.

Hindu Astronomy, Global Vision Publishing House, Delhi, 2004.

Shute, Charles

See Robins, Gay

Sluse, René François Walther de

Letter to Henry Oldenburg, December 17, 1671. In Turnbull, *The correspondence of Isaac Newton*, I, p. 71

Letter to Henry Oldenburg, January 17, 1673. Published with the title "An extract of a letter from the excellent *Renatus Franciscus Slusius*, Canon of *Liege* and

Counsellor to His Electoral Highness of *Collen*, written to the publisher in order to be communicated to the R. *Society*: concerning his short and easie *Method of drawing Tangents to all Geometrical Curves* without any labour of Calculation: Here inserted in the same language, in which it was written," *Philosophical Transactions*, **8** (1673) 5143–5147. The demonstration was given in **8** (1673) 6059 under the heading "Illustrissimi Slusiii modus, quo demonstrat methodum suam ducendi tangentes ad quaslibet curvas absqe calculo, antehac traditam in horum actorum N°. 90. The statement of the method, but not the entire letter, is reproduced in Gerhardt, *Der Briefwechsel von Gottfried Wilhelm Leibniz mit Mathematikern*, p. 232.

Mesolabvm, sev dvæ mediæ proportionales inter extremas datas per circvlvm et ellipsim vel hyperbolam infinitis modis exhibitæ. Accedit problematvm qvorvmlibet solidorvm effectio per easdem curvas, jisdem modis. & appendix de eorundem solutione per circulum & parabolam, Jan F. van Milst, Leodij Eburonum (Liège), 1659.

Smith, David Eugene

History of mathematics, 2 vols., Ginn and Company, Boston, 1923 and 1925; reprinted by Dover Publications, New York, 1958.

Portraits of Eminent Mathematicians, Scripta Mathematica, New York, Portfolio I, 1936; Portfolio II, 1938.

A source book in mathematics, McGraw-Hill Book Company, New York, 1929; reprinted by Dover Publications, New York, 1959, 1984.

Stedall, Jacqueline Anne

Mathematics emerging: A sourcebook 1540–1900, Oxford University Press, Oxford, 2008.

Stifel, Michael

Arithmetica integra, Iohannis Petreium, Nuremberg, 1544. German translation by Eberhard Knobloch and Otto Schönberger as *Michael Stifel vollständiger Lehrgang der Arithmetik*, Verlag Königshausen & Neumann, Würzburg, Germany, 2007.

Stokes, Sir George Gabriel

"On the critical values of the sums of periodic series," *Cambridge Philosophical Society Transactions*, **8** (1848) 533–583. Section III reproduced in Stedall, *Mathematics emerging*, pp. 527–528.

Struik, Dirk Jan

A source book in mathematics, 1200–1800, Harvard University Press, Cambridge, Massachusetts, 1969; reprinted by Princeton University Press, Princeton, New Jersey, 1986.

Suiseth, Richard Raymond

De motu, Cambridge University, Gonville and Caius College, 499/268, ff. 212^v–213^r. In Clagett, *The science of mechanics in the Middle Ages*, pp. 245–246.

Liber calculationum, Vatican Library 3095, ff. 113–114^r, 118^{r-v}. In Clagett, *The science of mechanics in the Middle Ages*, pp. 298–304.

Taylor, Brook
"De motu nervi tensi," *Philosophical Transactions of the Royal Society of London*, **28** (1713, pub. 1714) 26–32. Reproduced with English translation in *Philosophical Transactions of the Royal Society of London, Abridged*, **6** (1809) 7–12 and 14–17.

Methodus incrementorum directa, & inversa, printed by Pearson for William Innys at the Prince's Arms in Saint Paul's Church Yard (Typis Pearsonianis: prostant apud Gul. Innys ad Insignia Principis in Cœmeterio Paulino), London, 1715. It can be seen photographically reproduced, and also in English translation and updated notation (Taylor's use of dots and primes as subscripts and superscripts is too involved), in Struik, *A source book in mathematics, 1200–1800*, pp. 328–333; reprinted (without the photographic reproduction) in Calinger, *Classics of mathematics*, pp. 466–468. Page 23, containing Taylor's series, is reproduced with translation in Stedall, *Mathematics emerging*, pp. 204–206. Complete English translation by L. Feigenbaum, Ph.D. thesis, Yale University, 1981.

Thomas, Ivor, ed. and trans.
Selections illustrating the history of Greek mathematics, Vol. I *From Thales to Euclid*, Vol. II *From Aristarchus to Pappus*, William Heinemann Ltd., London and Harvard University Press, Cambridge, 1941, 1942; reprinted by Harvard University Press, 1980; reprinted with revisions 1993, 2000.

Thoren, Victor E.
The Lord of Uraniborg, Cambridge University Press, Cambridge, 1990, 2007.

Toomer, Gerald James
"The chord table of Hipparchus and the early history of Greek trigonometry," *Centaurus*, **18** (1973/74).

Turnbull, Herbert Westren, ed.
James Gregory tercentenary memorial volume, George Bell and Sons, Ltd., London, 1939.

The correspondence of Isaac Newton, Cambridge University Press, London and New York, Vol. I: 1661–1675, 1959; Vol. II: 1676–1687, 1960; Vol. III: 1688–1694, 1961. There are additional recent editions.

The mathematical works of Isaac Newton, I, Johnson Reprint Corporation, London and New York, 1964.

Viète, François
Ad angvlarivm sectionvm analyticen. Theoremata. καθολικώτερα. A Francisco Vieta Fontenaensi primum excogitata, et absque vlla demonstratione ad nos transmissa, iam tandem demonstrationibus confirmata. Operà & studio Alexandri Andersoni Scoti, Oliverivm de Varennes, Paris, 1615. It was included, with the title "Ad

angvlares sectiones theoremata καθολικώτερα demonstrata par Alexandrum Andersonum" in *Francisci Vietæ opera mathematica*, pp. 287–304. English translation as "Universal theorems on the analysis of singular sections with demonstrations by Alexander Anderson" in Witmer, *The analytic art by François Viète*, pp. 418–450.

Ad logisticem speciosam notæ priores, or more fully *In artem analyticem isagoge. Eiusdem ad logisticem speciosam notæ priores, nunc primum in lucem editæ. Recensuit, scholiisque illustravit J. D. B.* Published posthumously in Paris by Guillaume Baudry, with notes by Jean de Beaugrand, in 1631. It was included in *Francisci Vietæ opera mathematica* as "Ad logisticen speciosam, notæ priores," pp. 13–41. French translation by Fréderic Ritter as "Introduction à l'art analytique par François Viète," *Bollettino di Bibliografia e di Storia delle Scienze Matematiche e Fisiche* [Bollettino Boncompagni], **1** (1868) 223–276. English translation as "Preliminary notes on symbolic logistic" in Witmer, *The analytic art by François Viète*, pp. 33–82.

Canon mathematicvs sev ad triangvla cum adpendicibus, Ioannem Mettayer, Lvtetiæ (Paris), 1579.

De æqvationvm recognitione et emendatione tractatvs dvo. Quibus nihil in hoc genere simile aut secundum, huic auo hactenus visum (with an introduction by Alexander Anderson), Ioannis Laqvehay, Paris, 1615 (but probably written in the early 1590s). It was included in *Francisci Vietæ opera mathematica*, pp. 82–158. English translation by Witmer as "Two treatises on the understanding and amendment of equations," in *The analytic art by François Viète*, pp. 159–310.

De recognitione æquationvm, the first of the two treatises in *De æquationvm recognitione et emendatione tractatvs dvo*.

Francisci Vietæ opera mathematica, in unum volumen congesta, ac recognita, Operâ atque studio Francisci à Schooten Leydensis, Lvgdvni Batavorvm (Leiden), ex officinâ Bonaventuræ & Abrahami Elzeviriorum, 1646. Facsimile reproduction with an introduction by Joseph Ehrenfried Hofmann, Georg Olms Verlag, Hildesheim, 1970. French translation by Jean Peyroux as *Œuvres mathématiques; suivies du dénombrement des réalisations géométriques; et du supplément de géométrie*, Albert Blanchard, Paris, 1991.

Variorvm de rebvs mathematicis responsorvm, liber VIII: cuius præcipua capita sunt, de duplicatione cubi, & quadratione circuli; quæ claudit πρόχειρον seu, ad vsum mathematici canonis methodica, Iamettivm Mettayer, Tours, 1593. In *Francisci Vietæ opera mathematica*, pp. 347–436.

Vniversalivm inspectionvm ad canonem mathematicvm, liber singularis. This is the individually titled second book of the *Canon mathematicvs*.

Walker, Evelyn

A study of the Traité des Indivisibles of Gilles Persone de Roberval. With a view to answering, insofar as is possible, the two questions: which propositions contained therein are his own, and which are due to his predecessors or contemporaries? and what effect, if any, had this work on his successors?, Bureau of Publications,

Teachers College, Columbia University, New York, 1932; reprinted by the American Mathematical Society, Providence, Rhode Island, 1972.

Wallis, John

Arithmetica infinitorvm, sive nova methodus inquirendi in curvilineorum quadraturam, aliaq; difficiliora matheseos problemata, Leonard Lichfield for Thomas Robinson, Oxford, dated 1656, published 1655 = *Opera mathematica*, **1**, 1695, pp. 355–478. English translation by Stedall as *The arithmetic of infinitesimals. John Wallis 1656*, Springer-Verlag, New York, 2004.

De sectionibus conicis, nova methodus expositis, tractatvs, Leonard Lichfield for Thomas Robinson, Oxford, 1655 = *Opera mathematica*, **1**, 1695, pp. 291–296.

Johannis Wallis S.T.D. geometriæ professoris Saviliani, in celeberrima Academia Oxoniensi, Opera mathematica tribus voluminibus contenta, E Theatro Sheldoniano, Oxford, 1695–1699; reprinted by Georg Olms Verlag, Hildesheim, 1972 (originally published as *Operum mathematicorum* by Leonard Lichfield for Thomas Robinson, Oxford, 1656).

"Logarithmotechnia Nicolai Mercatoris," *Philosophical Transactions*, **3** (1668) 753–764.

Waring, Edward

Meditationes analyticæ, John Archdeacon, Cambridge University Press, Cambridge, 1776. Second ed., 1785.

Warren, John

A treatise on the geometrical representation of the square roots of negative quantities, John Smith, Cambridge University Press, Cambridge, 1828; reprinted in *Philosophical Transactions of the Royal Society of London*, **119** (1829) 241–254.

Weierstrass, Karl Theodor Wilhelm

Mathematische Werke. Herausgegeben unter Mitwirkung einer von der Königlich Preussischen Akademie der Wissenschaften eingesetzten Commission, 7 vols., Mayer & Müller, Berlin.

"Zur Functionenlehre," *Monatsberichte der Königlich Preussischen Akademie der Wissenschaften zu Berlin. Mathematische-Physikalische Klasse*, (August 12, 1880) 719–743= *Mathematische Werke*, Vol. 2, 1895, pp. 201–223, Art. 1, p. 202.

Zur Theorie der Potenzreihen, Münster, Fall 1841 = *Mathematische Werke*, Vol. 1, 1894, pp. 67–74.

Wessel, Caspar

"Om Direktionens analytiske Betegning, et Forsög anvendt fornemmelig til plane og sphæriske Polygoners Oplösning," *Nye Samling af det Kongelige Danske Videnskabernes Selskabs Skrifter*, **2**, Vol. V (1799) 469–518. French translation by Hieronymus Georg Zeuthen and others as *Essai sur la représentation analytique de la direction*, ed. by Herman Valentiner and Thorvald Nicolai Thiele, Copenhagen, 1897. English translation by Flemming Damhus as *On the analytical representation* *of direction. An attempt applied chiefly to solving plane and sphericial polygons*, with introductory chapters by Bodil Branner, Nils Voje Johansen, and Kirsti Andersen. Edited by Bodil Branner and Jesper Lützen. Matematisk-fysiske Meddelelser 46:1, Det Kongelige Danske Videnskabernes Selskab, Copenhagen, 1999.

Whiteside, Derek Thomas
"Henry Briggs: the binomial theorem anticipated," *The Mathematical Gazette*, **45** (1961) 9–12.

The mathematical papers of Isaac Newton, 8 vols., Vol. I: 1664–1666, 1967; Vol. II: 1667–1670, 1968; Vol. III: 1670–1673, 1969; Vol. IV: 1674–1684, 1971; Vol. V: 1683–1684, 1972; Vol. VI: 1684–1691, 1974; Vol. VII: 1691–1695, 1976; Vol. VIII: 1697–1722, 1981, Cambridge University Press; digitally printed paperback version, 2008.

Mathematical works of Isaac Newton, 2 vols., Johnson Reprint Corporation, New York and London, 1964, 1967.

Witmer, T. Richard
The analytic art by François Viète. Nine studies in algebra, geometry and trigonometry from the Opus restitutae mathematicae analyseos, seu algebrâ novâ [this was Viète's title for a collection of most of his mathematical treatises] *by François Viète*, The Kent State University Press, Kent, Ohio, 1983; reprinted by Dover Publications, New York, 2006.

Woepcke, Franz
"Recherches sur l'histoire des sciences mathématiques chez les orientaux, d'après des traités inédits arabes et persans. Deuxième article. Analyse et extrait d'un recueil de constructions géométriques par Aboûl Wafâ," *Journal Asiatique*, Ser. 5, **5** (February–March 1855) 218–256 and (April 1855) 309–359 = Woepcke, *Études sur des mathématiques arabo-islamiques*, Vol. 1, Institut für Geschichte der Arabisch-Islamischen Wissenschaften, Frankfurt am Main, 1986, pp. 483–572 = Sezgin, *Islamic mathematics and astronomy*, **61** II, pp. 84–173.

Wolf, Rudolph
"Notizen zur Geschichte der Mathematik und Physik in der Schweiz," *Mittheilungen der naturforschenden Gesellschaft in Bern*, **135** (1848). The fragments from Bernoulli's autobiography are reproduced under the title "Erinnerungen an Johann I Bernoulli aus Basel" in Literarische Berichte No. L, *Archiv der Mathematik und Physik mit besonderer Rücksicht auf die Bedürfnisse der Lehrer an höhern Unterrichtsanstalten*, **13**, ed. by Johann August Grunert, C. A. Koch's Separat-Conto, Greifswald, 1849, pp. 692–698.

Zeller, Mary Claudia
The development of trigonometry from Regiomontanus to Pitiscus, Ph.D. thesis, University of Michigan, Ann Arbor, Michigan, 1944. Lithoprinted by Edwards Brothers, Ann Arbor, Michigan, 1946.

INDEX

CPSIA information can be obtained
at www.ICGtesting.com
Printed in the USA
LVOW04*2050260216

476856LV00001B/1/P

9 780387 921532